# Evolution of the Pyrenees during the Variscan and Alpine Cycles 1

SCIENCES

*Geoscience*, Field Director – Yves Lagabrielle

*Dynamics of the Continental Lithosphere*,
Subject Head – Sylvie Leroy

# Evolution of the Pyrenees during the Variscan and Alpine Cycles 1

## *Variscan Cycle and Cretaceous Rifting*

*Coordinated by*
Nicolas Saspiturry
Jessica Uzel
Alexandre Ortiz

iSTE

WILEY

First published 2024 in Great Britain and the United States by ISTE Ltd and John Wiley & Sons, Inc.



ISTE Ltd
27-37 St George's Road
London SW19 4EU
UK

www.iste.co.uk

John Wiley & Sons, Inc.
111 River Street
Hoboken, NJ 07030
USA

www.wiley.com



Library of Congress Control Number: 2023942080

British Library Cataloguing-in-Publication Data
A CIP record for this book is available from the British Library
ISBN 978-1-78945-124-5

ERC code:
PE10 Earth System Science
*PE10_5 Geology, tectonics, volcanology*
*PE10_7 Physics of earth's interior, seismology, volcanology*

# Contents

# Preface

**Nicolas SASPITURRY[1], Jessica UZEL[2] and Alexandre ORTIZ[2]**

[1] *Géosciences Montpellier, CNRS, Université de Montpellier, France*

[2] *Géosciences Rennes, CNRS, Université de Rennes 1, France*

The Pyrenean Mountain belt is, without a doubt, one of the most emblematic of France's geological "objects". This mountain belt has long attracted many French and foreign scientists. Although its formation is relatively recent, the analysis of the innumerable variety of rocks composing it allows us to reconstitute the complex evolution of this area over the last 600 million years. The Pyrenees were studied extensively between 1960–1990 from a stratigraphic, sedimentological and tectonic point of view, through the realization of numerous PhD theses. The purpose of these theses was to produce the 1/50,000 scale geological maps published by the *Bureau de recherches géologiques et minières* (BRGM). The end of the geological BRGM mapping program caused geologists to lose interest in the Pyrenees in favor of the Alps. Indeed, the growth of knowledge concerning the formation of passive margins and sub-continental mantle exhumation at the ocean-continent transition has led many researchers to apply these concepts to the inverted Alpine passive margins, making them the emblematic fossil global standard for these objects. Nevertheless, the Pyrenees remain a first-class natural laboratory for better understanding of the Variscan and Alpine cycles.

Since the 2010s, Pyrenean research has experienced a significant boom, as evidenced by the many research projects that have emerged: (1) the ANR PYRAMID (2012–2016), (2) the Pyrenees project of the *Référentiel géologique de la France* (RGF) (2014-2019) and (3) the Convergence research program (2016–2020). The ANR PYRAMID, supported in part by the energy group TotalEnergies and BRGM, was entitled: "North PYRenees: Integrated assessment of fluid migration history, rift inversion, the role of surface processes and deformation in the evolution of an orogenic retrowedge and its foreland basin". The goals of this ANR grant were to (1) understand the overall 3D structural style and architecture of the North Pyrenean foreland system, and define the role of crustal inheritance within it; (2) link sediment transfer, landform formation and exhumation to the evolution of the foreland basin; and (3) quantify fluid-rock interactions from pre-orogenic extension to compression. Following the success of the ANR PYRAMID, BRGM has decided to continue by re-initiating scientific research on the Pyrenees.

Developed by BRGM, the RGF program, following the 1:50,000 geological map of France program, aimed to constitute the new foundation of French geological knowledge. Selected for their societal and academic attributes, the Pyrenees and their northern foreland constituted the first territory to apply the concepts of the RGF program. The RGF program consisted of reusing, reinterpreting and unifying different sources of surface and subsurface data on a single map, whether they came from geological maps, the subsurface data bank or oil and geothermal drilling. The objective of the RGF was, in the long term, the multi-scale representation of the geological knowledge of France in three dimensions, taking into account the physical properties of rocks. A true strategic project for BRGM and the entire scientific community, the RGF was part of a mission to support public policies by offering a response to the problems of land use planning, natural resource management and seismic and flood risk prevention. The RGF Pyrenees project, which lasted five years, revisited the 600 million years of geological history of this mountain belt. The goal of the RGF Pyrenees project was to (1) revisit the knowledge of the Paleozoic basement of the Pyrenees, exhumation and denudation during the tertiary orogeny and (2) constrain the processes of formation of the North Pyrenean rift and the exhumation of the continental mantle during the Lower Cretaceous, of which the lherzolites lining the belt are relics.

Finally, the Convergence research program (2016–2020) is the result of a close collaboration between TotalEnergies, BRGM and CNRS. This national research program aimed to structure research activities on mountain ranges and their adjacent sedimentary basins. This major scientific program included two projects: OROGEN and SOURCE TO SINK. The SOURCE TO SINK project was designed to better understand the processes of landform creation and erosion that lead to the formation of sedimentary basins from their source to the depositional zone. At a larger scale, OROGEN was an integrated holistic geosciences research project coordinated by INSU, with the objective of pushing back the limits of knowledge in terms of understanding the mechanisms behind mountain formation, as applied to the Pyrenean belt. The Pyrenees represented an exceptional study area due to its geological singularities: (1) the passive margins inherited from the Cretaceous rifting partially preserved during continental subduction and then collision, and (2) the variation of its deformation states, between the Bay of Biscay and the Gulf of Lion, allowing us to study, in detail, how this domain evolves during compression and postorogenic phases. The presence of preserved basins of Cretaceous age, accessible from the outcrop in the north of the axial zone of the chain, has allowed for the acquisition of a unique data set for this type of domain, in particular the continental crust thermal evolution during the rifting stage.

The works of the SCIENCES series by ISTE and Wiley on the Pyrenees thus report on the evolution of geological knowledge, from the Variscan cycle to the Alpine cycle, resulting from these three major scientific research programs: the ANR PYRAMID, the RGF Pyrenees project and the Convergence research program. These publications logically involve many of the leading scientific actors of these research programs. We would like to thank the authors of the chapters who participated in the construction of these works. We would also like to thank Yves Lagabrielle and Sylvie Leroy for trusting us to manage this edition on Pyrenean geology.

August 2023

# 1

# Tectonics and Geodynamics of the Variscan Cycle in the Pyrenees

**Bryan COCHELIN[1], Charles GUMIAUX[2,3], Benjamin LE BAYON[3], Yoann DENÈLE[4] and Thierry BAUDIN[3]**

[1] *State Key Laboratory for Mineral Deposits Research, University of Nankin, China*

[2] *ISTO, CNRS, Université d'Orléans, France*

[3] *BRGM, Orléans, France*

[4] *GET, CNRS, Université Toulouse-III-Paul-Sabatier, IRD, CNES, OMP, Toulouse, France*

## 1.1. Introduction

The Pyrenees have long been overshadowed by the enthusiasm of research on the Alps; nevertheless, they are a first-class natural laboratory that can help us to better understand the Earth system. Due to the great diversity and geological richness that the Pyrenean chain provides, as well as the long and complex tectonic history of which it is the result, the study of the Pyrenean chain allows us to address many fundamental scientific questions that once animated, and still animate, the scientific community. Triggered by the Alpine tectonic uplift, the erosion of the Mesozoic to Cenozoic cover during the Tertiary period denudated a large portion of the Paleozoic basement of the Pyrenean chain. At first sight, the tectonic structuring of this basement is largely inherited from an earlier orogenic cycle: the Variscan orogeny.

---

For a color version of all the figures in this chapter, see www.iste.co.uk/saspiturry/evolution1.zip.

*Evolution of the Pyrenees during the Variscan and Alpine Cycles 1*,
coordinated by Nicolas SASPITURRY, Jessica UZEL, Alexandre ORTIZ. 

Although the Pyrenean basement has been studied extensively since the 1970s, during the 2000s, the geological scientific community has focused on the study of other orogenic segments of the range, somewhat neglecting the progress of our knowledge and understanding of the structure of the Pyrenean basement. It is within this scientific framework that the *Bureau de recherches géologiques et minières* (BRGM) decided to initiate a major research program: the Geological Reference System of France (commonly referred to as RGF). This new program succeeded the 1:50,000 geological map program, while incorporating the advanced scientific methods and techniques of the 21st century, with the aim of restoring a continuous and coherent layer of geological knowledge throughout the country. The first of these RGF projects (2014–2019) focused on the Pyrenees, with the objective, in partnership with French research institutes, to propose a harmonized and customizable digital geological map, capable of integrating all new geological data acquisitions (cartographic, analytical, etc.). This chapter therefore aims to highlight the latest scientific advances concerning the Variscan cycle and its tectonic consequences in the Pyrenees, with regard to the recent work carried out ahead of and during the RGF Pyrenees project.

## 1.2. State of the knowledge at the start of the RGF Pyrenees project

### 1.2.1. *General*

In this section, the main characteristics of the Pyrenean basement inherited from the Variscan orogeny are summarized. For further information on works carried out before this date, an exhaustive synthesis is presented in Barnolas and Chiron (1996). The numerous cartographic surveys carried out over the course of the last century have found that the Paleozoic basement outcrops all along the Pyrenean range in the form of crustal blocks of variable size. The main block is located in the core of the Pyrenees and forms the Axial Zone (Figure 1.1(a)). The Axial Zone, which is about 330 km long and 60 km wide, is oriented parallel to the axis of the range and is bordered by the North Pyrenean Fault along its northern boundary (Figure 1.1(a)). To the north of this major fault, other much smaller basement blocks, called the North Pyrenean massifs, are found cartographically surrounded by the Mesozoic sedimentary series. The Variscan basement outcrops in the Basque massifs can be found in the western part of the Pyrenees (Figure 1.1(a)). All these massifs are constituted by metasedimentary rocks from the Ediacaran to Carboniferous age. The associated lithologies are essentially: (1) alternating sandstone and limestone, with rare conglomeratic levels, in the Cambro-Ordovician, (2) shales and black pelites associated with the Silurian, (3) alternating thick limestone series and sandstone and limestone levels in the Devonian and (4) silty and sandstone-conglomeratic series (turbidites) dating from the Carboniferous. The Lower Paleozoic metasedimentary series are sometimes intruded by Ordovician granitoids. Gneissic and/or migmatitic massifs from the Variscan orogeny are also observed locally. These gneisses, associated with migmatites, are evidence of Variscan high temperature metamorphism that affected the deepest rocks within the crust. These rocks, corresponding to the lower to middle crust, appear locally in the form of metamorphic domes, structurally below the weakly metamorphic sedimentary rocks from the upper Paleozoic (see section 1.2.2). The latter thus constituted the upper crust at the beginning of the Variscan orogeny. Numerous Variscan plutonic bodies of varying sizes and natures were emplaced within the formations described above (Figure 1.1(a)).

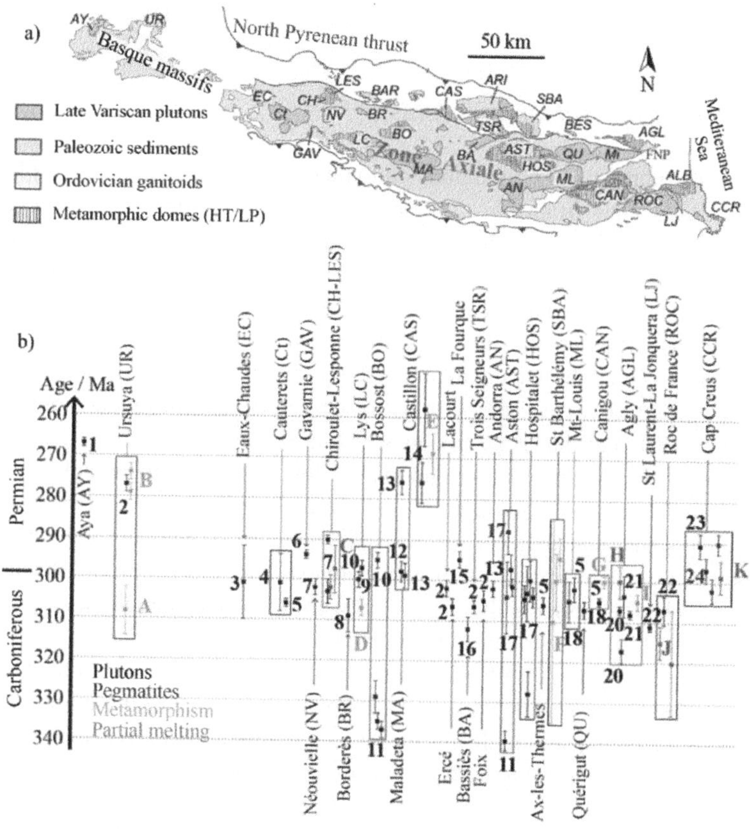


**Figure 1.1.** *(a) Geological map of the Pyrenees showing the outcroping zones of the basement. The main granitic plutons and metamorphic domes of the Axial Zone, the North Pyrenean massifs and the Basque massifs are highlighted (modified from Cochelin et al. (2017)); (b) compilation of available dating for magmatism and late Variscan metamorphism along the Pyrenees*

LEGEND FOR FIGURE 1.1. – *Dates from 1: Denèle et al. 2012, 2: Vacherat et al. 2017, 3: Ternet et al. 2003, 4: Guerrot 1998, 5: Denèle et al. 2014, 6: Kilzi et al. 2016, 7: Lemirre et al. 2019, 8: Gleizes et al. 2006, 9: Esteban et al. 2015, 10: Lopez-Sanchez et al. 2018, 11: Mezger and Gerdes 2016, 12: Evans 1993, 13: Pereira et al. 2014, 14: Kilzi 2014, 15: Poitrenaud et al. 2020, 16: Paquette et al. 1997, 17: Schnapperelle et al. 2020, 18: Maurel et al. 2004, 19: Roberts et al. 2000, 20: Olivier et al. 2004, 21: Olivier et al. 2008, 22: Aguilar et al. 2014, 23: Druguet et al. 2014, 24: Van Lichtervelde et al. 2014, A: Vacherat et al. 2017, B: Hart et al. 2016, C: Lemirre et al. 2019, D: Esteban et al. 2015, E: Kilzi 2014, F: Delaperriere et al. 1994, G: de Hoÿm de Marien et al. 2019, H: Tournaire Guille et al. 2019, I: Siron et al. 2020, J: Aguilar et al. 2014, K: Druguet et al. 2014.*

### 1.2.2. *Magmatism and metamorphism*

The most common and voluminous plutonic bodies were essentially emplaced in the upper crust. They often have a characteristic "mushroom-like" shape (Pouget et al. 1989). With a calc-alkaline geochemical signature, these plutons are most often composite in nature. Indeed, they are composed of both basic facies on their periphery (diorite, gabbro) and acid facies in their core (granite, rarely leucogranite). Similar plutons, albeit smaller in size (kilometers), are also frequently observed in the upper crust. When these calc-alkaline plutons are emplaced at the top of gneissic domes, they present a laccolitic form, as illustrated by the Mont-Louis pluton (Denèle et al. 2014) or the Lys-Caillaouas pluton (Pouget et al. 1989). The source of the magmas at the origin of these plutons has long remained uncertain, oscillating between purely crustal contribution and variable mantle involvement due to geochemical signatures that can be considered hybrid (Debon et al. 1996). The anatexis granites (leucogranites) of the Pyrenees are more modest in size, but widespread. They are essentially observable within metamorphic domes. They form decametric to plurikilometric lenticular bodies or appear as dykes. In migmatitic domains, they can be associated with small basic facies, as observed in the Gavarnie cirque or in the Aston massif (Pouget et al. 1989). The source of these basic rocks is described as mantelic, although the mechanism behind the partial melting of the mantle has long remained enigmatic and is still discussed to this day.

Until the democratization of U-Pb dating, established in the 1990s to 2000s, the emplacement of most plutonic massifs was interpreted as having taken place at the paroxysm of the Variscan orogeny (i.e. between 340 and 330 Ma), as described elsewhere in the internal zones of the chain (Massif Central, Bohemian Massif, etc.). However, new U-Pb dates have since revealed much younger emplacement ages than expected, between 310 and 290 Ma (Figure 1.1(b)), that is, at the transition between the Carboniferous and Permian. Nevertheless, some authors (Mezger 2010; Mezger and Gerdes 2016) maintain an older age, around 330–340 Ma, to some leucogranites. Still, the non-reproducibility of these results in more recent studies of the same magma bodies (Denèle et al. 2014; Lopez-Sanchez et al. 2018) (Figure 1.1(b)) leaves some doubt as to the significance of these older ages.

The partial melting of the lower Pyrenean crust is the result of a regional high temperature–low pressure (HT/LP) metamorphism, called "M2". In the Axial Zone, it is characterized in the middle and upper crust structural levels by, respectively, amphibolite and greenschist facies parageneses (Figures 1.2(a) and (b)). In the North Pyrenean massifs, the intensity of metamorphism seems more important than in the Axial Zone. Indeed, the latter are characterized by the occurrence of numerous outcrops of granulites, attesting to a temperature of the order of 800–850°C and a pressure of between 6 and 7 kbar. The main examples are located in the Ursuya, Castillon, Saint-Barthélémy and Agly massifs (Figure 1.2(c)). The apparent geothermal paleogradients, deduced from thermobarometric estimates made in different structural levels, appear anomalously high in the upper crust (up to 3–4 kbar), that is, between ~50 and 90°C/km (Figures 1.2(b) and (c)). However, they follow an adiabatic trend (apparently zero gradient) in the middle crust (700°C in the Axial Zone, 800°C in the North Pyrenean massifs) (Figures 1.2(b) and (c)). The pressure–temperature evolutions recorded by the rocks during the M2 metamorphic episode have remained very poorly constrained due to the small number of studies carried out. Most of the estimates obtained testify to a decompression synonymous with exhumation along either retrograde (Albères, Bossòst: Mezger et al. 2004; Vilà et al. 2007) or quasi-adiabatic (Roc-de-France: Aguilar et al. 2015) paths (Figure 1.2(a)). Therefore, during M2 metamorphism, high-grade metamorphic rocks derived from the middle crust were exhumed from depths ranging from about 20 km (North Pyrenean massifs) to about 12 km (Axial Zone) until they were found at the base of the upper crust, around 7–10 km deep, at the end of the Variscan orogeny (Figure 1.2(a)). Note, however, that in the Aston Massif, Mezger and Régnier (2016) conversely report that the M2 phase-related temperature rise could be accompanied by a moderate pressure rise or evolve isobarically (Figure 1.2(a)).

The age of M2 metamorphism has long remained poorly constrained, oscillating between the Middle Carboniferous (Mezger and Gerdes 2016) and Lower Permian (Delaperriere 1994; Kilzi 2014) (Figure 1.1(b)). In situ zircon dating within leucosomes in the Roc-de-France (Aguilar et al. 2014) or magmatic bodies described in the field as synchronous with migmatization, such as at Cap de Creus (Druguet et al. 2014), have refocused the M2 event between 315 and 290 Ma in the Axial Zone while evoking the possible existence of several partial melting pulses (Figure 1.1(b)). On the other hand, the evidence of zircons with inherited cores giving an age of 307 ± 3 Ma within the Lys-Caillaouas pluton dated at 300 ± 2 Ma (Esteban et al. 2015) has suggested a certain proximity or even synchronicity of M2 metamorphism with the main magmatism phase. This concomitance between migmatization and plutonism is also visible in the field, where the distinction and entanglement between the two facies is often confusing.

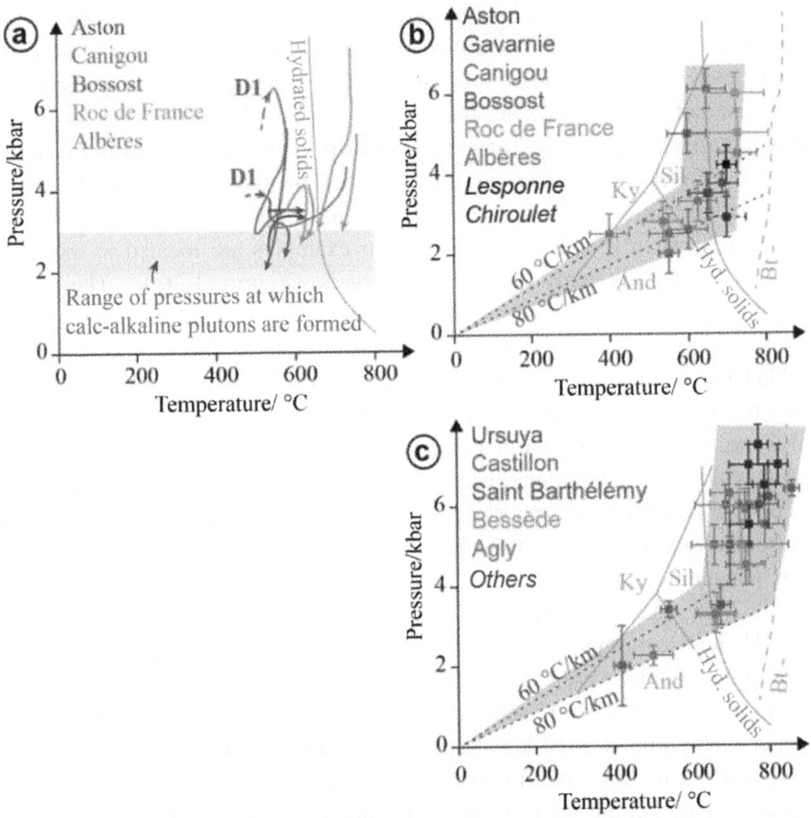


**Figure 1.2.** *P-T diagrams illustrating the metamorphic evolution*

LEGEND FOR FIGURE 1.2. *(a) From the Axial Zone. Aston: Mezger and Régnier 2016, Canigou: de Hoÿm de Marien et al. 2019, Bossòst: Mezger et al. 2004, Roc-de-France: Aguilar et al. 2015, Albères: Vilà et al. 2007, pluton emplacement estimates from Autran et al. 1970; Debon et al. 1996; Evans et al. 1998; Poitrenaud, 2018; (b) the thermal peak recorded in the Axial Zone; and (c) in the North Pyrenean massifs. Estimates of the metamorphic peak from Ayora et al. 1993; Guitard et al. 1996; Mezger et al. 2004; Triboulet et al. 2005; Vilà et al. 2007; Mezger and*

*Régnier 2016; Lemirre et al. 2019 for the Axial Zone and Andrieux 1982; Vielzeuf 1984; de Saint Blanquat et al. 1990; Mendia and Ibarguchi 1991; Giro and Alias 2007; Tournaire Guille 2017; Siron et al. 2020 for the North Pyrenean massifs.*

In the Pyrenees, the M2 metamorphic event follows an enigmatic event called M1, or "Barrovian", described as medium temperature–medium pressure (MT/MP). Geological evidence for such an early metamorphic event is rare, often found as isolated relics, found here and there in some metamorphic domes. For example, occurrences of ante-M2 kyanite have been described within the Canigou (Azambre and Guitard 2001). In a similar way, Mezger et al. (2004) described an early paragenesis attributed to M1 in the Bossòst dome. According to these authors, this M1 metamorphism would be associated with a prograde path, hypothetically related to an early thickening of the crust during the Variscan orogeny (see the following sections dedicated to tectonic events). M2 metamorphism and associated deformation have unfortunately largely erased most occurrences of the M1 event. Therefore, all attempts at an absolute dating of the M1 phase have as of yet been unsuccessful.

### 1.2.3. *Deformation and tectonics*

#### 1.2.3.1. *Main deformation*

In addition to the late Variscan magmatism and HT/LP metamorphism, the Pyrenean basement is unique in its structuring inherited from this orogenic cycle. A clear structural opposition has been demonstrated in the Axial Zone between the levels related to the upper and middle crust (de Sitter and Zwart 1960; Zwart 1979; Carreras and Capella 1994). These two domains have since been referred to as the "Superstructure" (i.e. upper crust) and "Infrastructure" (i.e. middle crust) (Figure 1.3). The Superstructure is described as a low-grade metamorphic domain, generally represented by the Upper Paleozoic series. These metasedimentary series are intensely folded and are affected by a subvertical penetrative cleavage, axial plane of the regional folds and striking parallel to the chain (Figure 1.3). Perpendicular to the chain, the Superstructure forms a half-fan of cleavage, with vertical structures to the north, and north dipping to the south (Zwart 1986). It is within the Superstructure that most of the calc-alkaline plutons described above are emplaced. The Infrastructure forms domes of high-grade rocks that appear as windows beneath the Superstructure. These domes are highlighted by a penetrative foliation and their stretching is parallel to the direction of the chain. The difference in deformation between a "ductile and rather flat" Infrastructure and a "brittle and vertical" Superstructure was conspicuous to researchers of the time (Figure 1.3).

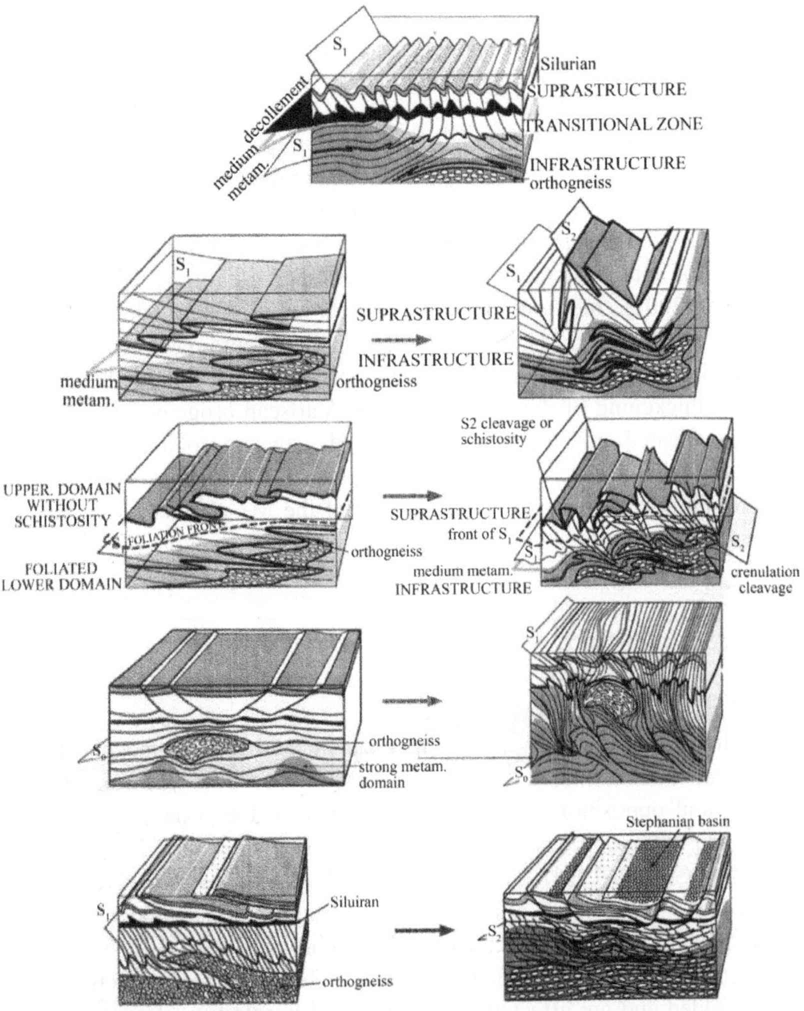


**Figure 1.3.** *Conceptual block diagrams reflecting the different formation models of the Superstructure and the Infrastructure (source: Carreras and Capella (1994))*

LEGEND FOR FIGURE 1.3. – *(a) Initial model of contemporaneous formation during the main Variscan orogenic phase (de Sitter and Zwart 1960); (b) Formation model for a single, shallowly dipping regional foliation, reworked by the late straight folds in the Superstructure (Séguret and Proust 1968; Matte 1969); (c) two-stage formation during which the foliation of the Infrastructure is early while the schistosity of the*

*Superstructure forms later (Carreras and Capella 1994); (d) early extensional model responsible for HT/LP metamorphism followed by a convergence phase associated with the formation of the Superstructure and the diapiric uplift of domes and plutons (Soula 1982; Soula et al. 1986a, 1986b). In this model, the Si and Ss foliations are formed during this last phase, within or without the domes, respectively; (e) early formation model for the shortening of the Superstructure followed by a late-orogenic extension phase responsible for the formation of the flat foliation in the Infrastructure (Verhoef et al. 1984; Van den Eeckhout, 1986).*

**Why are there so many contradictory theories about the genesis of the Suprastructure and Infrastructure of the Variscan basement of the Pyrenees?**

First, most of these studies were carried out at the local scale, that is, at the scale of a sub-massif, or even at the micro-local scale, that is, based on microscopic observations (e.g. Aerden 1994). It was therefore not always easy to place these scattered studies in a coherent regional geological framework.

Second, field analysis of the contact or transitional zone between the Infrastructure and the Superstructure is not always straightforward, which has often led to a multiplication of deformation phases to interpret the structures observed within these two structural levels. Only a detailed analysis of the structures, carried out systematically on all the domes of the Pyrenees, has made it possible to establish correlations and better interpret their different structures.

Third, it is important to place these studies and models in the scientific context of the time. In addition to the not-so-distant advent of plate tectonics theory (McKenzie and Parker 1967; Le Pichon 1968), many geologists were strongly influenced by earlier tectonic models that were very popular at the time. This is the case, for example, for the "Alpine" model on the formation of mountain ranges with "Pennine nappes" forming large crustal folds (Argand et al. 1911; Heim 1919; Ramsay 1963). This Alpine influence has been fully illustrated in the Pyrenees, as in the other ranges, through the publication of numerous studies illustrating the development of nappes and folds involving the Infrastructure (Guitard 1964, 1970; Lagarde 1978; Soliva et al. 1989; Laumonier and Autran 2001). It should be noted, however, that some of these basement folds were a "reality" at the time, as the Ordovian laccoliths were initially considered as the Cadomian basement of the Pyrenees. Their interstratification in the Lower Paleozoic sediments made them appear as basement folds.

**Box 1.1.** *Recalling the origins of the controversy concerning the formation of the different structural levels in the Axial Zone*

Indeed, the age and tectonic context for the formation of these two structural domains, as well as the relationship between deformation and metamorphism, have been important topics of debate, and even disagreement, between the 1970s and 2000 (Carreras and Cappela 1994; Carreras and Debat 1996). Historically, it was first proposed that these two superimposed domains formed synchronously, in a context of N-S convergence during the Variscan orogeny (de Sitter and Zwart 1960; Zwart 1979) (Figure 1.3(a)). Shortly thereafter, their formation was instead described as diachronous and resulting from interference between different successive tectonic phases. A first model, entirely in the context of convergence, evoked an early setting of folds – or even of basement nappes – subsequently reworked by a later folding phase (Figures 1.3(b) and (c)). In this case, regional metamorphism is described as either early (Séguret and Proust 1968; Matte 1969; Laumonier et al. 1984) or late (Carreras and Capella 1994). Other models including a phase of extension and then convergence have also been proposed (Figures 1.3(d) and (e)). For Soula (1982), Soula et al. (1986a, 1986b) and Pouget (1991), the thermal anomaly responsible for the HT/LP metamorphism results from an early rifting phase followed by a late N-S convergence phasc (Figure 1.3(d)). In this model, the emplacement of domes and plutons is explained by a strong viscosity contrast between low viscosity magmas from the lower crust and a highly viscous sedimentary host (diapir model). For Van den Eeckhout (1986), Van den Eeckhout and Zwart (1988) and Vissers (1992), on the contrary, it is a postorogenic extension that would explain the formation of the Infrastructure (Figure 1.3(e)).

The development of modern structural geology in the orogenic domain has allowed a quantum leap in the understanding of the deformation of the Pyrenean basement, in particular thanks to the systematic studies on all the plutons, and their country-rocks, since the 1990s. Therefore, the study of magmatic, gneissic fabrics and magnetic susceptibility anisotropy in the plutons has made it possible to highlight the synchronous character of their emplacement with the main deformation phase identified in the upper crust (Messaoudi 1990; Bouchez and Gleizes 1995; Leblanc et al. 1996; Evans et al. 1997; Gleizes et al. 1998a; Auréjac et al. 2004). Most of these authors concluded that the deformation was transpression with a dextral component, in a context of regional N-S shortening (Gleizes et al. 1998b). It has been shown that transpressive deformation is heterogeneously distributed via a process of strain localization within mylonitic corridors up to kilometer thick (Carreras 2001). In the Superstructure, this is reflected in the development of south-verging thrusts, which are mainly rooted in a decollement layer – mostly the Silurian black shales (García-Sansegundo 1991; Carreras and Debat 1996). These mylonitic zones have been described as having reverse-dextral kinematics. They form a longitudinal anastomosed network parallel to the regional schistosity, wrapping and even intersecting the metamorphic domes and granitic plutons (Carreras et al. 1980; Carreras and Druguet 1994; Carreras 2001). These appear to have been activated following a retrograde path from the amphibolite facies to the greenschist facies (Carreras et al. 1977; Bouchez and Gleizes 1995; Denèle et al. 2008). Where these regional mylonitic zones intersect the plutons, the result is a progressive strain localization from a magmatic to a solid state.

On the other hand, the study of the metamorphic domes of Aston, Hospitalet and Bossòst in the 2000s has allowed us to specify their mode of formation. The deformation into amphibolite facies is reflected in the formation of a subhorizontal penetrative gneissic foliation accompanied by a globally longitudinal, or WNW-ESE stretching (Mezger and Passchier 2003; Mezger 2005; Denèle et al. 2007, 2009), as suggested in older studies (Guitard 1960; Van den Eeckout 1986). It is illustrated by the recrystallization, reorientation or neoformation of the main mineral phases (alumina silicates, quartz, feldspars, calcite) parallel to this direction of stretching. The formation of these ductile fabrics reflects a vertical thinning of the crust and is accompanied by the formation of shallowly dipping shear bands (C' planes). From one massif to another, these shear zones may exhibit antithetic kinematics, that is, the displacement eastward or westward of the upper blocks on the C' planes (Van den Eeckout 1986; Mezger and Passchier 2003; Denèle et al. 2007, 2009; Mezger 2009). Locally, this vertical thinning can occur in pure shear (i.e. coaxial deformation) (Mezger and Passchier 2003; Mezger 2005, 2009). This global dynamic was interpreted by these authors as a lateral flow phase of the partially molten lower crust in response to regional N-S shortening.

These Pyrenean metamorphic domes have the peculiarity of having very steep septentrional and meridional borders. It was partly for this reason that they were interpreted by some geologists as diapiric structures (Soula et al. 1986a; Pouget et al. 1989). Most often, their southern flank appears in an overturned position (Denèle et al. 2007; Vilà et al. 2007). These regional E-W folds appear to be the result of a N-S shortening phase and are shown to be contemporary with the formation of the previously described reverse-dextral mylonitic zones (Carreras et al. 1980; Denèle et al. 2008, 2014). This folding phase was interpreted as responsible for the formation, or at least the amplification, of these dome-shaped structures (Denèle et al. 2014). This late phase of folding is referred to as the "doming" phase (or D3 and D2b depending on the author). Indeed, like middle crustal flow, the "doming" phase is also consistent with a dextral transpressive regime in a N-S convergence context. For this reason, for the majority of the scientific community, the deformation phase responsible for the major structuring of the two structural domains of the Pyrenean basement results from a late Variscan (310–300 Ma) dextral transpressional phase called D2.

The situation in the North Pyrenean massifs seems to be more complex than that described in the Axial Zone. Similarities between these different massifs have been highlighted, such as the structure and the emplacement mode of the main plutons. As with the Axial Zone, an emplacement under dextral transpression has been proposed (Leblanc et al. 1996; Olivier et al. 2008). The structural opposition between Infrastructure and Superstructure has also been documented (Soula 1982). Nevertheless, clear differences seem to emerge with respect to the Axial Zone, especially in the deepest crust structural levels. In addition to the fact that the intensity of HT/LP metamorphism appears to be greater there (Figure 1.3), the flow dynamics of the middle crust differs in part from that observed in the Axial Zone. Indeed, most of the known stretching lineations in the anatectic domains of the North Pyrenean massifs follow a N-S direction (Carreras and Debat 1996). The shear criteria are all compatible with N-S extension during which detachment zones were activated, allowing exhumation of the middle to lower crust (de Saint Blanquat et al. 1990; Bouhallier et al. 1991; Olivier et al. 2004), as in the Saint-Barthélémy massif (de Saint Blanquat 1993). The link between N-S extension and dextral transpression in response to N-S shortening is therefore not immediately obvious. This is partly related to the lack of geochronological and thermobarometric constraints on the extensional event. Rare studies would place it either synchronous with the transpression or slightly later (Olivier et al. 2008; Kilzi 2014). It should also be added that, for a number of authors, part of the extensive N-S ductile deformation observed in the North Pyrenean and Basque massifs could also be related to mantle exhumation during Cretaceous rifting (de Saint Blanquat et al. 1990; Jammes et al. 2009; Clerc and Lagabrielle 2014).

#### 1.2.3.2. *Early deformations*

The D2 phase follows an earlier deformation phase called D1. The latter remains poorly constrained due to the intensity of the D2 deformation and the associated M2 metamorphism. Indeed, the D2 deformation phase has largely obliterated the earlier structures from the D1 phase. Nevertheless, scatered evidence has been documented in both the Superstructure and Infrastructure. In the Superstructure, it is mainly evidence of folding prior to D2 that has been described. In this case, these folds have a WNW-ESE to N-S trace and may be open or recumbent (Müller and Roger 1977; Laumonier and Autran 2001). The P1 folds affect the Devonian formations up to the last turbiditic deposits of the Culm facies, dated to the Upper Carboniferous (325–316 Ma; see Delvolvé et al. (1998)). These folds therefore formed between the deposition of the Culm and the D2 deformation phase, that is, between 325 and 310 Ma. In contrast, Matte (2002) suggests that these N-S trending folds were generated during the D2 phase. In the Infrastructure, the attribution or not of certain structures to a D1 deformation phase depends intrinsically on the tectonic models historically proposed by certain authors. Indeed, the dominant foliation in the metamorphic domes was, for a long time, interpreted as the result of a D1 deformation phase that would have developed around 340 Ma. The latter would have been linked to the development of basement nappes (called "tangential tectonics") and pennic folds with a south or south-west vergence (see above). The recent assignment of this dominant foliation to the D2 episode, synchronous with the M2 magmatic and metamorphic event, has finally invalidated this initial interpretation. Nevertheless, recent observations have attributed local structures to a D1 phase of south-verging nappes, as with the Aston Dome (Denèle et al. 2009). Occurrences of kilometer-long recumbent folds in the Infrastructure were also associated with the D1 deformation phase, for example, in Hospitalet (Clariana and García-Sansegundo 2009), although the regional foliation is an axial plane of these folds. In the Infrastructure, evidence of early D1 deformation was always associated by default with the Barrovian M1-type metamorphism described above.

### 1.2.3.3. *Later Paleozoic deformations (Permian)*

The Paleozoic basement of the Pyrenees was rapidly dismantled as soon as the D2 transpressional phase ended. This resulted in the establishment of narrow continental sedimentary basins during the Gzhelian (formerly Stephanian), then the Cisuralian (formerly Autunian). The deposits are essentially detrital with an accumulation of thick conglomeratic and sandstone series at the main escarpment areas. All the deposits are clearly unconformable on the Paleozoic basement and are deposited on the folded and schistosed devono-carboniferous series (Mirouse 1962). Intense magmatic activity accompanied the formation of these basins and resulted in the emplacement of volcanics (basaltic, pyroclastic flows) or numerous dykes (Lago et al. 2004a; Rodríguez-Méndez et al. 2014). The most spectacular remnant of this magmatic event is undoubtedly the Ossau paleocaldeira, whose andesites and rhyolites currently form the Pic du Midi (Bixel and Lucas 1983; Bixel 1984; Ternet et al. 2003). Magmatic activity decreases in intensity during the Permian, with only episodic emplacement of a few basaltic lava flows (Pereira et al. 2014) and rare plutons (Aya pluton; see Denèle et al. (2012)). Detrital deposits also evolve into more pelitic terms and basins tend to widen during the Permian. The formation of all these basins results from intense tectonic activity materialized through the activation of numerous normal and/or strike-slip fault systems. The most numerous have an E-W and NW-SE strike, while others present a N-S to N30° strike (Soula et al. 1979; Lucas 1985; Speksnijder 1985; Lago et al. 2004a; Rodríguez-Méndez et al. 2016). While the basins bordering the Axial Zone are mainly oriented parallel to the longitudinal faults, those in the Basque Country instead seem to be controlled by the N-S fault system (Lucas 1985; Saspiturry et al. 2019a). These different, synchronously activated fault systems have been interpreted as a network of Riedel fractures resulting from a dextral transtensional regime.

### 1.2.4. *Location(s) of the Pyrenean segment in the Variscan chain*

Historically, the Paleozoic deformation phases described in the Pyrenean basement have always been linked to the geodynamic evolution of the internal zones of the West European Variscan range. The evidence of HT/LP metamorphism, probably Carboniferous, and the omnipresence of compressive structures, essentially southern vergents, have often led to the interpretation of the Pyrenees as the southern continuity of the nappes system described in the Massif Central (Faure et al. 2009). Therefore, for many authors, the Pyrenees occupied an internal position in the Variscan range (Zwart 1979) and also seemed to have been affected by a gravitational collapse event (Van den Eeckhout and Zwart 1988; Kriegsman et al. 1989; Bouhallier et al. 1991; Gibson 1991; Vissers 1992; de Saint Blanquat 1993; Aerden 1994, 1995), as described elsewhere in the internal zones (Vanderhaeghe et al. 1999; Martínez Catalán et al. 2014; Gapais et al. 2015). Nevertheless, the deposition of thick turbiditic sequences that are considered synorogenic, between the Namurian and the Westphalian, leads to the external placement of the Pyrenees in the Variscan range (Barnolas and Chiron 1996; Delvolvé et al. 1998). The deposition of these "Culm" series appears to have propagated from the northeast to the southwest. The Pyrenees have thus been reinterpreted as a transition zone between the internal zones, located in the northeast, and the external zones of the range, located in the southwest (García-Sansegundo 1991; Matte 2002; Clariana and García-Sansegundo 2009; Carreras and Druguet 2014). The eastern part of the Axial Zone and the North Pyrenean massifs would, according to them, form the inner zones of the Variscan chain, as evidenced by the presence of numerous metamorphic domes. Recent fundamental advances in our understanding of late Variscan tectonometamorphic events in Iberia have made it possible to propose an alternative view to those previously described (see below). Drawing parallels between late- to postorogenic Iberian and Pyrenean tectonometamorphic events, Denèle et al. (2014) proposed that the Pyrenees, as a whole, underwent the same process of abrupt temperature rise at the base of the continental crust in response to postorogenic delamination of the underlying lithospheric mantle. Since the Pyrenees should be considered as part of the foreland of the Variscan belt during the collisional events that affected the inner zones between 360 and 320 Ma (Franke 2014) (Figure 1.4). The Variscan Pyrenees would then be distinct from a classic late-orogenic evolutionary model because of the M2 metamorphism, magmatism, and deformation associated with the D2 deformation phase that characterize them.

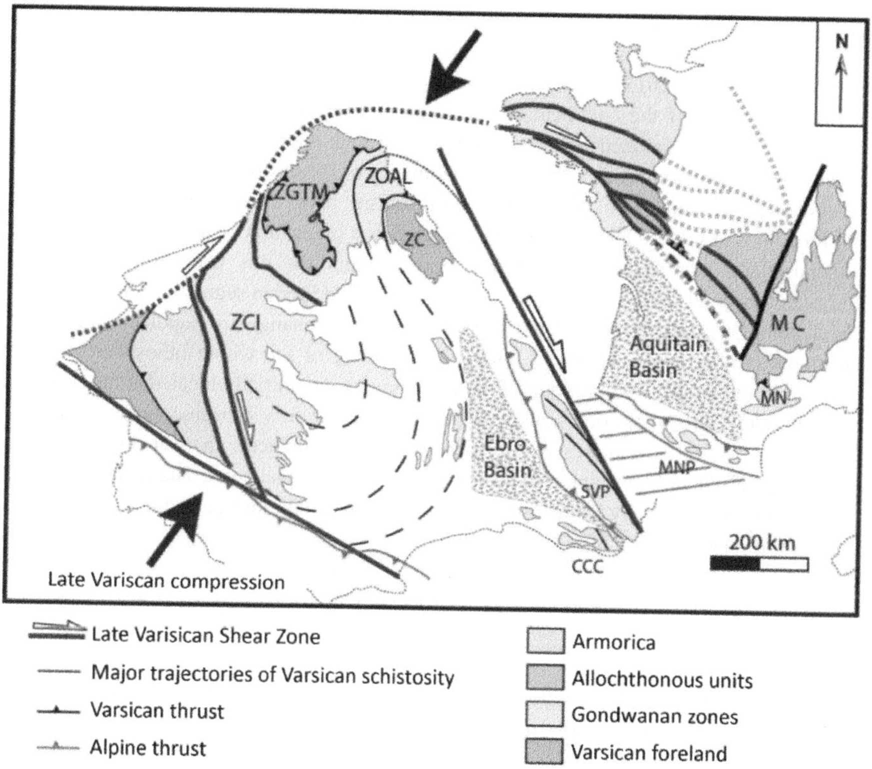


**Figure 1.4.** *Structural map of the Variscan chain in France and Spain at the end of the Carboniferous (source: Denèle et al. (2014))*

LEGEND FOR FIGURE 1.4.– *The main crustal structures formed at this time are the Cantabrian orocline and the main Late Variscan crustal-scale strike-slip faults. Abbreviations: CCC: Catalan Coastal Range, MC: Massif Central, MN: Montagne Noire, MNP: North Pyrenean Massifs, SVP: Variscan Pyrenean Segment, ZC: Cantabrian Zone, ZCI: Central Iberian Zone, ZGTM: Galicia-Trás-os-Montes Zone, ZOAL: West Asturias-León Zone.*

**How were the late-early oroclines of the West European domain formed?**

The Variscan chain results from the amalgamation of numerous continental blocks and microblocks from the Devonian to the Carboniferous. Following a phase of continental collision, the chain underwent, in its European part, a late reorganization of these orogenic structures around 310–290 Ma. This resulted in the activation of crustal strike-slip faults and intense folding of its structure, leading to the formation of oroclines

(Figure 1.4). The most spectacular example is undoubtedly the Cantabrian orocline, resulting from a cumulative 180° rotation of the Iberian-Armorican arc between 305 and 295 Ma (Weil et al. 2013). This late phase is also accompanied by active exhumation of the deep domains of the continental crust in an anomalously hot thermal context. This episode is clearly illustrated by the development of high-temperature metamorphism and intense bimodal magmatism in the southern domains of the chain (Iberia, southern Massif Central, among others). In Iberia, the synchronism between this thermal event and the formation of the Cantabrian orocline has led to the invocation of the major role of the asthenospheric mantle. A model proposed to account for this late evolution of the Variscan orogen at lithospheric scale is the folding of the Iberian segment of the Variscan chain along a vertical rotation axis in response to a change in regional shortening directions (Gutiérrez-Alonso et al. 2004). This folding of the whole lithosphere would have led to the intense compression of the Cantabrian zone, initially located at the edge of the belt. The Cantabrian zone now forms the core of the new orogenic syntaxis. In the Cantabrian zone, this compressive episode would have induced an over-thickening of the lithospheric mantle, forcing it to detach from the continental crust. The replacement of the lithospheric mantle by asthenospheric mantle would therefore be at the origin of the drastic heating of the base of the crust at this period.

**Box 1.2.** *Genesis of West European late Variscan oroclines*

### 1.2.5. *Line of scientific questioning guiding recent research in the framework of the RGF Pyrenees project*

At the end of the Variscan orogeny, the Pyrenees seem to have undergone a singular tectonometamorphic evolution as an external domain of a mountain range. Contradictory visions clash, oscillating between the first tectonic models that emphasize the construction through a Variscan orogenic perspective, and the more recent ones that take into account the late to postorogenic evolution of the Iberian basement, to which the Axial Zone is attached. These oppositions are essentially due to a lack of global vision concerning the deformations at the scale of the Pyrenees, mainly due to localized and fragmented studies. Certain massifs or domains have attracted the majority of scientific interest (e.g. the domes of the eastern Pyrenees, the plutons), while others have remained understudied (e.g. the Superstructure). In addition, the multitude of observations obtained, sometimes redundant, have been interpreted differently according to the sensitivity of the different authors (e.g. concerning the deformation phases). The inclusive attempts presented in Barnolas and Chiron (1996) did not lead to a scientific consensus because there was no alternative to the dominant models of the time. The lack of geochronological constraints on the age of the metamorphism was also a major hindrance to the development of a robust global deformation model. The uncertainties surrounding a

D1 tectonic event and M1 metamorphism associated with crustal thickening also contributed to confusion within the scientific community.

Moreover, the mechanical aspects of the deformation have often been neglected in the proposed tectonic models, while many studies have shown that the deformation styles recorded in the mountain ranges depend directly on the rheology of the lithosphere, contrasting the rigid lithosphere favorable to the emplacement of nappe sheets (Brun 2002) with the warm lithosphere, where the deformation is instead homogenous and distributed, as in the Archean (Chardon et al. 2009). The diapiric emplacement model of domes and plutons by Soula et al. (1986a) and the transpressional lateral flow model for domes proposed by Mezger and Passchier (2003) and Denèle et al. (2007) have allowed the impact of rheology on crustal structuring in the Pyrenees to be re-established, notably through the introduction of the notion of vertical strain partitioning, a mechanical principle that had never been taken into account before in the Pyrenees.

**How does deformation accommodation by partitioning work?**

As illustrated by plate tectonics, the continental lithospheres can slide at their boundaries, separate (divergence zone), or collide (convergence zone). In nature, the divergence or convergence of plates is very rarely perfect (orthogonal to the boundary zone between the two plates), but presents a more or less marked degree of obliquity. This is referred to as transpression in the case of oblique convergence and transtension for oblique divergence (Fossen et al. 1994). In such contexts, the induced deformation can be considered as the sum of two components: (1) a pure shear component (shortening or stretching) perpendicular to the plate boundary and (2) a simple shear component (strike-slip) parallel to the plate boundary.

These two components can be accommodated homogeneously by the entire lithosphere zone to which they are applied and are referred to as homogeneous deformation (Figure 1.5(a)). The deformation can also be distributed into different domains, each accommodating all or part of one of the components (Figure 1.5(b)). The deformation is then said to be partitioned into different domains (Teyssier et al. 1995). Taking into account the illustrations in Figure 1.5(b) as representing a plate boundary zone seen in map view, this corresponds to a case of lateral (or horizontal) partitioning of the deformation. Viewed in cross-section, deformation affects the lithosphere throughout its thickness, and not just within its superficial domains. The accommodation of deformation by partitioning can therefore also take place vertically, that is, between different structural levels. Of course, in a natural system, deformation can be accommodated by partitioning both laterally and vertically.

**Box 1.3.** *Recalling deformation mechanisms by partitioning*

Therefore, most of the recent work on the Pyrenean basement has focused on (1) the acquisition of new structural data in order to complete our clearly incomplete vision of the architecture of the Superstructure and the Infrastructure, (2) the structural relationship between these two domains, (3) the consideration of the rheology of the lithosphere and a potential partitioning of the deformation to be considered in tectonic models, and (4) the obtaining of new geochronological constraints and the estimation of the pressure–temperature conditions of the M1 and M2 metamorphic events. The main objective was to obtain an integrated view of the tectonometamorphic events that affected the Paleozoic basement of the Pyrenees and place them in a coherent geodynamic framework, taking into account those illustrated in the neighboring Variscan massifs.

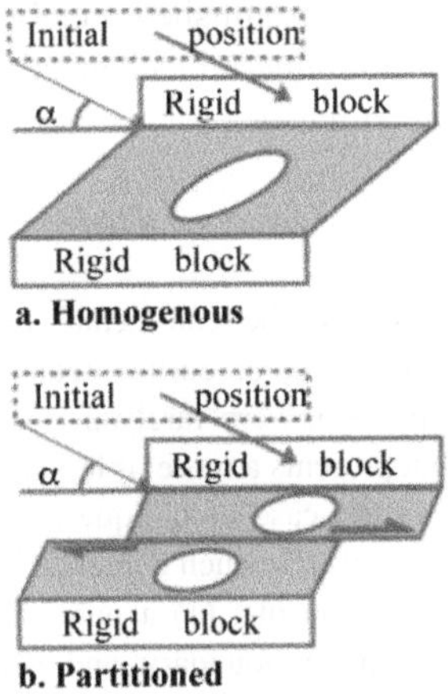


**Figure 1.5.** *Illustration of the theoretical transpressional deformation mode of a boundary zone between two rigid plates. (a) Example of homogeneous deformation and (b) partitioning of the deformation between domains of homogeneous deformation with almost pure-shear (shaded areas) and domains of localized simple shear (fault here). Modified from Tikoff and Teyssier (1994)*

The scientific implications of studies on the Pyrenean basement go far beyond our understanding of the Variscan chain and also touch on the Meso-Cenozoic history of the Pyrenees. Indeed, the Paleozoic layers structured during the Variscan cycle constitutes, as early as the Triassic, form the basement of the sediments linked to the Alpine cycle. This basement was dismantled during the Cretaceous rifting phase, then involved in the construction of the Pyrenean orogenic prism following the Eocene rift inversion. This last phase is supposed to have had a strong impact on the Pyrenean basement via the establishment of numerous nappes, the most famous of which is the Gavarnie nappe. Following the acquisition of the ECORS seismic profile in 1989, imaging the deep structure of the chain, the Pyrenees were described as a doubly vergent orogenic prism where numerous basement nappes are stacked (Roure et al. 1989; Muñoz 1992). These nappes are supposed to be folded in the form of anticlines, and this folding must be accompanied by penetrative deformation in the basement (Beaumont et al. 2000). On the other hand, nappe emplacement in the core of the Axial Zone must imply the presence of significant stratigraphic and/or metamorphic offsets. However, there has been no consensus on these points since the 1990s. Since then, studies of the Variscan cycle and the Alpine cycle have largely been detached from one another, and the study on the structuring of the Pyrenean chain has remained essentially limited to structural studies of the basement. Although the impact of the structural and thermal legacy of the Cretaceous rifting phase on the structuring of the Pyrenean orogenic prism was quickly recognized, the impact of Variscan structures remained under-documented. The recent studies carried out within the framework of the RGF Pyrenees project have therefore been concerned with placing the knowledge on the basement (which forms the major part of the crustal volume involved in the Pyrenean orogenic prism) back into the establishment of mountain belt formation models.

## 1.3. Variscan tectonics

In this section, the main structural characteristics of the Pyrenean basement are presented, based upon recent field studies. The following analysis is also based on an exhaustive synthesis of structural objects (foliations, schistosities, folds, lineations) from the bibliography and French and Spanish geological maps (Cochelin 2016). These data have been harmonized at the scale of the Pyrenees in order to define a coherent and global scheme. We will detail the structural characteristics of the main structural domains and illustrate the transition between the Infrastructure and the Superstructure through three examples: the Canigou, Bossòst and Chiroulet-Lesponne massifs. These massifs reflect the diversity of structures identified in the Axial Zone. For the sake of clarity, from here on we will use a new terminology concerning the fabrics resulting from the main deformation phase (i.e. D2) according to their respective structural level. Therefore, the main foliation in the gneissic domes forming the Infrastructure will be called Si and the associated stretching/mineral lineation Li. Similarly, we will refer to the regional schistosity in the Superstructure as Ss and the associated stretching lineation as Ls.

### 1.3.1. *Main deformation D2 phase: generalities*

#### 1.3.1.1. *Structure of the basement*

##### 1.3.1.1.1. Global architecture: the Superstructure

The regional Ss schistosity of the Superstructure and the regional mylonitic zones that control the overall architecture of the Axial Zone are striking between N90° and N100° (Figures 1.6 and 1.7(a)). The mylonitic zones outline an anastomosing network enveloping low-strain lens-like domains, as well as plutons and domes. These mylonitic zones are mainly oriented N100°, or even N130° when they develop at the eastern or western ends of a dome or pluton (Figure 1.6). They are essentially subvertical or dip about 50–60°, mainly to the north (Figures 1.7(a) and 1.8). The mylonitic foliations carry steeply west-plunging (or high pitch) stretching lineations, giving them a NW to N trend (Figure 1.7(b)). The kinematics associated with these mylonitic zones are systematically reverse with a dextral component.

This dextral component tends to increase as the mylonitic zones diverge from the dominant N100° strike, until it becomes dominant in N130° striking mylonitic zones (Figure 1.6). This tendency has been well described in studies of granitic plutons and their contact aureole (Gleizes et al. 1998b), and some domes in the eastern Pyrenees (Carreras et al. 1980). In this case, strain localization in the plutons results in the progressive acquisition of a magmatic fabric until the stage of mylonites forming N100 to N130° striking corridors. The retrogressive state of deformation from magmatic stages to greenschist facies is then illustrated by the progressive appearance of chlorite. It is illustrated in this case by the development of C/S fabrics, typical of coeval deformation and cooling of plutons since the magmatic stage (Gapais 1989). These mylonitic zones can be active until the brittle-ductile transition, with the development of tectonic breccias (cataclases). When this fracturing is accompanied by the circulation of fluids from the partially molten crust at depth, this can lead to the development of ore deposits rich in rare metals (rare earths, gold, among others), as observed at Salau in Ariège (Poitrenaud 2018). This late stage of fracturing in the late Variscan mylonitic zones has been estimated at 289 ± 2 Ma via the dating of hydro-thermal minerals (apatites) associated with tungsten mineralization (Poitrenaud et al. 2019).

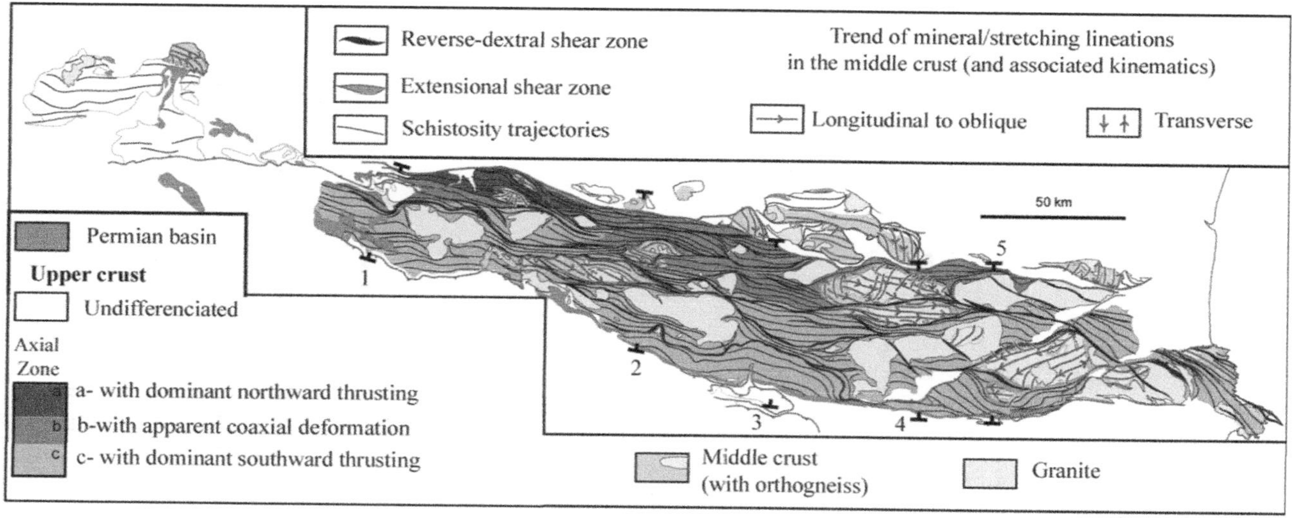


**Figure 1.6.** *Simplified strainfield of the Paleozoic basement of the Pyrenees, modified from Cochelin et al. (2017). The deformation dynamics of the crust are illustrated in the upper crust by the schistosity trajectories Ss and by Li stretching lineation trajectories in the middle crust. In this figure, the cross-sections are shown in Figure 1.8*

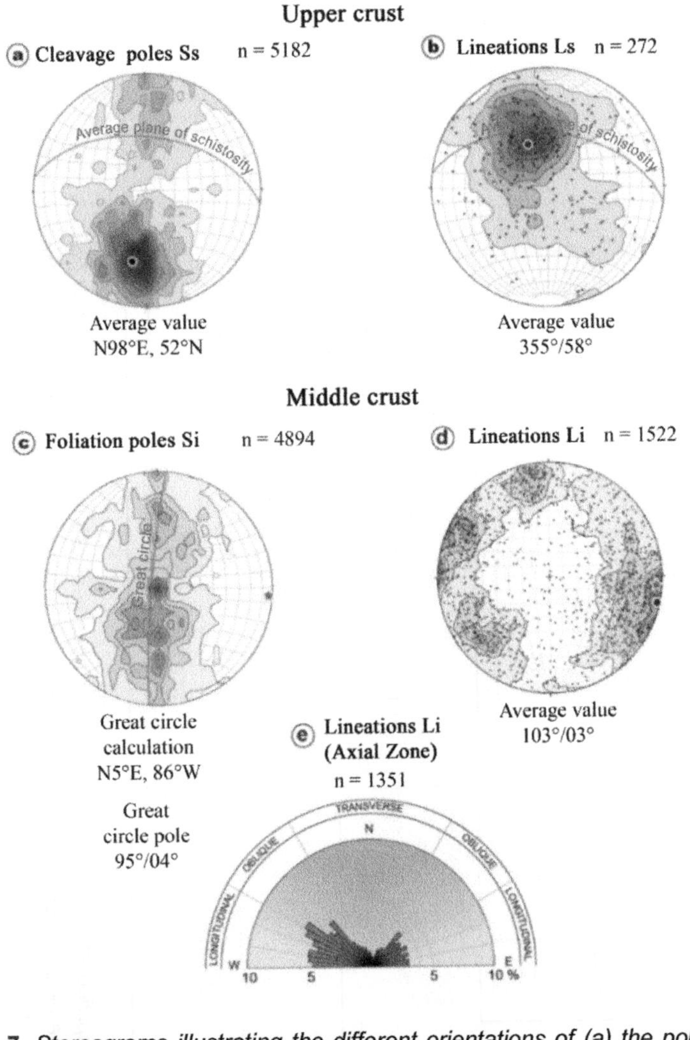


**Figure 1.7.** *Stereograms illustrating the different orientations of (a) the poles of the Ss schistosity planes; (b) the Ls stretching lineations; (c) the poles of the Si foliation planes; and (d) the Li stretching lineations in the Pyrenees basement. (e) Rose diagram illustrating the preferential orientation of the Li stretching lineations only in the Pyrenees Axial Zone. Figures modified from Cochelin et al. (2018b)*

The mylonitic foliation is often concordant with the regional Ss schistosity in areas of lower strain domains. The Ss schistosity paths (Figure 1.6) are actually parallel to the map traces of the P2 folds, forming the axial plane (Figure 1.9). This schistosity draws a double fan in the Axial Zone, with a northern domain where the schistosity dips to the south, and a southern domain where the schistosity dips to the north (Figure 1.6). The core of the Axial Zone has a predominantly vertical schistosity (Figure 1.8). The northern domain only appears in the northwestern part of the Axial Zone, because it is obliterated eastward by the North Pyrenean fault, partly oblique to the Variscan structure (Figure 1.6). The Ss schistosity can become subhorizontal in the southern border of the Axial Zone due to a generalized tilting of the basement involved in the emplacement of the Alpine nappes (Figure 1.8). This Alpine tilting is well illustrated by the Permo-Triassic series unconformably overlying the basement. Indeed, these series are tilted to the south and are almost verticalized at the front of the Alpine nappes in the Pyrenees. Their untilting (back to horizontal: Pyrenean ante-compression state) allows us to estimate that the dip of the Ss schistosity must have initially been about 50–60° toward the north in these southern sectors. These values are similar to the average value of the Ss schistosity dip in the rest of the Axial Zone (Figure 1.7(a)). The Ss schistosity yields a stretching lineation with similar characteristics in every respect to those observed in the mylonitic zones described above (Figure 1.7(b)). A well-marked NNW-SSE to N-S stretching direction is observed. In domains with vertical schistosity, stretching is also vertical. The movements associated with the development of shear zones are mainly vertical, with a minor dextral component. Upward movement of northern blocks is generally observed on these steep shear planes (Figure 1.8). In areas where the schistosity dips moderately, a divergence of shear directions is observed. The southern domains are characterized by southward thrusting movements, as historically described in the literature (García-Sansegundo 1991; Carreras and Debat 1996). On the other hand, in the northern domains, this trend is reversed, as evidenced by the presence of northward-verging thrusts (Figures 1.6 and 1.10). In these southern and northern domains, the regional folds form drag folds that are compatible with the vergence of the thrusts, that is, they dip outwards from the Axial Zone. In the central domain, the regional folds are often rather open and upright, suggesting a coaxial (non-rotational) deformation instead. The deformation is relatively homogeneous in the Superstructure domain, when observed on a large scale. In detail, it appears that the core of the Axial Zone is more deformed than its peripheral domains. This is reflected in frequent transposition of the bedding by the Ss schistosity, while the stratigraphy is clearly oblique to the schistosity in the south of the Axial Zone.

##### 1.3.1.1.2. Middle crust windows: the Infrastructure

By definition, metamorphic domes are deep crustal units that have been exhumed to the level of the Superstructure described above. Analysis of metamorphic domes thus give us valuable clues as to the deformation mode in the middle crust. From a geological point of view, these domes can be divided into two categories. Those of the eastern Pyrenees are the most voluminous (Aston, Hospitalet, Canigou). These are emplaced in the lower Paleozoic series and their roofs coincide more or less with those of the Ordovician granitoid laccoliths (Figure 1.6). The metamorphic domes of the central and western zones of the Axial Zone are smaller and incorporate a variable part of the upper Paleozoic series (Bossòst, Chiroulet and Lesponne). This difference illustrates a differential erosion of the Axial Zone between the eastern Pyrenees and the western Pyrenees, where the Axial Zone is buried under its well-preserved Mesozoic cover. The most extensive and developed domes observed in the eastern Pyrenees give us direct information on the mechanical behavior of the deepest domains of the Variscan continental crust. The metamorphic rock outcrops of the western zones of the Axial Zone give us precious clues on the dynamics of the transition level between the Infrastructure and the Superstructure. They all show a generally E-W map-view stretching, parallel to the dominant structural orientation of the Axial Zone described in the previous section (Figure 1.6). The Si foliation measurements are predominantly shallowly dipping. All their poles cluster along a N-S girdle, illustrating the folding of the domes along an E-W axis (Figure 1.7(c)).

About 90% of the stretching and mineral lineations Li have a trend sub-parallel to the elongation trend of the domes. Indeed, lineations strike between N70° and N120°E (Figure 1.7(e)). We will therefore describe them as longitudinal (Figures 1.6 and 1.7(e)). Most of these lineations are preferentially oriented N90° to N120°E in the Axial Zone, with an average of N98°E (Figure 1.7(e)). Those oriented N70-N90°E are restricted to the Canigou and Roc-de-France domes (Figure 1.6). Longitudinal stretching is also characterized in part of the North Pyrenean massifs, such as in the Arize, the Trois-Seigneurs and the Ursuya (Figure 1.6). This longitudinal stretching is accompanied by the formation of penetrative C' shear bands, characterized by directions of movement parallel to the stretching lineation Li. Quite often, one shear direction is dominant in each massif, but this varies from one massif to another. For example, shearing toward the E to ESE is located in the Aston, Hospitalet, Bossòst, Ursuya and, to a lesser extent, the Lys-Caillaouas massifs (Figure 1.6). Conversely, westward shearing is measured in the Canigou (and Nuría sub-dome, see next section), Chiroulet and Lesponne domes. Antithetic shears can be observed locally (e.g. at the eastern and western terminations of the domes) (Figure 1.6), while coaxial deformation can affect their cores (Cochelin et al. 2017). The C' shear planes are frequently injected with leucosomes from partially melted surrounding rocks. In this case, the leucosomes exhibit a magmatic fabric. These observations attest that shearing was synchronous with partial melting.

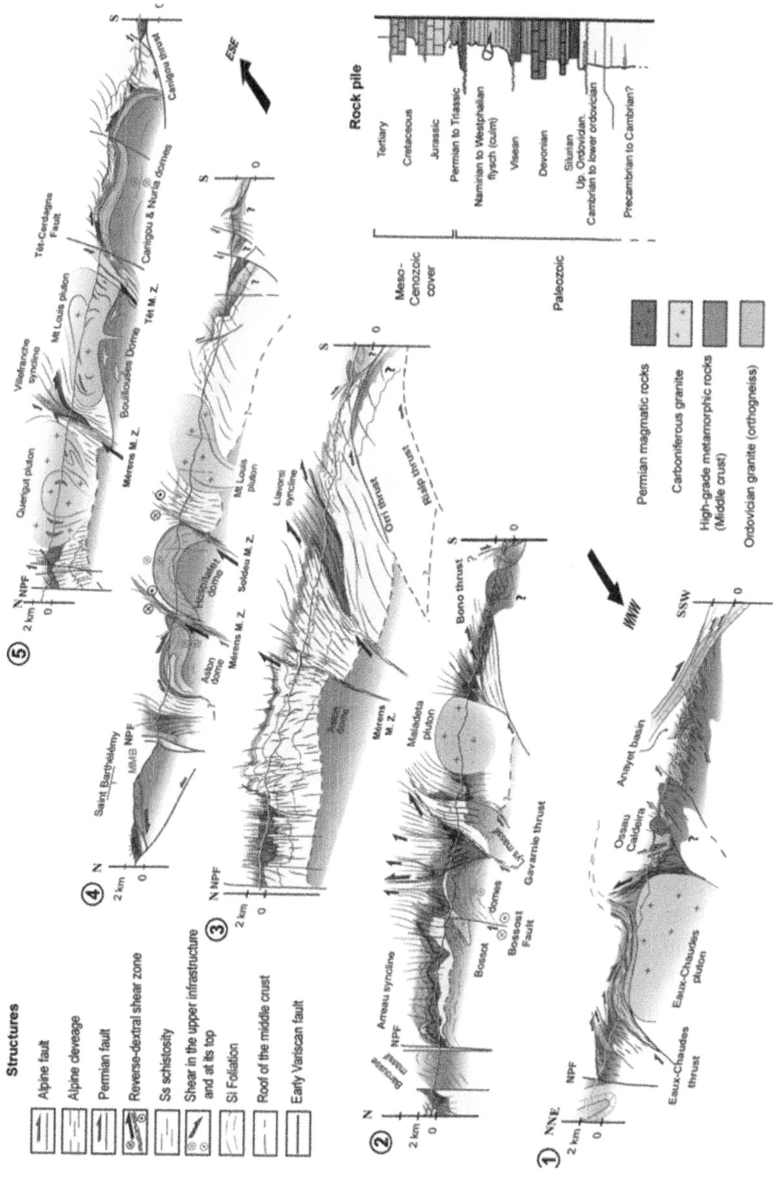


**Figure 1.8.** *Geological cross-sections drawn through the Axial Zone of the Pyrenees. See location in Figure 1.6. M. Z. = mylonitic zone. From Cochelin et al. (2017)*

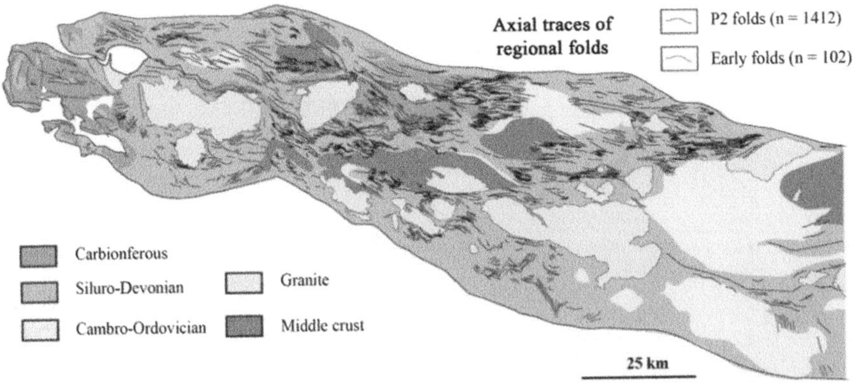


**Figure 1.9.** *Structural map of the western part of the Axial Zone, on which the axial traces of the regional folds observed in the Superstructure are represented*

Another family of stretching lineations can be observed in the Infrastructure. These are lineations with a very oblique or even orthogonal trend to the E-W global structure of the chain (Figures 1.6 and 1.7(e)). We will call these "transverse lineations". These are observed (1) at the top of many domes of the Axial Zone and (2) in the North Pyrenean massifs. With the exception of the very few areas affected by a predominantly coaxial deformation (e.g. the northern Bossòst dome), transverse and longitudinal lineations are not observed in the same structural levels. Therefore, in the Axial Zone, the transition between longitudinal and transverse Li-stretching lineations can occur in two ways. The first consists of a progressive evolution from the core to the roof of the dome via a gradual divergence of lineations. This is particularly the case in the Canigou, Hospitalet and Bossòst massifs (Figure 1.6). A similar trend can be observed in the Basque massif of Ursuya, which we will detail in section 1.3.3, due to an important imprint of Permian tectonics. On the other hand, the transition between longitudinal and transverse lineations can be abrupt. This is, for example, the case in the Aston or Chiroulet massifs, where the transition seems to be related to important rheological changes within the domes. In Aston, it seems to coincide with the accumulation of melt products at the base of the Ordovician laccolith. In the Chiroulet, it is at the transition between the migmatitic core of the dome and its very ductile envelope (marbles) that the change takes place (Figure 1.10(a)). The North Pyrenean massifs of the Castillon, Saint-Barthélémy and Agly have the particularity of presenting only transverse lineations, whatever their level in the Infrastructure (Figure 1.6). Areas with transverse lineations are affected by penetrative C' shear bands showing often extensive kinematics, with a clear dominance of southward shearing, both in the Axial Zone and in the North Pyrenean massifs. Nevertheless, northward shearing is observed, for example on the northern flank of the Canigou, in the Agly or in the Aston. These shear bands also develop in the amphibolite facies and can even operate in the greenschist facies. Indeed, leucosomes are sometimes observed as injected veins along these shear planes, attesting to their coeval activation with the partial melting phase. Sometimes, leucogranites show well-developed C/S fabrics, illustrating their deformation during cooling (Cochelin et al. 2017).

This dynamic is accompanied by significant strain localization at the top of the metamorphic domes, leading to the formation of extensional mylonitic zones of hectometric to kilometric thickness (Figure 1.6). These mylonitic bands are located at important rheological transitions, either related to lithological heterogeneities of the protolith (orthogneiss versus metasediments, sandstone versus pelites), or induced by the intensity of M2 metamorphism (difference in behavior of the main mineral phases). Therefore, these mylonitic zones are mainly observed in the stability domains of andalusite and biotite. In particular, they coincide with the attenuation of isogrades (Gibson and Bickle 1994; Mezger and Régnier 2016; Le Bayon and Cochelin 2020; Siron et al. 2020). All of this illustrates that the exhumation of domes is accommodated by extension. These extensional mylonitic zones can also be observed at the eastern and western termination of domes from the Axial Zone, where these domes show divergence of stretching lineations from their cores to their flanks (e.g. Hospitalet, Canigou, Bossòst) (Figure 1.6). In the Lys-Caillaouas massif, the deformation seems to have been localized in the southern edge of the massif, leading to the formation of an E-W mylonitic zone with a displacement of the upper block toward the SE (sinistral transtension). Indeed, the Lys-Caillaouas pluton appears to have used this mylonitic zone as a conduit to emplace atop the metamorphic dome (Cochelin et al. 2017). In the Axial Zone, the Infrastructure transverse lineations have often been interpreted as related to a D1 event of tangential tectonics (Soliva et al. 1989; Laumonier and Autran 2001; Denèle et al. 2009). Nevertheless, the fact that N-S shear events appear synchronous with partial melting and leucogranite emplacement has linked this deformation to the main D2 phase. These observations have been validated by the recent dating of migmatites, leucogranites and metamorphic rocks in Aston, Canigou, Bossòst and Chiroulet. The latter attest to a synchronous deformation of the M2 event, between 305 and 290 Ma (Denèle et al. 2014; Lemirre 2018; de Hoÿm de Marien et al. 2019; Lemirre et al. 2019). A similar pattern is found in the North Pyrenean massifs, where the activation of extensive mylonitic zones with N-S kinematics is now well characterized (Bouhallier et al. 1991; de Saint Blanquat 1993). However, the age of mylonitization in the Agly Massif is still debated to this day, as it is considered Variscan by some (Bouhallier et al. 1991; Olivier et al. 2004; Siron et al. 2020; Vanardois et al. 2020) and Cretaceous by others, as it is compatible with the ductile N-S stretching recorded in the pre-rift series (Clerc and Lagabrielle 2014; Odlum and Stockli 2019). The most recent work seems to demonstrate a Variscan impact, around 300 Ma, where thinning occurred in this massif (Tournaire-Guille et al. 2019; Vanardois et al. 2019). In contrast, recent dating in the Saint-Barthélémy, Castillon and Ursuya suggests that this extensive dynamic, initiated in the Late Carboniferous, continues until 280–270 Ma (Figure 1.1) (Kilzi 2014; Hart et al. 2016). This dynamic will be discussed further in the section dedicated to the D3 phase.

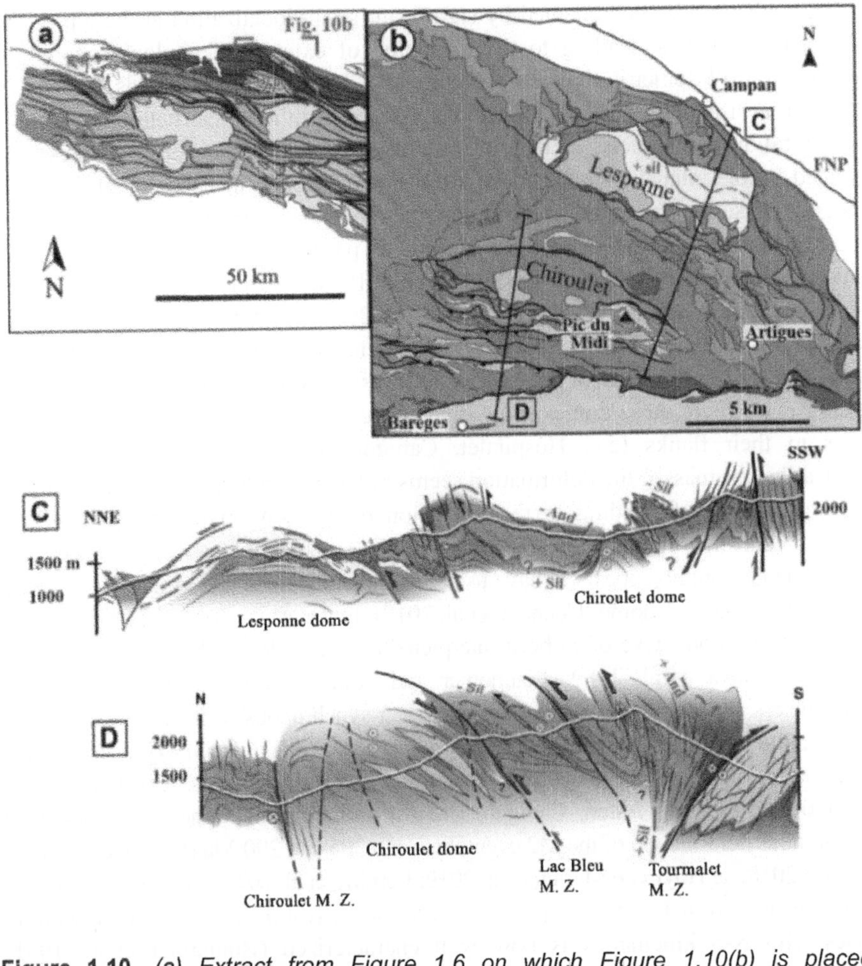


**Figure 1.10.** *(a) Extract from Figure 1.6 on which Figure 1.10(b) is placed; (b) simplified map of the Chiroulet and Lesponne massifs; (c) cross-section of the Lesponne massif and the eastern part of the Chiroulet; (d) geological cross-section of the Chiroulet massif*

#### 1.3.1.1.3. Examples of transition between Superstructure and Infrastructure

*Canigou-Nuría dome*

In the Canigou Massif (Figure 1.11), the transition between Infrastructure and Superstructure is abrupt, as is often the case in the domes of the eastern half of the Axial Zone. The typical deformation of the Infrastructure is clearly exposed, with the divergence of Li stretching lineations from a N70° direction at the apex of the dome to almost N-S on its northern and southern flanks (Figure 1.12).

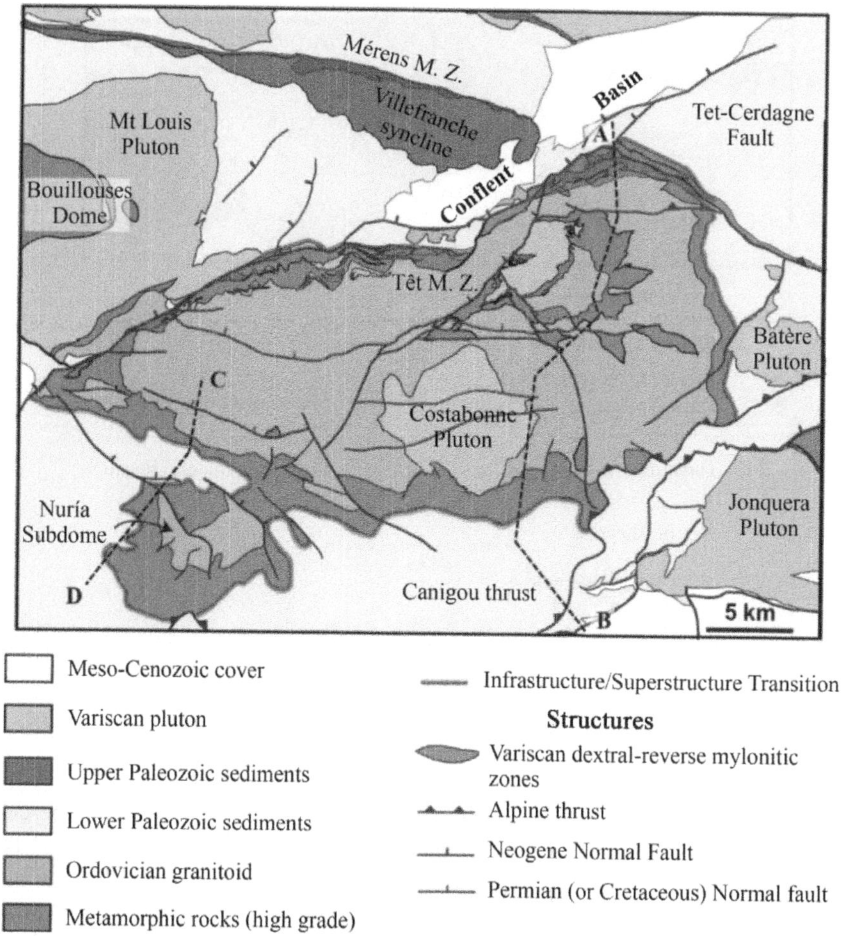


**Figure 1.11.** *Simplified geological map of the Canigou massif and Nuría massif modified from Cochelin et al. (2018b). The cross-sections in Figure 1.13 are located here*

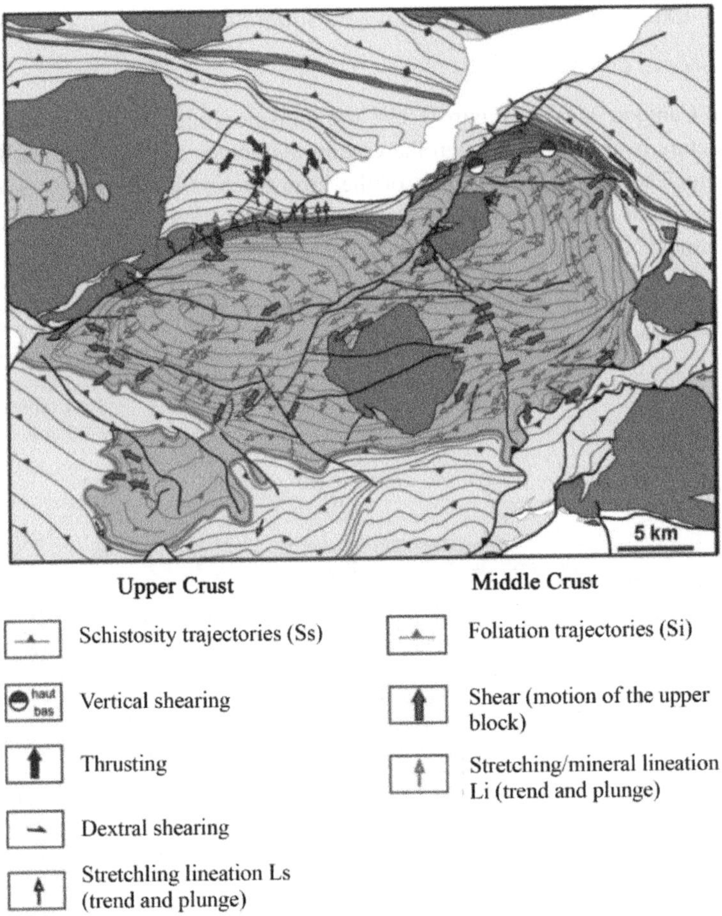


**Figure 1.12.** *Structural map of the Canigou and Nuría massifs, showing the trajectories of Ss schistosity, Si foliation, and the main directions of stretching and shearing*

The metamorphism associated with the formation of the Si foliation in the core of the massif has been dated at 300 Ma (de Hoÿm de Marien et al. 2019). The kinematics observed in the penetrative shear zones are essentially down-dip and thus extensional (Figure 1.12). In contrast, within the Superstructure, as elsewhere in the Axial Zone, a north-dipping schistosity is observed, bearing NNW-plunging lineations associated with southward to SSEward thrusting (i.e. perpendicular to dome stretching, Figure 1.12). The transition from one structural level to another, and thus from one tectonic style to another, is less than 1 km thick and coincides with the zone of tightening of the metamorphic isograds (andalusite, biotite). In this dome, this transition zone corresponds everywhere to a mylonitic zone, whose kinematic characteristics vary according to their location around the dome. The relationship between the two domains is most clearly seen in the Nuría massif, southwest of the Canigou dome (Figure 1.11). In the Nuría Massif, this transition is materialized by a deformation gradient that can be observed over a thickness of a few hundred meters (Figures 1.13(a) and (b)). In the biotite orthogneiss, a simple gneissic foliation (Si) is observed, whereas a strain localization is seen at the top of the orthogneiss, in the paragneiss, micaschists and Ediacaran to Cambrian marbles. The Si foliation appears mylonitic, forming what is called the South Canigou mylonitic band (Figures 1.12 and 1.13). This foliation seems to have developed along a retrograde path, from amphibolite facies (andalusite, stable biotite) to the beginning of greenschist facies (appearance of chlorite). The mylonitic band of the South Canigou has extensional kinematics, top to the west at Nuría, and top to the SW on the southern flank of the Canigou (Figures 1.12 and 1.13). The intensity of the deformation decreases upwards, while a new upright schistosity (crenulation) appears, N100°E striking and affecting the Si foliation in places. The appearance of this crenulation schistosity also coincides with a decrease in the intensity of HT metamorphism. Therefore, in the space of a few hundred meters, we go from a metamorphic domain to metasediments showing relics of sedimentary structures (bedding) and “evidence” of low-grade metamorphism (i.e. maximum chlorite greenschist facies). The schistosity becomes penetrative and is, as elsewhere, north dipping, forming the axial plane of longitudinal regional folds. Stretching lineations on the schistosity planes are N-S to NW-SE trending, which is also typical of the Superstructure (Figure 1.12). The development of this schistosity at the top of the Infrastructure is accompanied by the activation of south to SSE vergent shallowly dipping thrusts, reactivating the mylonitic Si foliation planes (Le Bayon and Cochelin 2020).

On the northern flank of the Canigou dome, a rather similar geometry can be observed. The extensional dynamics related to the emplacement of the Canigou dome could be clearly observed. It is characterized by the development of extensional penetrative shear bands (top to N or NW) synchronous with the emplacement of leucogranites dated at 305 Ma by Denèle et al. (2014) (Figure 1.12) (Cochelin et al. 2017). This trend is quickly obliterated by the development of thick greenschist facies mylonites over 500 m–1 km in thickness, and whose kinematics are south-verging (Figure 1.13; North Canigou or Têt mylonitic zone; see also Laumonier and Autran (2001)). Obviously, this thrust system seems to run along the roof of the Canigou dome, using the orthogneiss as a decollement layer, and thus tends to propagate southward (Figure 1.13(a)). Toward the east, this mylonitic zone follows the shape of the dome and strikes N130° while becoming vertical (Figure

1.13(b)). It extends to the massif of the Roc-de-France and the Albères (Figure 1.6). The dextral component becomes dominant, whereas it was non-existent along the northern flank of Canigou (Figures 1.12 and 1.13).

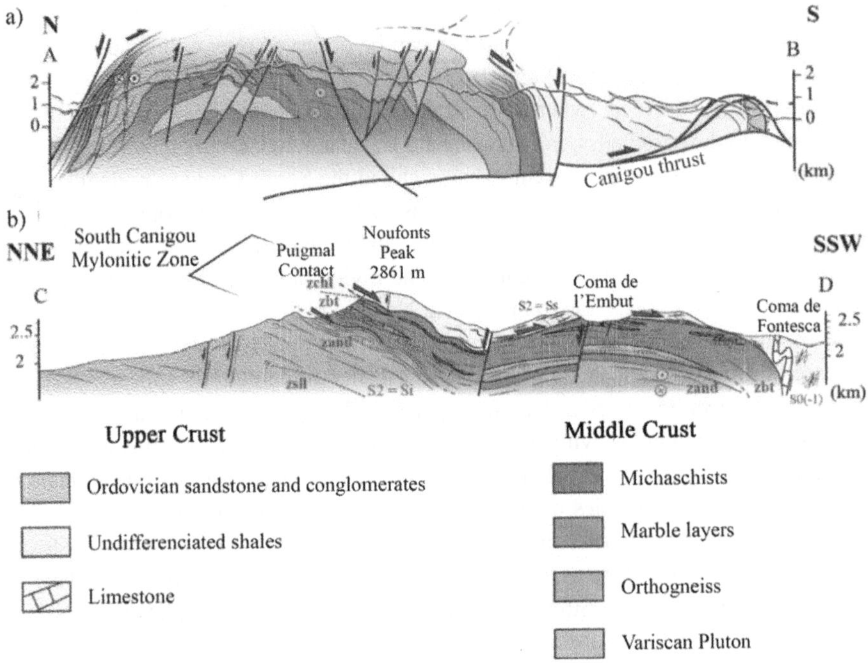


**Figure 1.13.** *Cross-sections: (a) of the Canigou massif, modified from Casas et al. (2010) and (b) of the southern flank of Canigou and the Nuría massif, adapted from Le Bayon and Cochelin (2020)*

### *Bossòst dome*

The Bossòst dome, otherwise known as the "Garonne dome", is a perfect example of the structural relationships between Infrastructure and Superstructure. Its original structure has been very well preserved, since this dome has only been cut into two sub-domes by the Bossòst fault, which is probable alpine in age (Figures 1.8 and 1.14(a)). Here, the outer envelope of the dome coincides with the andalusite-in isograd (Figure 1.14(a)). The dome consists mainly of Cambrian to Ordovician metasediments, except in its southeastern part, where the andalusite stability domain encompasses a portion of the upper Paleozoic. The core of the dome displays sillimanite paragneisses and numerous leucogranite sills, dated at 295 Ma (Lopez-Sanchez et al. 2018). The southern dome is stretched along an E-W direction, while only the northern half of the dome is observed. The stretching lineations Li are longitudinal (Figure 1.14(b)), although in the northern part of the dome, the coexistence of longitudinal and transverse lineations has been described (Mezger and Passchier 2003). This emphasizes the development of a flattening regime during the formation of the dome. In the southern part of the dome, the stretching lineations are sigmoidal, from N100°E at the core to N160–170°E at the edge.

The leucogranites systematically intersect the Si foliation and do not show any particular fabric. They therefore appear to be postkinematic. The observed shear criteria are mostly directed SE to SSE, although opposite kinematics were observed in the northern part of the dome (Figure 1.14(b)). In all cases, the shear direction is in the same direction as the plunge of the Li stretching lineation. It developed in the upper part of the Cambro-Ordovician, whose transition to the Silurian corresponds to marble levels (Figures 1.15(b) and (c)). This transition locally reaches the Devonian, which also consists of a thick marble series intercalated with andalusite micaschists. Once again, non-coaxial top SE shearing developed from amphibolite facies (sillimanite stability) into greenschist facies, suggesting exhumation of the dome related to strong vertical thinning for this portion of the crust. The Si factories appear strongly disturbed in the Upper Paleozoic series within the Bossòst dome structure. Indeed, the southern mylonitic zone is intersected by another mylonitic zone located in the Silurian black shales (Figures 1.7 and 1.15(a)). This mylonitic zone, developed in greenschist facies, thickens in the upper domains of the crust and wraps the inner envelope of the dome defined by the paragneiss (Figures 1.15 (a) and (b)). It is materialized in the Silurian black shales by an anastomosed network of metric to decametric shear bands (Figure 1.15(b)).

Stretching in these shear zones is vertical or N-S. The kinematics of the shear zones are mainly north-verging on the southern flank of the dome, while south-verging (or northern block uplift) is observed on the northern flank of the dome (Figures 1.15(a) and (c)). Above the Silurian black shales, the Devonian series are affected by steep thrusts rooted in the Silurian. The Devonian is also characterized by tight E-W folds accompanied by a vertical penetrative Ss schistosity (Figures 1.8 and 1.15(a)). The lineations observed on this schistosity are systematically vertical. Therefore, in the south-eastern part of the Bossòst dome, the Si foliation is largely affected by the folds observed in the Superstructure. The Si-foliation is even totally transposed by the Ss-schistosity at the approach of the dome edge. The overall structure of the Superstructure and the kinematics on the flanks of the Bossòst dome thus appear symmetrical, with northward thrusting on the southern flank and southward thrusting on the northern flank (Figure 1.15(a)). This pattern, as elsewhere in the Pyrenees, is compatible with a regional horizontal N-S striking shortening. Here, the main thrusts are located in the Silurian black shales, acting as a preferential decollement level. At the apex of the dome, that is, at the convergence of the various thrusts, the Upper Paleozoic is intensely flattened (Figures 1.8 and 1.15(a)).

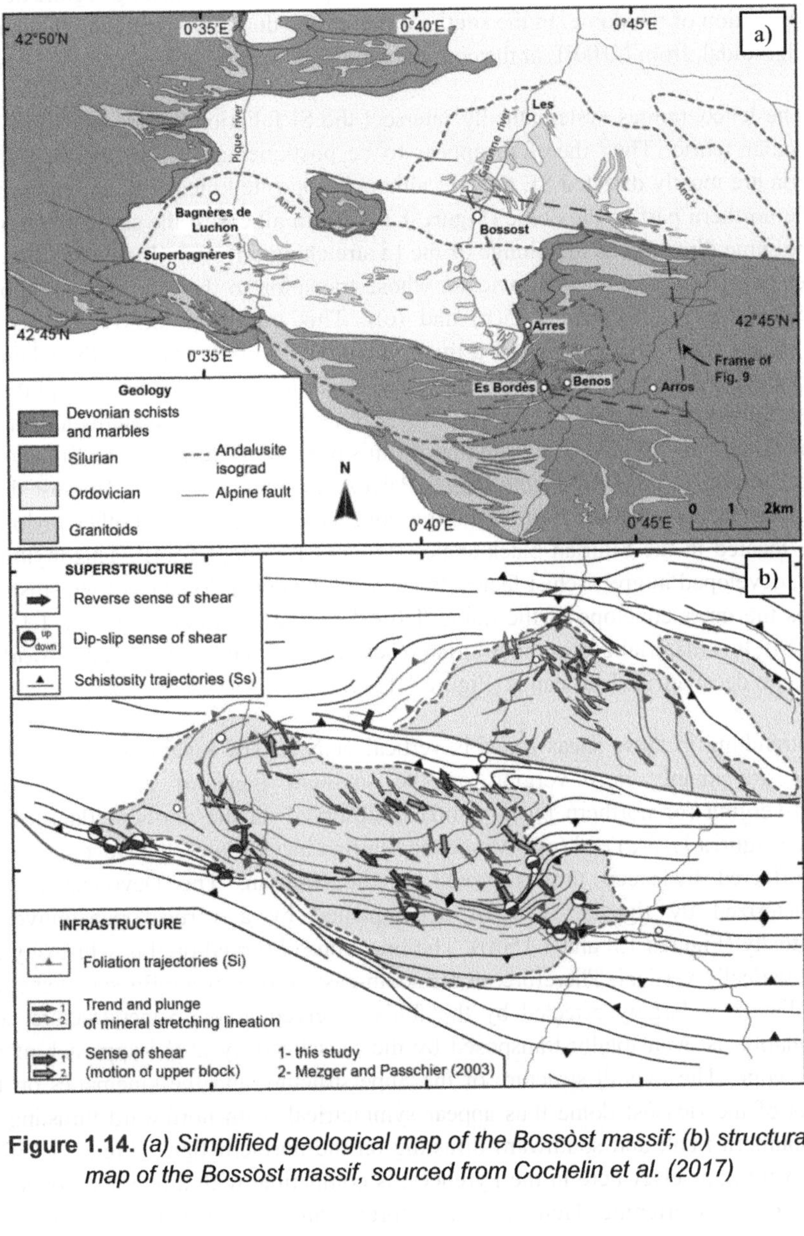


**Figure 1.14.** *(a) Simplified geological map of the Bossòst massif; (b) structural map of the Bossòst massif, sourced from Cochelin et al. (2017)*

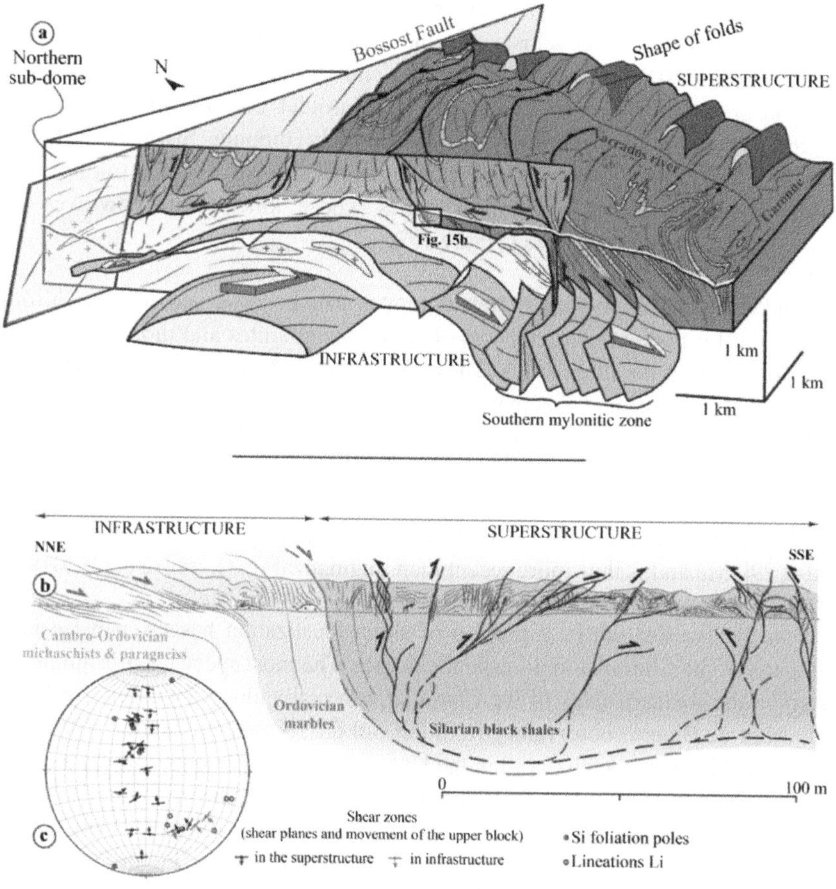


**Figure 1.15.** *(a) 3D block illustrating the structural and metamorphic features of the southern Bossòst dome; (b) geological section illustrating the transition from Si foliation in the Bossòst dome to the reactivation of Silurian black shales in the transpressive mylonitic zone; (c) stereogram representing shear plane measurements and their kinematics in the Infrastructure and Superstructure along the section (Figure 1.15(b)). Figures modified from Cochelin et al. (2017)*

### *Chiroulet and Lesponne domes*

The Chiroulet and Lesponne massifs form the westernmost domes of the Axial Zone (Figure 1.10(a)) and show characteristics very similar to the Bossòst dome. In fact, the sillimanite and andalusite isograds incorporate a very large part of the Upper Paleozoic metasedimentary series (Figure 1.10(b)). Unlike the southern

Bossòst dome, where Silurian schists mark the contact between sillimanite paragneiss and andalusite schists, in the Chiroulet dome these same schists constitute the transition between migmatites and sillimanite paragneiss. As in the other domes described in this chapter, the Chiroulet and Lesponne domes are characterized by the development of a shallowly dipping Si-foliation bearing essentially longitudinal Li-stretching lineations (N120° to 130°E striking). This deformation is associated with shear zones whose kinematics show westward displacement of the upper blocks (Figure 1.6).

Numerous sills or pockets of leucogranites can be observed in the Chiroulet dome. In the Lesponne, kilometer-long sills of leucogranites and diorites have been emplaced on top of the migmatites. These magmatic rocks have the same fabrics (Si foliation, Li lineation) and kinematic characteristics as their metasedimentary host. Although geochemical analyses of leucogranites show that they result from partial melting of metasediments, diorites appear to be derived from partial melting of the mantle (Lemirre et al. 2019). In these massifs, migmatization has been dated at 300 Ma (Lemirre et al. 2019). Crystallization of leucogranites and diorites has been dated at 290 Ma and is thus more recent than migmatization (Lemirre et al. 2019).

As in the Bossòst dome, a process of strain localization has taken place in the upper parts of the Chiroulet and Lesponne domes. The most spectacular examples can be seen on the southern flank of the Chiroulet, especially along the access road to the Pic du Midi de Bigorre observatory from the Col du Tourmalet (Figure 1.10(d)). The Si foliation, which can be observed in the Upper Paleozoic, presents a mylonitic fabric over a thickness of about 1 km. Similar to the Bossòst dome, the mylonitic Si foliation is very intensely folded. These folds are tight on the southern flank of the Chiroulet, while their reverse flanks are sheared into north-verging thrusts (Figures 1.10(b) and (d)). These thrusts are all located in the marble levels. The presence of numerous tight folds on the southern flank of the Chiroulet leads to multiple repetitions of the Devonian metamorphic series, which can be observed over nearly 4 km. Between the Chiroulet and Lesponne, the folds are more open, but their vergence and that of the associated thrusts remain directed towards the north. Note also that the shape of the Chiroulet dome is asymmetrical. Its northern flank is vertical, even dipping to the north, while its southern flank dips at about 50°S. This configuration is compatible with a folding phase associated with the development of northward thrusts. Approaching the andalusite-out stability domain, a steeply dipping Ss schistosity develops as the axial plane of regional folds. As elsewhere, these schistosity planes carry vertical or N-S stretching lineations. The Ss schistosity, eventually, totally erases the Si foliation as it approaches the roof of the domes and the Superstructure. North of the Tourmalet mylonitic zone, the observed penetrative shear zones are characterized by top to the north reverse kinematics. The latter, as well as the more southerly domains, are instead characterized by southward thrusting (Figure 1.10(a)). At the Chiroulet mylonitic zone, a dextral component could be observed on the northern flank of the dome (Figures 1.10(b) and (d)). In contrast to the Chiroulet, the andalusite isograd remains essentially confined to the Cambro-Ordovician series in the Lesponne Dome. Therefore, the transition between the Infrastructure and Superstructure generally coincides with the presence of Silurian black shales (Figure 1.10(c)). As in the Bossòst Dome, the thrusts seem to be localized into the Silurian pelitic series, following the shape of the dome. The presence of this hard core of crystalline rocks surmounted by weakely metamorphic metasediments led to the reactivation of the Variscan thrusts during the Alpine orogeny (Figures 1.10(b) and (c)).

#### 1.3.1.2. *Deformation model of dextral transpression*

Recent studies carried out within the framework of the RGF Pyrenees project have allowed us to characterize the two structural domains observed in the Variscan basement more precisely, while clarifying their structural relationship and the age of their formation, that is, between 310 and 290 Ma. The synthesis of these studies allows us to draw a unique model (Figure 1.16) of the deformation of the Variscan continental crust that takes into account both the atypical rheology of the crust at the Carboniferous-Permian transition and the structural and kinematic contrasts between the Infrastructure and Superstructure.

From 310 Ma onwards, the Variscan crust of the Pyrenees was affected by a HT/LP metamorphism, leading to a drastic drop in its overall strength. The rise of isotherms in the crust led to the formation of anomalous geothermal gradients in the crust. Indeed, the estimated thermal paleogradients in the upper part of the crust are relatively high to the extent that they were estimated to be between 45 and 80°C/km (Cochelin et al. 2018a; Lemirre 2019). As for the middle crust, it was characterized by a near constant temperature between 750°C and 850°C (Lemirre 2019; Siron et al. 2020). This drastic change in the thermal regime coincides with the transition between Infrastructure and Superstructure and leads to an atypical horizontal layering (Figure 1.16). Due to the presence of a large amount of melt (migmatites, granites), the middle crust has become much less resistant than the upper crust. This led it to flow laterally and thus to thin homogeneously in response to N-S compression. The storage of melt at a depth of 10–12 km (i.e. near the brittle-ductile transition) has in fact amplified the drastic drop in and viscosity of the first few kilometers of the middle crust. This had the effect of accentuating the ascent of these low viscosity rocks within the upper crust in the form of gneissic domes (Figure 1.16). The exhumation of these domes was favored by the activation of extensional shear zones, often located in rheological weak layers, such as Silurian black shales or marble levels. The directions of movement of these extensional mylonitic zones may be at a high angle to the mean stretching direction observed in the middle crust (Figure 1.16). This local tendency can be interpreted as a gravity component induced by viscosity contrasts in this portion of the middle crust. Indeed, apart from these local anomalies, most of the middle crust tends to stretch horizontally in a longitudinal manner, that is, perpendicular to the N-S regional shortening direction (Figure 1.16).

Simultaneously, the upper crust belonging to the future Axial Zone is intensely shortened, inducing folding and the formation of a steep longitudinal schistosity. In response to the regional shortening, a thrust system is developed, showing opposite vergence (Figure 1.16). In this pattern, the core of the Axial Zone is the most compressed zone, with vertical penetrative schistosity and vertical stretching. In the northern and southern domains, the schistosity dips about 60° toward the core of the Axial Zone. It thus forms a double-vergence orogenic prism in which the middle and lower crust is laterally extruded, while the upper crust is thrusted and extruded parallel to the direction of shortening, illustrating vertical strain partitioning (Figure 1.16). Indeed, it is to the front of these thrusts that narrow transpressive sedimentary basins are developed in the form of andesite-like volcanic flows interbedded with argillites, coals and limestones or conglomerates between 307 and 290 Ma (Pereira et al. 2014; Gretter et al. 2015; Lloret et al. 2018). These basins would later become the main Permian depositional areas in the Axial Zone (see section 3.3). Crystallization of anatectic melts induced by crustal cooling around 290 Ma generates an increase in domes' roof strength. This allows reverse dextral mylonitic zones, developed in the upper crust, to propagate deep into the middle crust, which leads to the amplification by folding of the already formed domes. Their edges underwent a localized deformation, marked by the formation of transpressive mylonitic zones. In the upper crust, these mylonitic layers eventually operate in a brittle regime (Poitrenaud et al. 2019).

In contrast to the Axial Zone, the North Pyrenean massifs showed little or no evidence of shortening over the period 310–290 Ma. Instead, they were affected by a global thinning of the crust combining both lateral flow and transverse extension (Figure 1.6). This extension favored the exhumation of granulites, contrary to the Axial Zone where only amphibolite facies rocks appear within the domes. Activation of mylonitic zones involving N-S extension appears to be partly contemporaneous with transpression in the Axial Zone (Lemirre 2018; Siron et al. 2020; Vanardois et al. 2020). A dextral component, similar to that described in the Axial Zone, has been described in the Arize, Trois-Seigneurs and Agly massifs (Leblanc et al. 1996; Olivier et al. 2004; Vanardois et al. 2020). Furthermore, some authors would be tempted to propose that the North Pyrenean massifs may not have been located in front of the Axial Zone around 300 Ma due to hypothetical strike-slip movements along the present North Pyrenean fault (Arthaud and Matte 1975; Burg et al. 1994; Matte 2001; Carreras and Druguet 2014). These late movements, often hypothetical, now appear to be widely questioned (Jammes et al. 2009; Tugend et al. 2014; Saspiturry et al. 2019a; Angrand et al. 2020; Asti et al. 2022). We therefore suggest that the North Pyrenean massifs were indeed in direct proximity to the Axial Zone, but at the edge of this transpressive orogenic miniprism (Figure 1.16). The deformation recorded in these massifs would thus be similar to that observed in the neighboring Variscan massifs, such as the Armorican Massif or the Massif Central, that is, combining extension and wrenching (Gumiaux et al. 2004; Faure et al. 2009; Gapais et al. 2015). Therefore, the structural contrast between the Axial Zone and the North Pyrenean massifs can instead be explained by horizontal strain partitioning between shortened and horizontally stretched domains at the end of the Variscan orogeny.

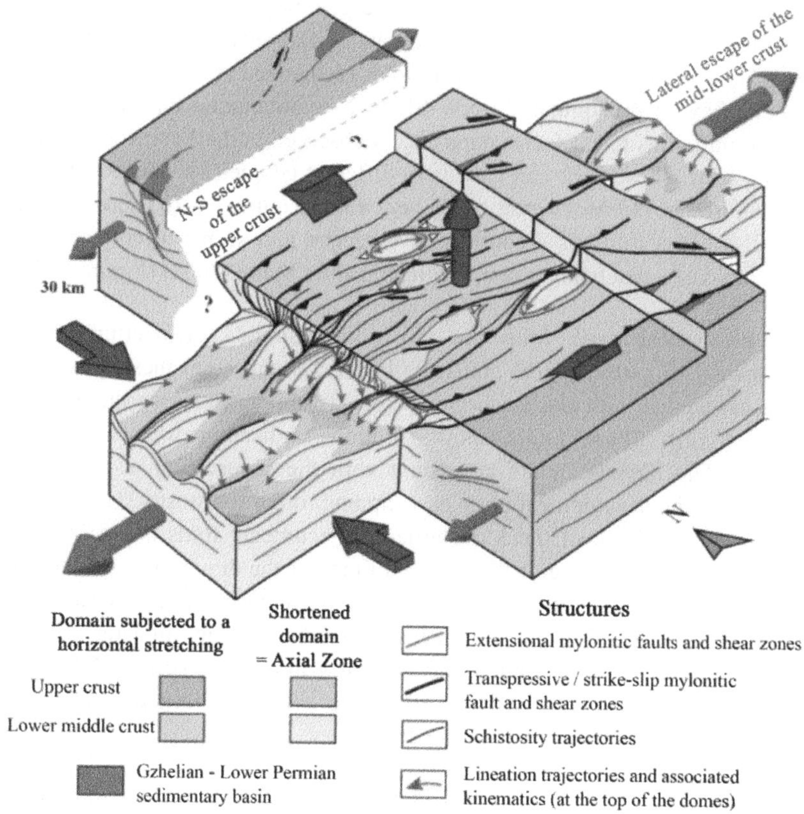


**Figure 1.16.** *Block-diagram illustrating the late Variscan vertical (in the Axial Zone) Axial Zone and horizontal (between the Axial Zone and the North Pyrenean massifs) strain partitioning in response to N-S shortening*

### 1.3.2. *Early phase D1*

Early structures in the Variscan history of the Pyrenees remain difficult to identify due to the intensity of M2 metamorphism and/or D2 deformation. Nevertheless, cartographic analysis of regional folds from geological maps can distinguish folds earlier than those related to the D2 deformation phase (Figure 1.9). The numerous alternations of detrital and calcareous deposits in the Upper Paleozoic favor their recognition, in contrast to the Cambro-Ordovician, where contrasting stratigraphic markers are rarer. It is therefore normal to find traces of these early deformations mainly in the western part of the Pyrenees, where the main Devonian and Carboniferous basins can be observed (Figure 1.9). Indeed, successive uplift from the Upper Jurassic to the Tertiary led to the erosion of the main Upper Paleozoic series in the central and eastern Pyrenees. Some relics of this erosion can be found in the synclines of Couflens, Sault and Villefranche (Delvolvé et al. 1998; Laumonier and Autran 2001). These early folds recognized in the western part of the Axial Zone have a predominantly NW-SE to N-S orientation and may affect the Upper Carboniferous Culm facies deposits, as observed in the Pyrénées-Atlantiques (Figure 1.17(a)). These folds can have very different geometries. The best known of these structures is the Moustardé recumbent fold (Figure 1.17(b)), located in the Ossau valley. This fold has a south-west vergence and seems to correspond to a secondary fold of a much larger one, as illustrated by Matte (2002) (Figure 1.18(a)).

Elsewhere, these early folds are most often straight or overturned to the west/southwest (Figures 1.18(b) and (c)). In the Lescun area, these west-verging folds are associated with the development of east-dipping reverse faults (Figures 1.18(b) and (c)). These faults show striations indicating westward thrusting movements. The kinematics are consistent with the observation of west-verging microfolds and the development of an east-dipping rough schistosity (Cochelin et al. 2017). The development of the Moustarde-like recumbent folds is also associated with southwestward shearing of the inverted limbs of the thrust folds (Matte 2002). Such early west-verging thrust systems have also been observed in the Ordovician to Devonian series of the Basque Country, despite the presence of later D2-related schistosity (Figure 1.6). In a similar way, an early thrust now striking N120°E and reactivated during D2 by the Mérens mylonitic zone has been described in the eastern Pyrenees (Laumonier and Autran 2001).

In the Infrastructure, recent works have clarified the age of certain structures historically associated with D1. The formation of the Infrastructure around 350–340 Ma proposed by some authors had already been invalidated by the dating of magmatism in the Canigou, Roc-de-France or Aston massifs at 310–300 Ma (Figure 1.1). However, there were still occurrences of transverse lineations in the Aston massif, interpreted as being related to a D1 tangent-tidal tectonic event (Denèle et al. 2009). New observations in this massif have shown that this transverse stretching was contemporaneous with partial melting and thus associated with extensional shearing (Cochelin et al. 2017), as evidenced by in situ dating of leucosomes, between 310 and 295 Ma (Lemirre 2019), that is, synchronous with the D2 event. Similarly, most of the recumbent folds described in the Infrastructure (Clariana and García-Sansegundo 2009) can be related to the D2 event, since the regional Si foliation is the axial plane of these folds. In contrast, in the Canigou Massif, relics of pre-D2 schistosity have been demonstrated beneath the Ordovician laccolith by de Hoÿm de Marien et al. (2019). The formation of this S1 schistosity, associated with staurolite and garnet crystallization, appears to reflect a higher pressure than that associated with M2, by about 1–2 kbar, consistent with Barrovian-type metamorphism. Unfortunately, this M1 event could not be dated, nevertheless this Barrovian facies seems to suggest a phase of crustal thickening that would be manifested through the emplacement of the early thrusts described above in the upper crust between 325 and 310 Ma.

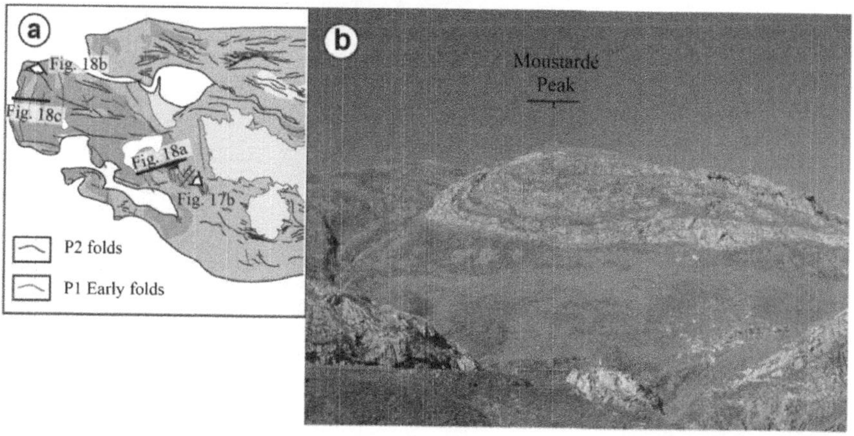


**Figure 1.17.** *(a) Excerpt from Figure 1.9 representing the axial traces of regional folds in the western part of the Axial Zone. The location of the cross-sections and viewpoints in Figures 1.17(b) and 1.18 are indicated. (b) Viewpoint from the SE of the Moustarde peak and its famous recumbent fold*

The deformation style observed in the upper crust is a fold-and-thrust belt type. It is therefore consistent with the deformation of the Variscan southern foreland in Pennsylvania (Cochelin et al. 2017), recorded at the surface by the deposition of the Culm conglomerates. No major stratigraphic jumps or early thrusting involving the Lower Paleozoic/Proterozoic (i.e. Carboniferous *substratum*) have been clearly identified. This rejects the existence of early basement nappes in the Pyrenees. Therefore, the only structures that could lead to a thickening of the crust, albeit relative, are the multi-kilometer recumbent folds and their associated minor thrusts (Figure 1.18(a)). This relegates the rooting of these nappes to a basement domain (thick skin) more to the NE of the present Pyrenean chain. The magnitude of the D1 deformation and associated crustal thickening thus remain unconstrained, but can reasonably be considered minor in view of the structures known to date. The relics of this first event found in the Infrastructure could thus reflect the beginnings of a shortening of the middle crust, already at high temperature, prohibiting the formation of thrust sheets in a deformation regime that is already of too low viscosity.

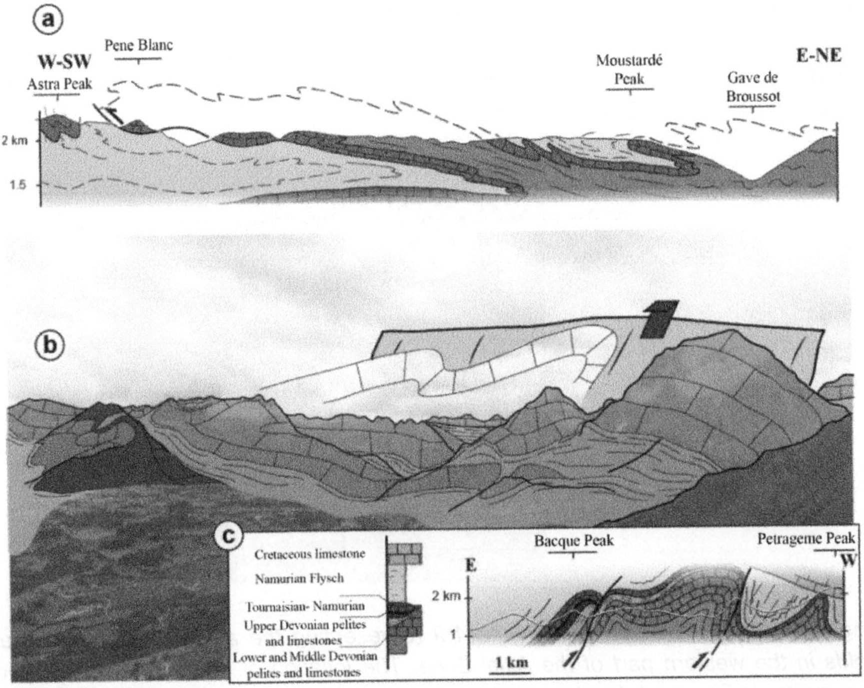


**Figure 1.18.** *a) Cross-section illustrating the recumbent fold system developed in the western part of the Axial Zone, modified from Matte (2002); (b) view southward of the Lescun area from the Ourtasse ridge, illustrating the west-verging fold and reverse fault system; (c) geologic section of the Lescun area, shown in Figure 1.18(b), taken from Cochelin et al. (2017)*

### 1.3.3. *Late Permian D3 phase*

#### 1.3.3.1. *Upper crust*

While evidence of syn-D2 sedimentation, that is, up to about 290 Ma, is particularly rare in the Pyrenees (see Stephanian basins), tectonic evolution in the Permian favored the formation of numerous sedimentary depositional zones. The so-called “gray” and “transitional” units deposited between 307 and 290 Ma were succeeded by two important sedimentation phases, leading to the deposition of the “red” series so recognizable in the Pyrenean landscape (Figure 1.19(a)). These red series have been subdivided into the “lower red unit” and the “upper red unit” (Mey et al. 1968; Gisbert Aguilar 1981). In the Axial Zone, these two units generally present the same characteristics, that is, the emplacement of conglomeratic series evolving laterally and vertically toward finer series (sandstone) up to the deposition of red mudstones and limestone layers, locally. Recent sedimentological and stratigraphic work by Gretter et al. (2015) and Lloret et al. (2018) have made the depositional environments of these basins clear and described their overall geometry in great detail. Indeed, these authors were able to show that basins on the southern edge of the Axial Zone were characterized by the development, in the first instance, of braided river systems, and in the second instance, of meandering river systems. While the braided systems appear to develop more obliquely to the E-W strike of the sedimentary basins, the meandering system is longitudinal (Gretter et al. 2015). This cycle is followed by the development of shallow, ephemeral lake systems (floodplain, carbonate production). Within these series, pyroclastic or basaltic flows of calc-alkaline and then alkaline affinities may be intercalated. These volcanic series have allowed for the deposition of the lower red unit to be precisely dated between 290 and 280 Ma, while the upper red unit was deposited between 280 and 266 Ma (Pereira et al. 2014). The location of the depositional zones appears to be primarily controlled by tectonics. Indeed, the base of the conglomeratic series is characterized by the development of breccias, interpreted as debris flows located at fault escarpments. In detail, the structural analysis of the Permian basins observed in the Axial Zone has highlighted several synsedimentary fault sets (Bixel and Lucas 1987; Gretter et al. 2015; Rodriguez-Mendes et al. 2016; Lloret et al. 2018) (Figure 1.19(b)). The main faults have E-W to NW-SE strike, that is, parallel to the Variscan basement structures. These faults were already active during the emplacement of the basal gray and transitional series. A second set of faults is oriented N-S to NE-SW. Kinematic analysis of these faults remains complicated due to frequent reactivation during the Alpine cycle (e.g. for the Anayet basin, Figures 1.19(b) and (c)). Nevertheless, the spatial distribution of these faults and their geometrical relationship seem to indicate a dextral strike-slip transtensional tectonic context. In this pull-apart basin formation model, activation of longitudinal dextral strike-slip faults led to the formation of depression zones between them and to the activation of NE-SW normal faults. The activation and subsequent reactivation of these faults during the Permian is highlighted by a clear angular unconformity between the lower red unit and the Paleozoic basement (or transitional gray unit), but also between the lower and upper red units.

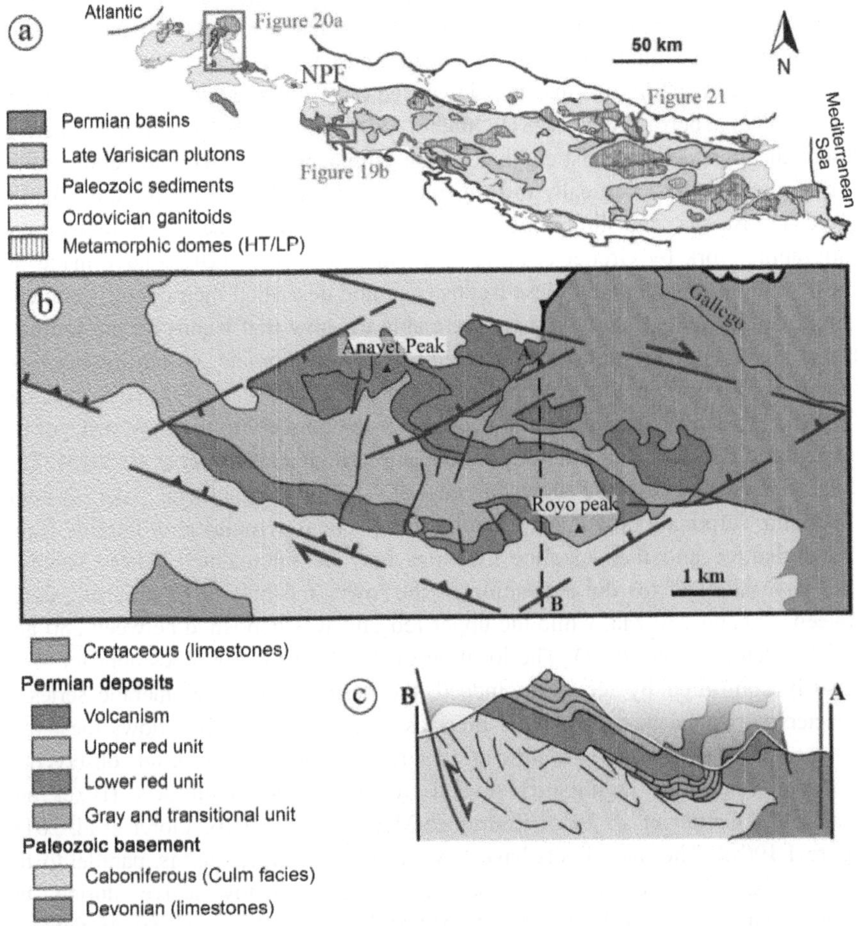


**Figure 1.19.** *(a) Simplified geological map of the Paleozoic basement of the Pyrenees and its Permian cover. Figures 1.19(b), 1.20(a) and 1.21 are located here; (b) simplified geological map; (c) cross-section of the Permian Anayet Basin, modified from Rodríguez-Méndez et al. (2016)*

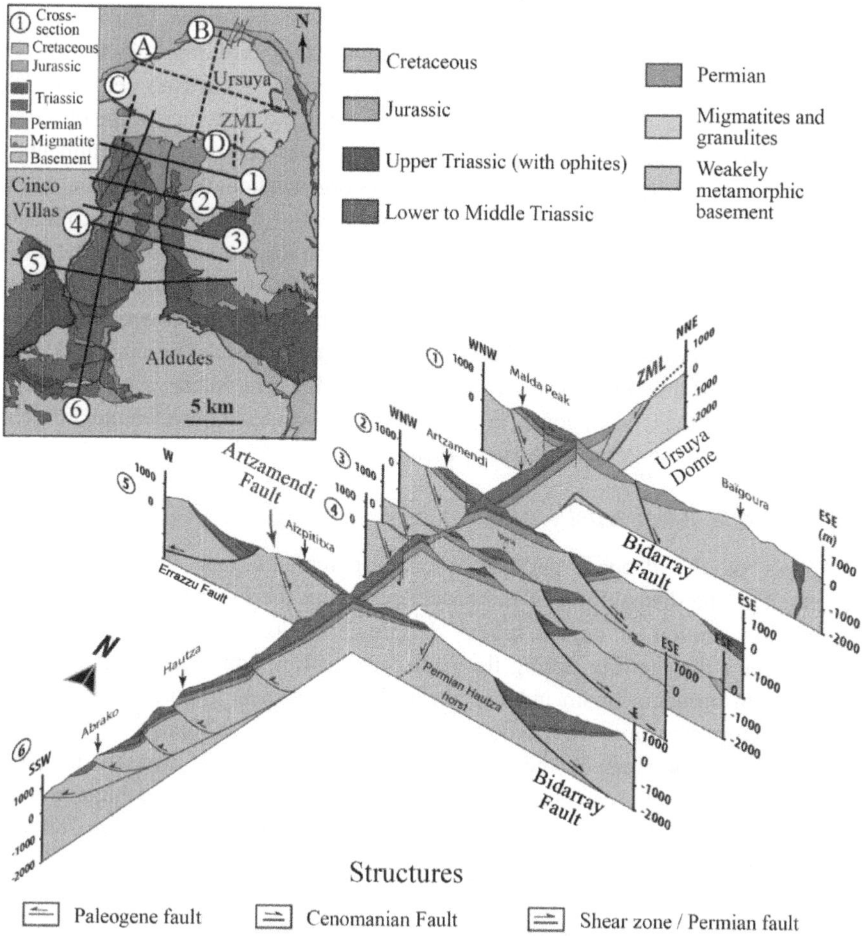


**Figure 1.20.** *Simplified geologic map and cross-sections of the Permian Bidarray Basin, modified from Saspiturry et al. (2019a)*

In the Basque massifs, only the upper red unit has been recognized (Lucas 1985). The main Permian outcrops in these massifs are in the Bidarray Basin, bordered by the Ursuya granulite massif to the north, the Cinco Villas to the west, and the Aldudes to the east (Figure 1.20(a)). Sedimentologically, the Bidarray Upper Red Unit has many similarities to that of the Axial Zone (Saspiturry et al. 2019a). Indeed, basin filling is accompanied by the emplacement of thick breccia and conglomeratic series along the major faults bordering the basin. These series pass laterally to finer terms toward the center of the basin (siltstones, argillites). Rare volcanic lava-flows are intercalated in the upper levels. As in the Axial Zone, the observed facies illustrate the development of alluvial fans evolving toward a fluvial system and the floodplain without evidence of carbonate lacustrine deposits. The originality of the Bidarray Basin lies in its geometry and the dynamics of the observed fluvial network. Indeed, the main alluvial fans show an eastward sedimentary discharge, while the fluvial network follows a N-S organization, that is, parallel to the basin strike (Saspiturry et al. 2019a). The establishment of the alluvial fans coincides with the Artzamendi normal fault trace striking N20° on the western edge of the basin (Figures 1.20(a) and (b)). It thus appears to have a hemigraben and, locally, a graben geometry, via activation of a conjugate normal fault bordering the eastern part of the basin (Figure 1.20(b), section 5). The basin appears to be distalizing and thickening northward toward the Louhossoa Mylonitic Zone (LMZ). This shear zone borders the southern flank of the Ursuya dome (Figure 1.20(b), section 6). It played a major role in the construction of the Permian Bidarray Basin.

#### 1.3.3.2. *Middle-lower crust*

Access to the deepest parts of the crust remains rare in the Pyrenees. Nevertheless, the dismantling of the basement during the Cretaceous rifting led to the individualization of crustal blocks defining the North Pyrenean and Basque massifs. Their dismantling during the Cretaceous rifting and their integration into the Pyrenean orogenic prism induced their tilting and their important denudation, allowing the exposure of portions of middle crust. Although the HT/LP metamorphism observed in the North Pyrenean and Basque massifs is partly inherited from the late Variscan D2 phase, recent dating works have highlighted a recurrent influence of a metamorphic and magmatic episode between 280 and 260 Ma (Denèle et al. 2012; Kilzi 2014; Hart et al. 2016; Vacherat et al. 2017; Lemirre 2018) (Figure 1.1). This Permian HT/LP metamorphism (M3) appears to occur continuously with the M2 episode. It seems that the granulite stage was reached around 300 Ma, while migmatization in some massifs would have continued until 280 or even 260 Ma. This is notably the case in the Saint-Barthélémy massif, but also in the Bessède de Sault and Castillon massifs in the northern Pyrenees (Figure 1.1) (Kilzi 2014; Lemirre 2018) and the Ursuya massif in the Basque Country (Hart et al. 2016; Vacherat et al. 2017; Lemirre 2018). This M3 metamorphism is associated with significant exhumation of the middle crust from granulitic stages, around 6–7 kbar and 800–850°C to conditions at 2–4 kbar and 550–650°C (de Saint Blanquat et al. 1993; Lemirre 2018). This M3 episode is accompanied by non-negligible magmatism, as evidenced by the emplacement of calc-alkaline plutons in the Basque massifs, such as the Aya granite around 267 Ma (Figure 1.1) (Denèle et al. 2012) or Ixtassou around 277 Ma (Vacherat et al. 2017), the latter bordering the Ursuya dome. Rare peraluminous granitic dykes also appear to be emplaced in the Axial Zone around 280 Ma (Lemirre 2018). Therefore, Permian magmatic activity results in volcanic and pyroclastic flows at the surface that originate in plutonism of predominantly calc-alkaline and then alkaline affinity in the upper crust, and locally by migmatization in the middle crust.

The analysis of ductile deformations in the North Pyrenean and Basque massifs, often interpreted as extensional, has led to the attribution of either a late Variscan or Cretaceous age. Recent petrological and microstructural analyses and dating of minerals associated with HT metamorphism in the main massifs have led to the interpretation of these basement ductile fabrics as essentially Paleozoic in age (Lemirre 2018; Saspiturry et al. 2019a; Siron et al. 2020; Vanardois et al. 2020). We will therefore follow this interpretation here. As for metamorphism and magmatism, deformation in the middle crust during D3 thus seems to evolve in continuity with the dynamics initiated during D2/M2 in the North Pyrenean and Basque massifs. This results, around 280 Ma, in a N-S extensional dynamic in some North Pyrenean massifs. The best example of this Permian N-S extension is found in the Saint-Barthélémy massif (Figures 1.1, 1.6 and 1.21). As initially described by de Saint Blanquat (1993), this massif is characterized, from base to top, by the presence of gneisses having recorded granulitic metamorphism, migmatites, a series of micaschists assimilated to the Cambro-Ordovician, and then the Upper Paleozoic series (Figure 1.21). This massif is affected by two major ductile tectonic contacts. The shallowest is located at the interface between the Ordovician and the Silurian black shales, that is, the most widespread decollement layer in the Pyrenees. The second is the transition zone between the granulitic gneisses and the migmatites. It is materialized by a mylonitic zone about 2 km thick, called the major mylonitic band (MMB) (Figure 1.21). Strain localization is progressive and is materialized in the gneisses and migmatites by the transposition of a steeply dipping early foliation by a secondary foliation with a shallow northward dipping (Lemirre 2018). This penetrative foliation is associated with the formation of shear zones with "top to south" kinematics (de Saint Blanquat 1993). The core of the MMB is characterized by the development of ultramylonites over several meters in thickness. Recent thermobarometry work carried out on the hanging-wall and foot-wall of the BMM has made it possible to quantify the thinning necessary to about 300%, via its activation, to bring the lower unit into contact with the upper unit at a depth of less than 10 km (Lemirre 2018). Of course, the current subhorizontal position of the MMB does not reflect its original orientation, as the entire massif underwent a northward Alpine tilt. The MMB was moderately dipping to the south during its formation in the Early Permian (Lemirre 2018), forming a major extensional mylonitic zone in the crust (detachment). Recent dating of hydrothermal apatites within the MMB suggests that it was reactivated during the Cretaceous rifting phase (Odlum and Stockli 2020). It is also likely that the decollement, localized in the Silurian black shales, also accommodated some of the extension into the upper parts of the crust, both during the Permian and Cretaceous. However, poor outcropping conditions make structural analysis of this area difficult.

In the Basque massifs, the Ursuya massif (Figure 1.20(a)) is the only witness of the deformation dynamics in the Permian middle crust. This massif was relatively unaffected by the Tertiary collision and still retains its late Variscan original position (Figures 1.20(a) and 1.22). Therefore, the preserved part of the Ursuya massif forms a foliation half-dome of which only the south-southeast part is outcropping today (Figure 1.22). Granulite relics (Vielzeuf 1984) now occupy the core and the northern part of the dome (Figure 1.22). The rest of the massif is mainly made up of migmatites and paragneisses in which rare dioritic and granitic sills are emplaced. The edges of the dome are characterized by the transition from migmatites to micaschists and paragneisses in which marble levels, as well as pegmatitic veins can be observed (Figure 1.22). This transition coincides with the disappearance of sillimanite in favor of andalusite. As in the Saint-Barthélémy described above, metamorphic studies attest to a granulite peak (6 kbar and 800°C) around 300 Ma, followed by a decrease in pressure and temperature to reach about 3 kbar and 650°C around 280 Ma (Vielzeuf 1984; Lemirre 2018).

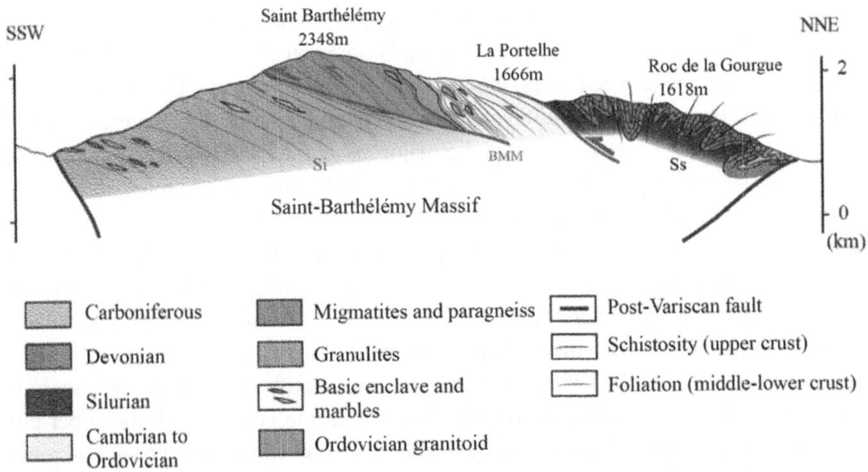


**Figure 1.21.** *SSW-NNE cross-section of the Saint-Barthélémy massif showing the tectonic relationships between the Saint-Barthélémy metamorphic dome, the main mylonitic zones recognized in the massif and the portion of the upper crust located at the level of the Roc de la Gourgue. Modified from Soula (1982)*

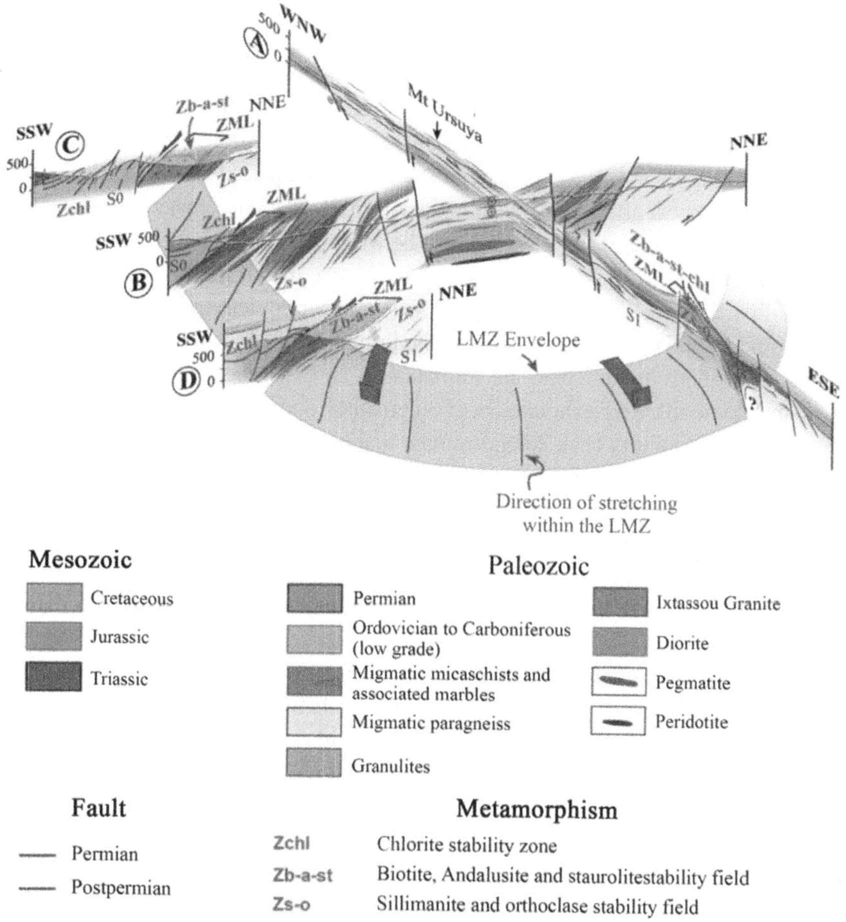


**Figure 1.22.** *Geological cross-sections through the dome of Ursuya, located in the north of the Bidarray Basin. Modified from Saspiturry et al. (2019a)*

Structurally, the core of the Ursuya Dome is characterized by subhorizontal foliations, becoming progressively south-dipping in the southern half of the massif and east-dipping in its eastern termination (Figure 1.22). These foliations yield longitudinal (N100°–110°E) stretching and mineral lineations throughout most of the massif (Figure 1.6). Approaching its southern edge, these lineations tend to diverge outward from the massif until they reach a near N-S trend (Figure 1.6). This type of geometry is similar to that described in some domes of the Axial Zone (section 3.1). This divergence in stretching direction coincides with the sillimanite-andalusite transition zone and is accompanied by a tightening of the isograds (andalusite, biotite and chlorite zones) and strain localization forming a mylonitic zone about 500 m thick (Figure 1.22). The LMZ appears to have operated along a retrograde path from amphibolite facies (stable sillimanite) to greenschist facies (chlorite zone) (Saspiturry et al. 2019a). Within it, mylonitization indiscriminately affects quartzites, micaschists and pegmatitic veins. In contrast, the Ixtassou pluton dated at 277 Ma (Vacherat et al. 2017), emplaced in the upper part of the LMZ, does not appear affected by this ductile deformation. It thus appears post-kinematic, constraining the activity of the LMZ at most between 300 Ma (assumed age of granulite metamorphism) and 277 Ma. Therefore, the Ursuya massif recorded a lateral flow episode of partially melted domains coupled with a localized deformation at the transition between middle and upper crust, leading to the formation of a metamorphic dome with an N110°E elongation. In particular, the formation of an extensional mylonitic zone at the top of the dome allowed exhumation of migmatites and granulites for nearly 10 km (Lemirre 2018; Saspiturry et al. 2019a), similar to the Saint-Barthélémy massif. A trace of the LMZ is also observable at the eastern end of the Ursuya massif. Its roof is materialized by massive, chaotic basement-element breccia deposits with a red sandstone matrix (Saspiturry et al. 2019a), typical of Permian fault scarp deposits. The Permian Bidarray Basin appears to deepen and thicken toward the LMZ, suggesting synchronicity between the formation of the Bidarray graben and the extension recorded in the middle crust of the Ursuya massif. Furthermore, the opening of the basin via N20° faults and the dominant N110°E stretching direction in the dome both appear compatible with a WNW-ESE regional extension. It therefore appears reasonable to consider these two phenomena as contemporary and reflecting the common response of both the upper and lower crust to an extensional regime during the Permian in the Basque Country.

#### 1.3.3.3. *Tectonic model of horizontal strain partitioning under extension/transtension*

The tectonic event D3 was set up in a context where the crust was already strongly structured because of the intensity of the late Variscan D2 event (Figure 1.23(a)). At the end of D2, a major distinction could already be made between the shortened (or even thickened?) domain, materialized by the Axial Zone, and the thinned domain corresponding to the various North Pyrenean massifs. In addition to this structural distinction, thermal conditions persist in the lower crust of the North Pyrenean massifs (Figures 1.3(b) and (c)), as shown by a metamorphism into granulite facies in the north and amphibolite facies in the south. The relaxation of late Variscan constraints seems to have simply led to an accentuation of this singularity during the D3 phase. Considering the recent work done in the Axial Zone, the North Pyrenean massifs of the central and eastern Pyrenees and the Basque massifs, three distinct domains can be identified, with their own dynamics (Figure 1.23). On the one hand, we have the North Pyrenean massifs of the central and eastern Pyrenees, in which the N-S extension, recorded during phase D2, seems to continue and even be accentuated via the activation of crustal detachments during phase D3. This tectonic regime is notably responsible for the exhumation of the middle crust to depths of about 6 km. In these massifs, the main detachments seem to be south-vergent, although a symmetrical system cannot be excluded, via the activation of conjugate extensional shear zones (Figure 1.23(a)). The exhumation of the middle crust, thanks to the activation of these shear zones, leads to the formation of longitudinal metamorphic domes, that is, perpendicular to the dominant stretching direction within these massifs. The thinning of the crust, on either side of the domes, must necessarily have caused the formation of subsidence zones at the surface, and their filling by the erosion products of the newly denudated basement (Figure 1.23(a)). Still at the surface, this extension had to be accommodated via longitudinal normal faults in the North Pyrenean massifs. Unfortunately, this system of basins and normal faults has partly been obliterated by the various erosions that have taken place since the Late Jurassic. The last remnants of these basins are now found in the northern part of the Axial Zone, or cover some of the North Pyrenean massifs. To the west, in the Basque massifs, phase D3 also seems to have amplified the deformation of the middle-lower crust developed during phase D2, namely a longitudinal horizontal stretching (N100°–110°E) (Figure 1.23(a)). HT/LP metamorphism in the middle crust, post 290 Ma, appears to have enhanced mid-crustal exhumation within the Ursuya Dome via activation of major detachment at the dome roof and margins in the Permian. The amount of exhumation recorded in this case (in the order of about 10 km) is comparable to that calculated for the North Pyrenean massifs, where a Permian age M3 metamorphic event has been found. This longitudinal stretching dynamic was also expressed in the upper crust of the Basque Country during the D3 phase via the formation of grabens and hemigrabens, perpendicular to the regional stretching direction (Figure 1.23(a)). The maximum depositional zones then coincide with the scarp zones of the main normal faults, N20°E or N100°E along the dome. Finally, recent work on the Permian bordering the southern parts of the Axial Zone has made generalizations of the model established in the 1990s, namely the emplacement of longitudinal transtensional (pull-apart) basins formed in the context of dextral transtension (Figure 1.23(a)). Magmatic activity is limited to the emplacement of andesitic and more rarely granitic dykes, as well as volcanic flows at the surface. This pull-apart basins system can be locally disturbed in the vicinity of the main magma chambers and leads to the formation of calderas, such as the Ossau one. This difference in crustal dynamics is probably due to the colder thermal state of the crust of the Axial Zone since 290 Ma, as compared to that of the other massifs that remain in extension. This transtensional geometry in the Axial Zone is consistent with a generally WNW-ESE strike-slip direction. This direction is also consistent with magnetic lineations measured in the rare post-D2 granite veins described in some domes of the Axial Zone.

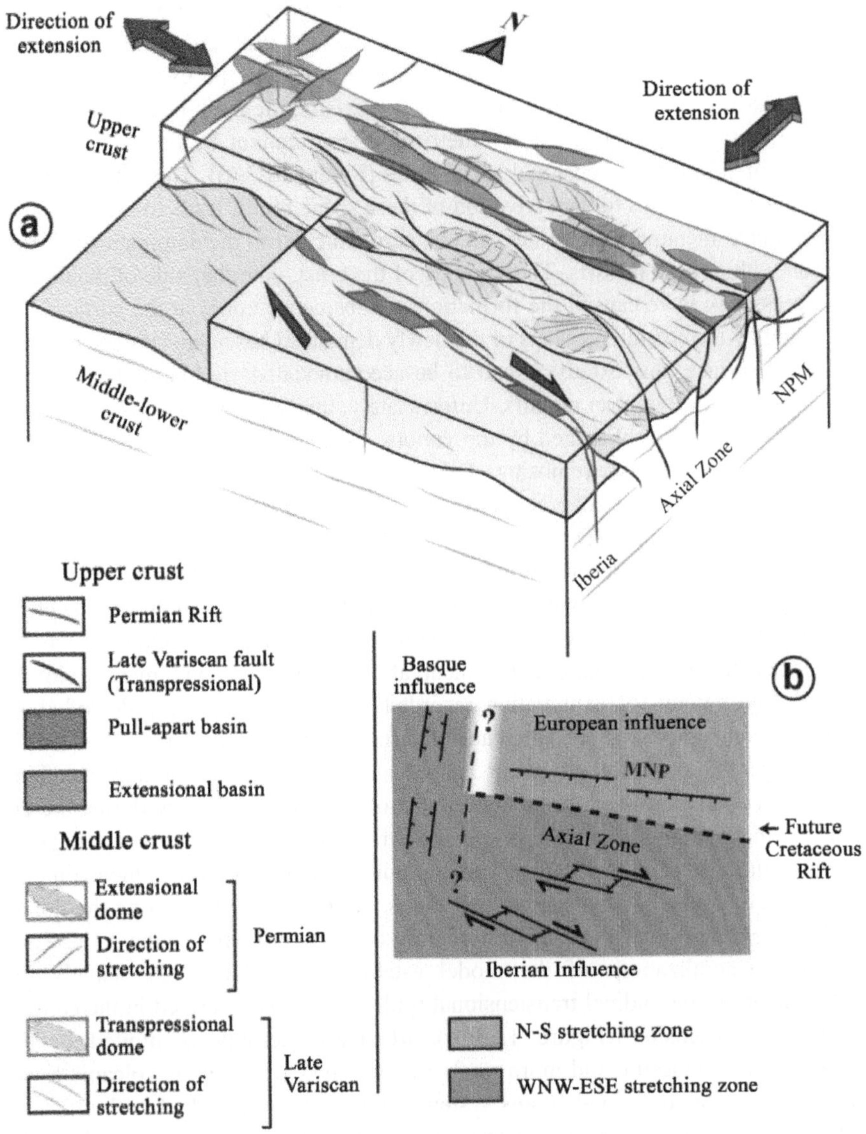


**Figure 1.23.** *(a) Block diagram illustrating the crustal architecture around 280 Ma in the Pyrenees. The structural legacy of the D2 transpressional phase is also shown. (b) Schematic map of the different tectonic influences recognized during D3 in the Pyrenees*

Therefore, the Pyrenean domain around 280 Ma is characterized by three domains with distinct deformation styles, thus defining three regional influence zones (Figure 1.23(b)). The Basque massifs and the Axial Zone share the same WNW-ESE stretching dynamics, but these dynamics are expressed differently in these different massifs. In this scheme, the Basque massifs have an atypical WNW-ESE stretching behavior, while the Axial Zone is characterized by a dextral transtensional system. The latter can be easily related to the tectonic context recognized in Iberia during the Permian (Gretter et al. 2015). The shift from a "Basque" to an "Iberian" type of influence (Figure 1.23(b)) can be explained by the rheological contrast between an area where the lower crust is still partially molten and highly favorable to horizontal flow (Basque massifs) and an area that is already rigid (Axial Zone). These two responses to the same apparent stresses thus reflect a lateral strain partitioning, directly dependent on the thermal state of the crust. The extent of the Basque influence zone, along the Atlantic coast, remains difficult to ascertain. Nevertheless, it does not seem to extend further south into Spain, and thus appears to be secondary at a regional scale.

The last zone of influence corresponds to the area affected by a N-S extension, well characterized in the north Pyrenean massifs of the central and eastern Pyrenees. This regime appears to be almost orthogonal to that of the other two zones. However, it is not so aberrant on a regional scale, since most of the French Permian basins are developed according to an extensional regime with global N-S stretching (Burg et al. 1994; Pochat and Van Den Driessche 2011) (see section 1.4), as in other European countries (e.g. Italy). The transition between this "European" zone of influence and the Axial Zone, reflecting more "Iberian" dynamics, appears to be quite abrupt and is obviously parallel to the Axial Zone (Figure 1.23(b)). Although the Cretaceous rifting followed by the construction of the Pyrenean orogenic prism deprives us of access to the portion of crust that laid between the Axial Zone and the North Pyrenean massifs, it seems that their radically different late Variscan dynamics related to the D2 and then D3 phases signal the existence of a major discontinuity between these two domains. Indeed, in contrast to the Axial Zone, the European domain was homogeneously thinned in a N-S direction for nearly 30 Ma during the Permian. This probable narrow WNW-ESE transition zone is certainly delimited to the north by the Permian detachments observed in the North Pyrenean massifs and to the south by the relics of Permian basins bordering the northern part of the Axial Zone (Figure 1.23(a)). The transition between the European and Basque influence zones was most likely accommodated in a progressive manner by the major N20°E faults (Figure 1.23(b)) recognized in the basement beneath the Aquitaine Basin (and probably Permian) in the vicinity of the Basque massifs

(Saspiturry 2019). In the latter case, the N20°E faults certainly acted as transfer faults, accommodating both extension and strike-slip displacement on both sides.

## 1.4. Geodynamic integration

In this section, we will attempt to relate the different tectonic events described to their West European geodynamic context in the late Variscan period (330–280 Ma) (Figures 1.24 and 1.25). We will therefore endeavor to relate the main phase (D2) to the late-orogenic formation of the Cantabrian orocline. In particular, we will use the early structures associated with D1 as passive markers to quantify late deformation. We will also discuss the extensional Permian D3 dynamics, described as the beginnings of the Alpine cycle in Europe. Therefore, we will not develop the complete tectonometamorphic history of the Variscan chain.

### 1.4.1. *Early D1 phase*

Around 320–310 Ma, the propagation of the Variscan front reached the Pyrenees, accompanied by the development of folds and thrusts. Their present vergence, toward the west or southwest, is similar to that described in the southern areas of the Variscan chain (Matte 2001). In Iberia, that is, on the southern branch of the Iberian-Armorican-Canadian arc, it is, on the contrary, toward the east. The structural affinities of the Pyrenees with Europe seem to link the Pyrenean segment to the northern branch of the arc. When the front of the chain reached the Pyrenees, around 320 Ma, it is important to specify that the internal domains of the Variscan chain had already reached an advanced stage of thinning, since 30 Ma: (1) the high-pressure units, formed by continental subduction between 380 and 360 Ma, had already been exhumed (Figure 1.24(a)) (Lotout et al. 2020) and (2) the inner zones were affected by synchronous extension and HT metamorphism, materialized by the emplacement of numerous plutons and migmatitic domes between 340 and 310 Ma (Figures 1.24 (a) and (b)) (Burg et al. 1994).

This synorogenic extension is characterized by stretching directions parallel to the chain and perpendicular to the convergence direction. The coevally emplacement of thrusts to the south or south-west (in France) also reflects a lateral strain partitioning here. The partial melting and synorogenic extension dynamic developed toward the southern parts of the inner zones (or rather toward the southeast after restoration, see Figure 1.24(a)) via the propagation of thrust systems (Matte 2007; Martínez Catalán et al. 2009). Note that around 330 Ma, the Variscan Range had an almost north-south strike (Figure 1.24(a)), and its known arcuate structure is the result of the late-orogenic events described below (Weil et al. 2010).

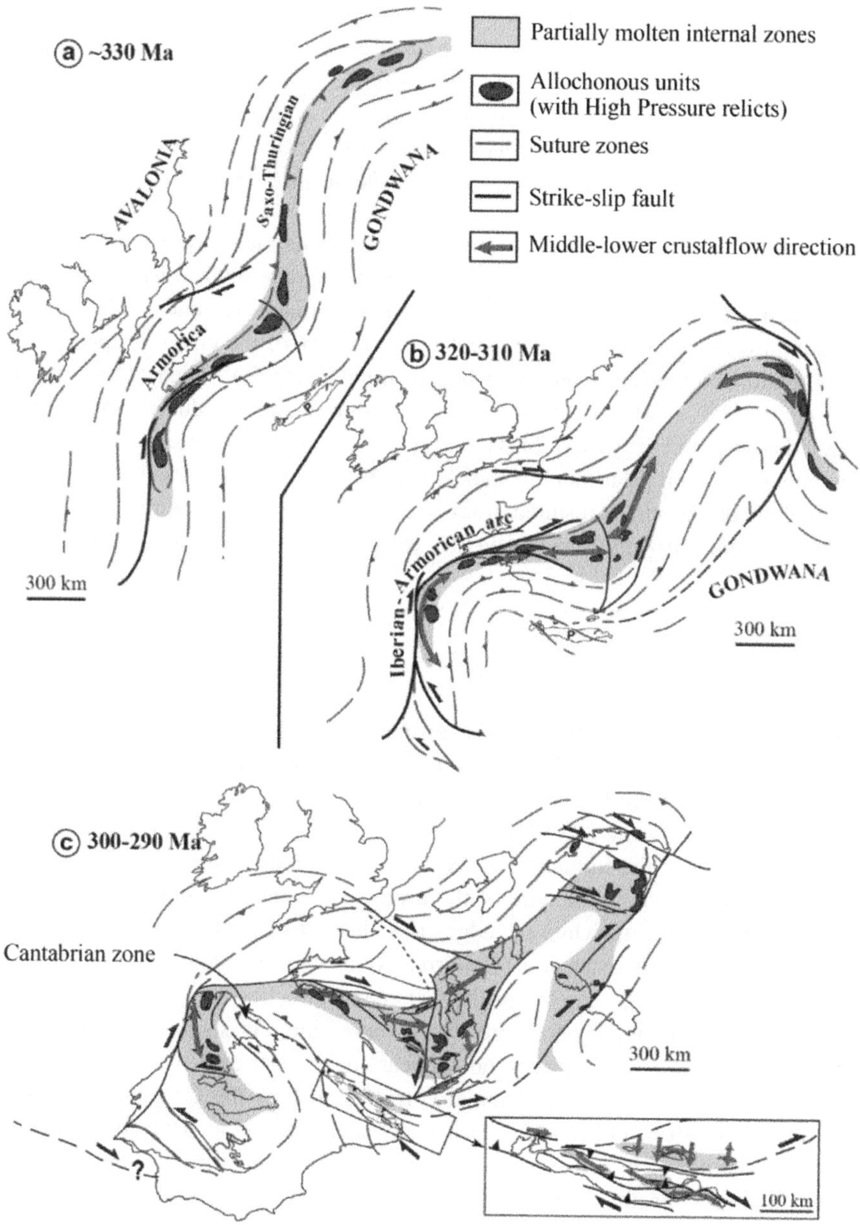


**Figure 1.24.** *Reconstruction of the Iberian-Armorican arc in the Carboniferous and Early Permian*

LEGEND FOR FIGURE 1.24 – *(a) Late phase of continental collision, ca. 330 Ma; (b) phase of late-orogenic extension in the inner zones and thrust propagation in the outer zones of the belt; (c) end of the closure of the Iberian-Armorican arc leading to the formation of the Cantabrian orocline, contemporary with the D2 transpressional event in the Pyrenees and HT/LP metamorphism in the southern domain of the Variscan belt. Modified from Cochelin et al. (2017).*

The origin of the curvature of the Variscan chain has always been greatly debated amongst the scientific community. Since the year 2000, it has been clearly established that the formation of the Iberian-Armorican arc, and at its core the Cantabrian orocline, is a very late event in the history of the chain. It appears that this virgation of the Variscan chain began around 330 Ma, when (1) the partial melting of the crust in the inner areas significantly reduced its strength (Vanderhaeghe and Teyssier 2001) and (2) dextral shear zones were forming throughout the chain (Martínez Catalán 2011). Available paleomagnetic data suggest that folding of the chain resulted from a change in shortening directions by 90° around 330 Ma, evolving from WNW-ESE to NNE-SSW (Edel et al. 2018).

### 1.4.2. *Late Variscan D2 phase*

The strain partitioning recognized in the Pyrenees between 310 and 290 Ma appears to be synchronous with important transfers of material within the Variscan chain during its dismantling (Figure 1.24(c)). Indeed, amplification of orogenic arcs led to the closure of the Iberian-Armorican arc and the formation of the Cantabrian orocline between 305 and 295 Ma (Weil et al. 2010; Edel et al. 2018). The formation of oroclines leads to the formation of atypical local strainfields leading to horizontal strain partitioning, where the core of the orocline is intensely compressed (folding and thrust development), while the edges (i.e. the rest of the chain) are affected by wrenching and lateral extension (Gutiérrez-Alonso et al. 2004) (Figure 1.24(c)). Therefore, it seems that the core of the Cantabrian orocline and the Pyrenees were both coevally structured by transpression/compression. As such, the formation of highly compressed domains does not appear to be restricted to the core of the oroclines, but rather draws longitudinal corridors such as the Cantabrian-Pyrenees axis (Figure 1.24(c)). Other transpressive corridors appear to have formed in the Variscan chain, such as in the outer crystalline massifs of the Alps, in Corsica-Sardinia, or in North Africa (Michard et al. 2010; Casini et al. 2015; Fréville et al. 2018). The difference in dynamics between the Axial Zone and the North Pyrenean massifs, that is, between transpressional and extensional dynamics, reflects this lateral strain partitioning on a belt scale (Figure 1.24(c)).

Therefore, the entire West European Variscan chain was subjected to intense structural reorganization, either by lateral extension and wrenching, or locally by transpression. This reorganization took place in an abnormally hot thermal context, marked by intense magmatism and widespread HT/LP metamorphism in both the internal zones and the Gondwanan margin external ones. Indeed, the intense folding of the lithosphere into multiple oroclines can only take place under abnormally hot thermal conditions, because otherwise it would deform into blocks (rigid plates). This anomalously hot thermal regime could result from a delamination process of the lithospheric mantle, as illustrated in the eastern Massif Central (Laurent et al. 2017), but also in Iberia (Gutiérrez-Alonso et al. 2004) or in southeastern Europe (Buzzi and Gaggero, 2008). Although no migmatitic domes similar to the Pyrenean ones appear in the Cantabrian area due to a lower Cenozoic denudation of the basement than in the Pyrenees, it still seems to have experienced the same thermal event marked by anomalously high geothermal gradients and magmatism (Blanco-Ferrera et al. 2017; García-López et al. 2018).

### 1.4.3. *Late Permian D3 phase*

Another turning point in the history of the Variscan chain took place around 280 Ma, when the Variscan convergence that led to the formation of Pangea ceased. In most of the inner areas of the chain, this phase was accompanied by thermal relaxation and deepening of the narrow Gzhelian-Autunian basins formed during phase D2 (Figure 1.25). Intense volcanism continues in northern Europe (Oslo rift), at the apex of a mantle plume (Lundmark et al. 2018). In this area, rift deformation occurs along an E-W direction of extension (Figure 1.25). In contrast, the southern part of the former Variscan range is still affected by bimodal magmatism and HT/LP metamorphism that is accompanied by generalized extension (Galli et al. 2012; Kunz et al. 2018; Pohl et al. 2018) (Figure 1.25), as observed in the Pyrenees (Figure 1.23). A mantle contribution has been clearly recognized for the plutonism and volcanism of this period (Buzzi and Gaggero 2008; Spalla et al. 2014). This extension is characterized, as in the Pyrenees, by the exhumation of granulitic crust via the activation of detachments and the formation of Permian basins, globally due to N-S extension (Figure 1.23(b)) (Pohl et al. 2018). Therefore, the direction of extension seems to point irrevocably in the direction of the hypothetical subduction zone of the Paleotethyan Ocean (Figure 1.25). This Permian extension is also accompanied by the rotation of some continental blocks belonging, today, to the Mediterranean-Alpine domain (e.g. Corsica, Sardinia, Maures and Apulia) (Edel et al. 2018), coupled with magmatic and metamorphic events. It could result from a back-arc extension process, in response to the subduction dynamics of the Paleotethys Ocean. This subduction, initiated around 330 Ma, is supposed to end around 240 Ma with the complete consumption of the subducting ocean (Bussy et al. 2000). The segmentation of Pangea during Permian rifting thus outlines the main continental blocks that will be involved in the Alpine cycle (Iberian plate, Corsican-Sardinian block, Apulia). The Permian rifting phase thus seems to have played a major role in the location of future plate boundaries and suggests a major impact of the Permian structural inheritance on the future tectonics of the European Alpine cycle.

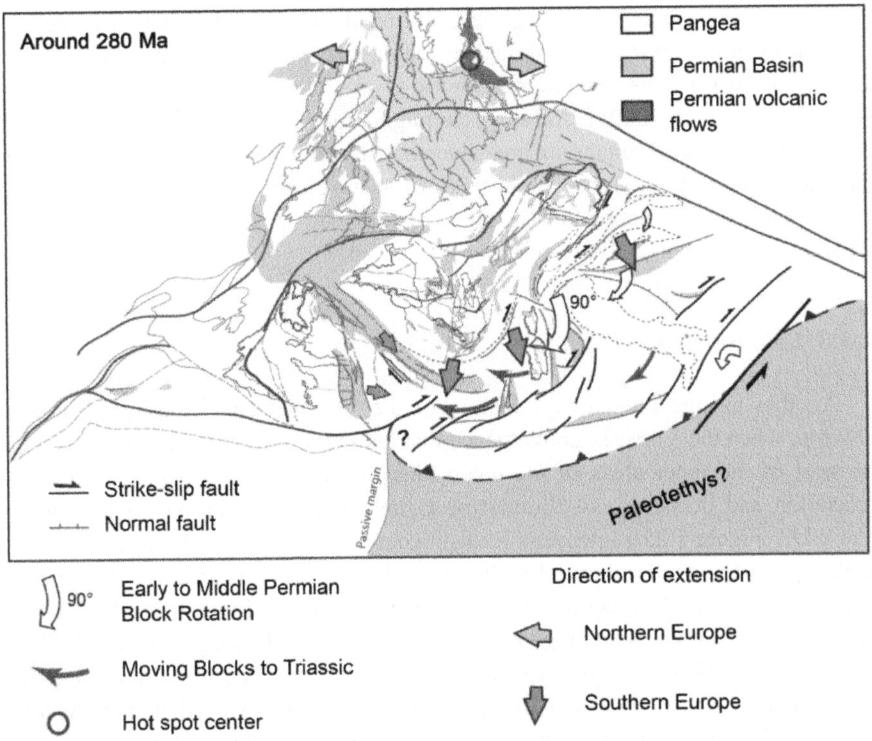


**Figure 1.25.** *Structural map of western Europe around 280 Ma, illustrating the Permian rifting dynamics and its probable relation in the southern part of Europe with the subduction dynamics of the Paleotethyan Ocean*

LEGEND FOR FIGURE 1.25.– *Reconstructions from Stampfli et al. (2013); Franke (2014); Cochelin et al. (2017). Location of Permian basins and magmatism from Lago et al. (2004b); Neumann et al. (2004); Lundmark et al. (2018); Ducassou et al. (2019).*

## 1.5. Implications of Paleozoic basement structuring on Alpine phases

As the Meso-Cenozoic tectonic history of the Pyrenees is extensively developed in great detail in the following chapters and in the second book on the Pyrenees, we will focus here only on the implications of our new knowledge of the basement on the formation models of the Pyrenean orogenic prism and briefly on the Cretaceous rifting phase. Regarding the Cretaceous rifting phase, it appears that the location of the rift is centered on the North Pyrenean massifs (Clerc and Lagabrielle 2014) and thus coincides with domains affected by a Permian extensional phase, bounded to the south by the remnant of Permian basins bordering the Axial Zone. This suggests that the Permian rift was preferentially reactivated in the Cretaceous, thus conducive to preserving the future Axial Zone (i.e. the Iberian margin) and forming a rather asymmetric Cretaceous rift in the central and eastern Pyrenees (Cochelin et al. 2018a; Ford and Vergés 2021). This Cretaceous thinning phase led to the dismantling of the Paleozoic basement of the European margin, individualizing the North Pyrenean massifs as tilted blocks, resulting in local mantle exhumation (Lagabrielle et al. 2019, 2020; Saspiturry et al. 2020a, 2020b). This mantle exhumation was accompanied by intense ductile deformation in the overlying continental crust (Asti et al. 2019), but also by the reactivation of inherited Paleozoic structures (Odlum and Stockli 2020). In the Axial Zone, Cretaceous extension is thus limited to the activation of rare normal faults, mostly localized along its northern edge, or to the reactivation of Permian or late Variscan normal faults (Figure 1.23). The Cretaceous rift tends to become symmetrical westward toward the Basque massifs (Teixell et al. 2016; Saspiturry et al. 2019b, 2021a, 2021b). Note that this western segment also corresponds to the segment where the Permian extensional dynamics were different at that time (Figure 1.23), so the influence of Paleozoic inheritance can also be considered.

Regarding the Cretaceous rift inversion and the formation of the orogenic prism during the Alpine orogeny, recent works have shown that the localization of the main Alpine thrusts in the Axial Zone corresponds to the reactivation of former contacts of (1) the Cretaceous rifting phase (Mencos et al. 2015) and/or (2) the Permian rifting phase (Saura and Teixell 2006; Izquierdo-Llavall et al. 2014), but also (3) areas of strong rheological contrasts in the crust, that is, at the inter-faces between crystalline rocks (plutons, gneissic domes) and poorly metamorphosed Upper Paleozoic series (Superstructure). These zones correspond in most cases to late Variscan mylonitic zones (Figures 1.6 and 1.8). In the former case, an example is the Lakhora-Eaux-Chaudes nappe (Figure 1.26(a)), located in the western part of the Axial Zone, which is located in an ancient Cretaceous normal fault, illustrated by the presence of synrift deposits north of this contact (Saspiturry et al. 2019b, 2021a). In the later case, we can mention the Gavarnie thrust and the Esera-Gistain fault, reactivating the Variscan (intra-Silurian) tectonic contacts located between the Lys-Caillaouas gneissic dome and the Devono-Carboniferous (Figure 1.8).

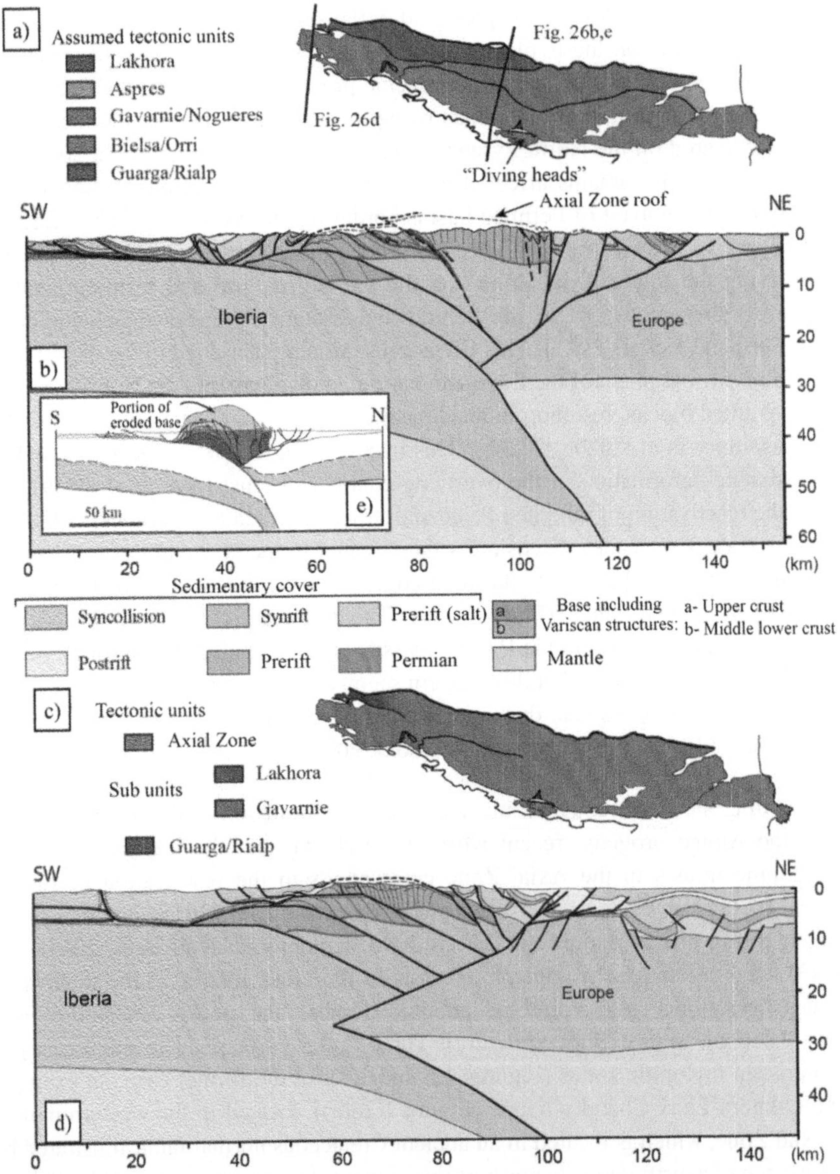


**Figure 1.26.** *Cross-sections of the central and eastern Pyrenees highlighting the structures related to the Variscan and Alpine orogenies*

LEGEND FOR FIGURE 1.26.– *(a) Location map of the different alpine nappes recognized in the Axial Zone, based on the model of Muñoz (1992), modified from Mouthereau et al. (2014). Sections b and d are located there; (b) interpretation of the ECORS profile in the central Pyrenees, modified from Beaumont et al. (2000) and taking into account structural and metamorphic markers in the basement; (c) new structural map unifying all alpine units in map (a) as a single "Axial Zone" structural unit, following the model of Laumonier (2015) and Cochelin et al. (2018a); (d) section in the western part of the Axial Zone, modified from Teixell et al. (2016) to add basement structural and metamorphic markers; (e) dated interpretation of the ECORS profile (Beaumont et al. 2000) following the anticlinal nappe-stack model.*

This chapter has illustrated that the continuity of Variscan structures (Figures 1.6 and 1.8) occurs on either side of Alpine faults, assumed to be major (Cochelin et al. 2017, 2018b). This is notably the case for both sides of the Nogueres thrust, located in the central Pyrenees and forming the formerly assumed continuity of the Gavarnie thrust (Figure 1.26(a)). In the central Pyrenees, the Nogueres thrust corresponds to the continuity of the Mérens late Variscan mylonitic zone (Figures 1.6 and 1.26(b)). A 1–2 km shift in Variscan paleo-isotherms could be demonstrated on either side of this mylonitic zone (Cochelin et al. 2018a) (Figure 1.26(b)). Whether this shift is late Variscan or related to a hypothetical alpine reactivation, it confirms the idea of a rapid vanishing of the Gavarnie thrust toward the east (Soler et al. 1998). In detail, we observe an almost perfect structural and metamorphic continuity of the Axial Zone in the central and eastern Pyrenees. The Axial Zone forms a single crustal block bounded to the north by the NPF and to the south by the Orri thrust (Figures 1.26(a) and (b)). Therefore, the amount of basement denudation and orogenic prism architecture in the central and eastern Pyrenees appear very similar to those described in the western part of the Axial Zone (Figure 1.26(d)) (Teixell 1996; Teixell et al. 2016; Bellahsen et al. 2019; Espurt et al. 2019). It appears that this part of the Axial Zone cannot be considered as a stack of anticlinally folded basement sheets, as previously proposed by Roure et al. (1989) and Muñoz (1992) (Figures 1.26(a) and (e)) following the ECORS seismic profile (Choukroune et al. 1989). This model implied (1) significant block tilting on both sides of the main thrusts (Beaumont et al. 2000), (2) displacement in the order of 20 km along the Nogueres thrust (Berastegui et al. 1993; Mouthereau et al. (2014), and (3) significant denudation of the basement, which was found to be inconsistent with the regional geology (compare Figures 1.26(d) and (e)). The Nogueres (or "plunging noses") unit (Séguret 1972; Choukroune and Séguret, 1973) (Figure 1.26(a)), located on the southern flank of the Axial Zone, is thought to correspond to thin pieces from the top of the Axial Zone (Figure 1.26(c)), ripped off during the activation of superficial thrusts at the basement-cover interface (Laumonier 2015; Cochelin et al. 2018a), and not to a unit rooted within the Axial Zone.

The formation of the orogenic prism appears to have occurred first by inversion of the Cretaceous rift, located in the North Pyrenean zone (i.e. north of the North Pyrenean fault). This inversion leads to the detachment of the synrift Cretaceous basins from the basement and the progressive accretion of the North Pyrenean massifs (Ford et al. 2016; Vacherat et al. 2016). Accretion, by underplating the Iberian margin beneath the European margin, led, in a second stage, to the formation of the basement nappes (Figures 1.26(b) and (d)). In this scheme, the Axial Zone thus forms a pop-up structure, as previously suggested by Brun (2002). This second phase is also accompanied by discrete alpine penetrative deformation within the Axial Zone, in the form of (1) new schistosity, (2) shear zones, or (3) by the partial reactivation of some late Varsican structures (faults, Ss schistosity) (Cochelin 2016). This penetrative deformation forms at low temperatures (250–300 °C), that is, in ranges below the temperatures recorded by the crust in the Carboniferous and Permian (Bons 1988; Soler et al. 1998; Cochelin et al. 2018a; Airaghi et al. 2020).

Therefore, the present architecture of the Pyrenean chain depends intrinsically on the structuring inherited from the Upper Paleozoic. In the first order, it is a probable structuring inherited from the Permian rifting that controlled the Cretaceous thinning zones and therefore the preferential inversion zones (i.e. zones showing the maximum shortening amount) during the Pyrenean orogeny. Finally, the internal heterogeneities of the Axial Zone inherited from the late Variscan transpressional phase played only a second-order role in the process of localization of the Alpine deformation in the crust. Most of the shortening was accommodated along a few main thrusts located at the edge of the Axial Zone and by intracover (or supra-axial zone) decollements. With respect to the alternative model proposed, it appears that the deformation models prior to the realization of the ECORS profile (Mattauer 1968; Choukroune and Séguret 1973) were far more faithful to the geological reality of the Pyrenees than those developed afterward. Clearly, the coupling between the different state-of-the-art analytical methods (geophysics, thermochronology, etc.) and our complete geological knowledge (i.e. of both the basement and its cover) will undoubtedly allow us to propose new innovative models on the formation and evolution of the Pyrenees (e.g. Teixell et al. 2018; Calvet et al. 2020).

The RGF Pyrenees project has thus made it possible to make an inventory of existing knowledge and data on the chain, and to provide new data, particularly on the entire Variscan basement of the Pyrenees, both in the Axial Zone and North Pyrenean massifs. These cartographic and analytical data, integrated into the RGF geological information system, have been linked in order to propose an integrated vision of the major geological events identified within the Pyrenean basement. This approach has allowed an unexpected meeting between different thematicians, leading to major scientific advances on the most recent geological events, namely the Cretaceous extension and the Pyrenean inversion.

## 1.6. References

Aerden, D.G.A.M. (1994). Late orogenic extension kinematics of orogenic collapse in the Variscan Pyrenees deduced from microstructures in porphyroblastic rocks from the Lys-Caillaouas massif. *Tectonophysics*, 238, 139–160 [Online]. Available at: https://doi.org/10.1016/0040-1951(94)90053-1.

Aerden, D.G.A.M. (1995). Porphyroblast non-rotation during crustal extension in the Variscan Lys-Caillaouas massif, Pyrenees. *Journal of Structural Geology*, 17, 709–725 [Online]. Available at: https://doi.org/10.1016/0191-8141(94)00090-M.

Aguilar, C., Liesa, M., Castiñeiras, P., Navidad, M. (2014). Late Variscan metamorphic and magmatic evolution in the eastern Pyrenees revealed by U-Pb age zircon dating. *J. Geol. Soc. London*, 171, 181–192 [Online]. Available at: https://doi.org/10.1144/jgs2012-086.

Aguilar, C., Liesa, M., Štípská, P., Schulmann, K., Muñoz, J.A., Casas, J.M. (2015). P–T–t–d evolution of orogenic middle crust of the Roc de Frausa massif (Eastern Pyrenees): A result of horizontal crustal flow and Carboniferous doming? *J. Meta. Geol.*, 33, 273–294 [Online]. Available at: https://doi.org/10.1111/jmg.12120.

Airaghi, L., Bellahsen, N., Dubacq, B., Chew, D., Rosenberg, C., Janots, E., Waldner, M., Magnin, V. (2020). Pre-orogenic upper crustal softening by lower greenschist facies metamorphic reactions in granites of the central Pyrenees. *Journal of Metamorphic Geology*, 38, 183–204 [Online]. Available at: https://doi.org/10.1111/jmg.12520.

Andrieux, P. (1982). Conditions de cristallisation et évolution paragénétique d'une charnockite hercynienne : le complexe granulitique d'Ansignan (Massif de l'Agly, Pyrénées-Orientales). *Bulletin de Minéralogie*, 105, 253–266 [Online]. Available at: https://doi.org/10.3406/bulmi.1982.7613.

Angrand, P., Mouthereau, F., Masini, E., Asti, R. (2020). A reconstruction of Iberia accounting for Western Tethys-North Atlantic kinematics since the late-Permian-Triassic. *Solid Earth*, 11, 1313–1332 [Online]. Available at: https://doi.org/10.5194/se-11-1313-2020.

Argand, E. (1911). Les nappes de recouvrement des Alpes Pennines et leurs prolongements structuraux. *Bull. Lab. Géol. Univ. Lausanne*, 14, 1–25.

Arthaud, F. and Matte, P. (1975). Les décrochements tardi-hercyniens du sud-ouest de l'Europe. Géométrie et essai de reconstitution des conditions de la déformation. *Tectono-physics*, 25, 139–171 [Online]. Available at: https://doi.org/10.1016/0040-1951(75)90014-1.

Asti, R., Lagabrielle, Y., Fourcade, S., Corre, B., Monié, P. (2019). How do continents deform during mantle exhumation? Insights from the Northern Iberia inverted paleopassive margin, Western Pyrenees (France). *Tectonics*, 38, 1666–1693 [Online]. Available at: https://doi.org/10.1029/2018TC005428.

Asti, R., Saspiturry, N., Angrand, P. (2022). The Mesozoic Iberia-Eurasia plate boundary: A wide domain of distributed transtensional deformation rather than a single wrench fault. *Earth-Science Reviews*, 230, 104040 [Online]. Available at: https://doi.org/10.1016/j.earscirev.2022.104040.

Aurejac, J.B., Gleizes, G., Diot, H., Bouchez, J.L. (2004). Le complexe granitique de Quérigut (Pyrénées, France) réexaminé par la technique de l'ASM : un pluton syntectonique de la transpression dextre hercynienne. *Bull. Soc. Géol. France*, 175, 157–174.

Autran, A., Fonteilles, M., Guitard, G. (1970). Relations entre les intrusions de granitoïdes, l'anatexie et le métamorphisme régional considérées principalement du point de vue du rôle de l'eau ; cas de la chaîne hercynienne des Pyrénées orientales. *Bulletin de la Société Géologique de France*, S7-XII, 673–731 [Online]. Available at: https://doi.org/10.2113/gssgfbull.S7-XII.4.673.

Ayora, C., Liesa, M., Delgado, J. (1993). Low-thermal-gradient Hercynian metamorphism in the eastern Pyrenees. *Journal of Metamorphic Geology*, 11, 49–58 [Online]. Available at: https://doi.org/10.1111/j.1525-1314.1993.tb00130.x.

Azambre, B. and Guitard, G. (2001). Disthène et staurotide reliques dans les métapelites du Canigou (Pyrénées orientales). Relations avec les épisodes hercyniens de basse et moyenne pressions. *Comptes Rendus de l'Académie des Sciences – Series IIA – Earth and Planetary Science*, 333, 601–609 [Online]. Available at: https://doi.org/10.1016/S1251-8050(01)01670-6.

Barnolas, A. and Chiron, J.C. (1996). *Synthèse géologique et géophysique des Pyrénées. Volume 1 : Introduction géophysique. Cycle Hercynien*. BRGM-ITGE, Orléans/Madrid.

Beaumont, C., Muñoz, J.A., Hamilton, J., Fullsack, P. (2000). Factors controlling the Alpine evolution of the central Pyrenees inferred from a comparison of observations and geodynamical models. *J. Geophys. Res.*, 105, 8121–8145 [Online]. Available at: https://doi.org/10.1029/1999JB900390.

Bellahsen, N., Bayet, L., Denele, Y., Waldner, M., Airaghi, L., Rosenberg, C., Dubacq, B., Mouthereau, F., Bernet, M., Pik, R. et al. (2019). Shortening of the Axial Zone, pyrenees: Shortening sequence, upper crustal mylonites and crustal strength. *Tectonophysics*, 766, 433–452 [Online]. Available at: https://doi.org/10.1016/j.tecto.2019.06.002.

Berastegui, X., Losantos, M., Munoz, J.A., Puigdefàbregas, C. (1993). Tall geologic del Pirineu central 1/200.000. *Publ. Geol. Catalunya*.

Bixel, F. (1984). Le volcanisme stéphano-permien des Pyrénées. PhD Thesis, Université Paul Sabatier, Toulouse.

Bixel, F. and Lucas, C. (1983). Magmatisme, tectonique et sédimentation dans les fossés stéphano-permiens des Pyrénées occidentales. *Revue de géologie dynamique et de géographie physique*, 24, 329–342.

Blanco-Ferrera, S., Sanz-López, J., García-López, S., Bastida, F. (2017). Tectonothermal evolution of the northeastern Cantabrian zone (Spain). *Int. J. Earth Sci. (Geol. Rundsch)*, 106, 1539–1555 [Online]. Available at: https://doi.org/10.1007/s00531-016-1365-5.

Bons, A.J. (1988). Intracrystalline deformation and slaty cleavage development in very low-grade slates from the Central Pyrenees. *Geologica Ultraiectina*, 56.

Bouchez, J.L. and Gleizes, G. (1995). Two-stage deformation of the Mont-Louis-Andorra granite pluton (Variscan Pyrenees) inferred from magnetic susceptibility anisotropy. *Journal of the Geological Society*, 152, 669–679 [Online]. Available at: https://doi.org/10.1144/gsjgs.152.4.0669.

Bouhallier, H., Choukroune, P., Ballèvre, M. (1991). Évolution structurale de la croûte profonde Hercynienne : exemple du massif de l'Agly (Pyrénées Orientales, France). *C. R. Acad. Sci. Paris*, 312, 647–654.

Brun, J.-P. (2002). Deformation of the continental lithosphere: Insights from brittle-ductile models. *Geological Society, London, Special Publications*, 200, 355–370 [Online]. Available at: https://doi.org/10.1144/GSL.SP.2001.200.01.20.

Burg, J.-P., Van Den Driessche, J., Brun, J.-P. (1994). Syn- to post-thickening extension: Mode and consequences. *Comptes rendus de l'Académie des sciences. Série 2. Sciences de la terre et des planètes*, 319, 1019–1032.

Bussy, F., Hernandez, J., Raumer, J.V. (2000). Bimodal magmatism as a consequence of the post-collisional readjustment of the thickened Variscan continental lithosphere (Aiguilles Rouges-Mont Blanc massifs, Western Alps). *Geological Society of America Special Papers*, 350, 221–233 [Online]. Available at: https://doi.org/10.1130/0-8137-2350-7.221.

Buzzi, L. and Gaggero, L. (2008). Petrogenesis of post-orogenic Lower Permian andesites in southern Europe: Insights into the collapse of the Variscan range. *Geodinamica Acta*, 21, 273–290 [Online]. Available at: https://doi.org/10.3166/ga.21.273-290.

Calvet, M., Gunnell, Y., Laumonier, B. (2020). Denudation history and palaeogeography of the Pyrenees and their peripheral basins: An 84-million-year geomorphological perspective. *Earth-Science Reviews*, 103436 [Online]. Available at: https://doi.org/10.1016/j.earscirev.2020.103436.

Carreras, J. (2001). Zooming on Northern Cap de Creus shear zones. *Journal of Structural Geology*, 23, 1457–1486 [Online]. Available at: https://doi.org/10.1016/S0191-8141(01)00011-6.

Carreras, J. and Capella, I. (1994). Tectonic levels in the palaeozoic basement of the Pyrenees: A review and a new interpretation. *Journal of Structural Geology*, 16, 1509–1524 [Online]. Available at: https://doi.org/10.1016/0191-8141(94)90029-9.

Carreras, J. and Debat, P. (1996). Tectonique Hercynienne. In *Synthèse géologique et géophysique des Pyrénées. Volume 1*, Barnolas, A. and Chiron, J.C. (eds). BRGM-ITGE, Orléans/Madrid.

Carreras, J. and Druguet, E. (1994). Structural zonation as a result of inhomogeneous non-coaxial deformation and its control on syntectonic intrusions: An example from the Cap de Creus area, eastern-Pyrenees. *Journal of Structural Geology, Structures and Tectonics at Different Lithospheric Levels*, 16, 1525–1534 [Online]. Available at: https://doi.org/10.1016/0191-8141(94)90030-2.

Carreras, J. and Druguet, E. (2014). Framing the tectonic regime of the NE Iberian Variscan segment. *Geological Society, London, Special Publications*, 405, 249–264 [Online]. Available at: https://doi.org/10.1144/SP405.7.

Carreras, J., Estrada, A., White, S. (1977). The effects of folding on the C-axis fabrics of a quartz mylonite. *Tectonophysics*, 39, 3–24 [Online]. Available at: https://doi.org/10.1016/0040-1951(77)90085-3.

Carreras, J., Julivert, M., Santanach, P. (1980). Hercynian mylonite belts in the eastern Pyrenees: An example of shear zones associated with late folding. *Journal of Structural Geology, Shear Zones in Rocks*, 2, 5–9 [Online]. Available at: https://doi.org/10.1016/0191-8141(80)90028-0.

Casas, J.M., Castiñeiras, P., Navidad, M., Liesa, M., Carreras, J. (2010). New insights into the Late Ordovician magmatism in the Eastern Pyrenees: U–Pb SHRIMP zircon data from the Canigó massif. *Gondwana Research*, 17, 317–324 [Online]. Available at: https://doi.org/10.1016/j.gr.2009.10.006.

Casini, L., Cuccuru, S., Puccini, A., Oggiano, G., Rossi, P. (2015). Evolution of the Corsica-Sardinia Batholith and late-orogenic shearing of the Variscides. *Tectonophysics*, 646, 65–78 [Online]. Available at: https://doi.org/10.1016/j.tecto.2015.01.017.

Chardon, D., Gapais, D., Cagnard, F. (2009). Flow of ultra-hot orogens: A view from the Precambrian, clues for the Phanerozoic. *Tectonophysics*, 477, 105–118 [Online]. Available at: https://doi.org/10.1016/j.tecto.2009.03.008.

Choukroune, P. (1989). The Ecors Pyrenean deep seismic profile reflection data and the overall structure of an orogenic belt. *Tectonics*, 8, 23–39 [Online]. Available at: https://doi.org/10.1029/TC008i001p00023.

Choukroune, P. and Séguret, M. (1973). Tectonics of the Pyrenees: Role of compression and gravity. *Gravity and Tectonics*, 141–156.

Clariana, P. and García-sansegundo, J. (2009). Variscan structure in the eastern part of the Pallaresa massif, Axial Zone of the Pyrenees (NW Andorra). Tectonic implications. *Bulletin de la Société Géologique de France*, 180, 501–511 [Online]. Available at: https://doi.org/10.2113/gssgfbull.180.6.501.

Clerc, C. and Lagabrielle, Y. (2014). Thermal control on the modes of crustal thinning leading to mantle exhumation: Insights from the Cretaceous Pyrenean hot paleomargins. *Tectonics*, 33, 2013TC003471 [Online]. Available at: https://doi.org/10.1002/2013TC003471.

Cochelin, B. (2016). Champ de déformation du socle paléozoïque des Pyrénées. PhD Thesis, Université Toulouse 3 Paul Sabatier.

Cochelin, B., Chardon, D., Denèle, Y., Gumiaux, C., Le Bayon, B. (2017). Vertical strain partitioning in hot Variscan crust: Syn-convergence escape of the Pyrenees in the Iberian-Armorican syntax. *Bull. Soc. Géol. Fr.*, 188(39) [Online]. Available at: https://doi.org/10.1051/bsgf/2017206.

Cochelin, B., Lemirre, B., Denèle, Y., de Saint Blanquat, M., Lahfid, A., Duchêne, S. (2018a). Structural inheritance in the Central Pyrenees: The Variscan to Alpine tecto-nometamorphic evolution of the Axial Zone. *Journal of the Geological Society*, 175, 336–351 [Online]. Available at: https://doi.org/10.1144/jgs2017-066.

Cochelin, B., Gumiaux, C., Chardon, D., Denèle, Y., Le Bayon, B. (2018b). Multi-scale strainfield analysis using geostatistics: Investigating the rheological behavior of the hot Variscan crust of the Pyrenees (Axial Zone). *Journal of Structural Geology*, 116, 114–130 [Online]. Available at: https://doi.org/10.1016/j.jsg.2018.07.024.

De Hoÿm de Marien, L., Le Bayon, B., Pitra, P., Van den Driessche, J., Poujol, M., Cagnard, F. (2019). Two-stage Variscan metamorphism in the Canigou massif: Evidence for crustal thickening in the Pyrenees. *J. Metamorph. Geol.*, 37, 863–888 [Online]. Available at: https://doi.org/10.1111/jmg.12487.

Debon, F., Enrique, P., Autran, A. (1996). Magmatisme Hercynien. In *Synthèse géologique et géophysique des Pyrénées. Volume 1*, Barnolas, A. and Chiron, J.C. (eds). BRGM-ITGE, Orléans/Madrid.

Delaperriere, E., de Saint Blanquat, M., Brunel, M., Lancelot, J.R. (1994). Géochronologie U-Pb sur zircons et monazites dans le massif du Saint Barthélémy (Pyrénées, France) ; discussion des âges des événements varisques et pré-varisques. *Bulletin de la Société Géologique de France*, 165, 101–112.

Delvolvé, J.-J., Vachard, D., Souquet, P. (1998). Stratigraphic record of thrust propagation, Carboniferous foreland basin, Pyrenees, with emphasis on Pays-de-Sault (France/Spain). *Geol. Rundsch.*, 87, 363–372 [Online]. Available at: https://doi.org/10.1007/s005310050215.

Denèle, Y., Olivier, P., Gleizes, G., Barbey, P. (2007). The Hospitalet gneiss dome (Pyrenees) revisited: Lateral flow during Variscan transpression in the middle crust. *Terra Nova*, 19, 445–453 [Online]. Available at: https://doi.org/10.1111/j.1365-3121.2007.00770.x.

Denèle, Y., Olivier, P., Gleizes, G. (2008). Progressive deformation of a zone of magma transfer in a transpressional regime: The Variscan Mérens shear zone (Pyrenees, France). *Journal of Structural Geology*, 30, 1138–1149 [Online]. Available at: https://doi.org/10.1016/j.jsg.2008.05.006.

Denèle, Y., Olivier, P., Gleizes, G., Barbey, P. (2009). Decoupling between the middle and upper crust during transpression-related lateral flow: Variscan evolution of the Aston gneiss dome (Pyrenees, France). *Tectonophysics*, 477, 244–261 [Online]. Available at: https://doi.org/10.1016/j.tecto.2009.04.033.

Denèle, Y., Paquette, J.-L., Olivier, P., Barbey, P. (2012). Permian granites in the Pyrenees: The Aya pluton (Basque Country). *Terra Nova*, 24, 105–113 [Online]. Available at: https://doi.org/10.1111/j.1365-3121.2011.01043.x.

Denèle, Y., Laumonier, B., Paquette, J.-L., Olivier, P., Gleizes, G., Barbey, P. (2014). Timing of granite emplacement, crustal flow and gneiss dome formation in the Variscan segment of the Pyrenees. *Geological Society, London, Special Publications*, 405, 265–287 [Online]. Available at: https://doi.org/10.1144/SP405.5.

Druguet, E., Castro, A., Chichorro, M., Pereira, M.F., Fernández, C. (2014). Zircon geochronology of intrusive rocks from Cap de Creus, Eastern Pyrenees. *Geological Magazine*, 151, 1095–1114 [Online]. Available at: https://doi.org/10.1017/S0016756814000041.

Ducassou, C., Mercuzot, M., Bourquin, S., Rossignol, C., Pellenard, P., Beccaletto, L., Poujol, M., Hallot, E., Pierson-Wickmann, A.C., Hue, C. et al. (2019). Sedimentology and U-Pb dating of Carboniferous to Permian continental series of the northern Massif Central (France): Local palaeogeographic evolution and larger scale correlations. *Palaeogeography, Palaeoclimatology, Palaeoecology*, 533, 109228 [Online]. Available at: https://doi.org/10.1016/j.palaeo.2019.06.001.

Edel, J.B., Schulmann, K., Lexa, O., Lardeaux, J.M. (2018). Late Palaeozoic palaeomagnetic and tectonic constraints for amalgamation of Pangea supercontinent in the European Variscan belt. *Earth-Science Reviews*, 177, 589–612 [Online]. Available at: https://doi.org/10.1016/j.earscirev.2017.12.007.

Espurt, N., Angrand, P., Teixell, A., Labaume, P., Ford, M., de Saint Blanquat, M., Chevrot, S. (2019). Crustal-scale balanced cross-section and restorations of the Central Pyrenean belt (Nestes-Cinca transect): Highlighting the structural control of Variscan belt and Permian-Mesozoic rift systems on mountain building. *Tectonophysics*, 764, 25–45 [Online]. Available at: https://doi.org/10.1016/j.tecto.2019.04.026.

Esteban, J.J., Aranguren, A., Cuevas, J., Hilario, A., Tubía, J.M., Larionov, A., Sergeev, S. (2015). Is there a time lag between the metamorphism and emplacement of plutons in the Axial Zone of the Pyrenees? *Geological Magazine*, 152, 935–941 [Online]. Available at: https://doi.org/10.1017/S001675681500014X.

Evans, N.G. (1993). Deformation during the emplacement of the Maladeta granodiorite, Spanish Pyrenees. PhD Thesis, University of Leeds.

Evans, N.G., Gleizes, G., Leblanc, D., Bouchez, J.-L. (1997). Hercynian tectonics in the Pyrenees: A new view based on structural observations around the Bassie's granite pluton. *Journal of Structural Geology*, 19, 195–208 [Online]. Available at: https://doi.org/10.1016/S0191-8141(96)00080-6.

Evans, N.G., Gleizes, G., Leblanc, D., Bouchez, J.L. (1998). Syntectonic emplacement of the Maladeta granite (Pyrenees) deduced from relationships between Hercynian deformation and contact metamorphism. *Journal of the Geological Society*, 155, 209–216 [Online]. Available at: https://doi.org/10.1144/gsjgs.155.1.0209.

Faure, M., Lardeaux, J.-M., Ledru, P. (2009). A review of the pre-Permian geology of the Variscan French Massif Central. *Comptes Rendus Geoscience*, 341, 202–213 [Online]. Available at: https://doi.org/10.1016/j.crte.2008.12.001.

Ford, M. and Vergés, J. (2021). Evolution of a salt-rich transtensional rifted margin, eastern North Pyrenees, France. *Journal of the Geological Society*, 178 [Online]. Available at: https://doi.org/10.1144/jgs2019-157.

Ford, M., Hemmer, L., Vacherat, A., Gallagher, K., Christophoul, F. (2016). Retro-wedge foreland basin evolution along the ECORS line, eastern Pyrenees, France. *Journal of the Geological Society*, 2015–129 [Online]. Available at: https://doi.org/10.1144/jgs2015-129.

Fossen, H., Tikoff, B., Teyssier, C. (1994). Strain modeling of transpressional and transtensional deformation. *Norsk Geologisk Tidsskrift*, 74, 134–145.

Franke, W. (2014). Topography of the Variscan orogen in Europe: Failed-not collapsed. *Int. J. Earth Sci. (Geol. Rundsch.)*, 103, 1471–1499 [Online]. Available at: https://doi.org/10.1007/s00531-014-1014-9.

Fréville, K., Trap, P., Faure, M., Melleton, J., Li, X.-H., Lin, W., Blein, O., Bruguier, O., Poujol, M. (2018). Structural, metamorphic and geochronological insights on the Variscan evolution of the Alpine basement in the Belledonne massif (France). *Tectonophysics*, 726, 14–42 [Online]. Available at: https://doi.org/10.1016/j.tecto.2018.01.017.

Galli, A., Le Bayon, B., Schmidt, M.W., Burg, J.-P., Reusser, E., Sergeev, S.A., Larionov, A. (2012). U-Pb zircon dating of the Gruf complex: Disclosing the late Variscan granulitic lower crust of Europe stranded in the Central Alps. *Contrib. Mineral. Petrol.*, 163, 353–378 [Online]. Available at: https://doi.org/10.1007/s00410-011-0676-6.

Gapais, D. (1989). Shear structures within deformed granites: Mechanical and thermal indicators. *Geology*, 17, 1144–1147 [Online]. Available at: https://doi.org/10.1130/0091-7613(1989)017<1144:SSWDGM>2.3.CO;2.

Gapais, D., Brun, J.-P., Gumiaux, C., Cagnard, F., Ruffet, G., Le Carlier de Veslud, C. (2015). Extensional tectonics in the Hercynian Armorican belt (France). An overview. *Bulletin de la Société Géologique de France*, 186, 117–129 [Online]. Available at: https://doi.org/10.2113/gssgfbull.186.2-3.117.

García-López, S., Voldman, G.G., Bastida, F., Aller, J. (2018). Tectonothermal analysis of a major unit of the Cantabrian zone: The Ponga unit (Variscan belt, NW Spain). *Int. J. Earth Sci. (Geol. Rundsch.)*, 107, 2727–2740 [Online]. Available at: https://doi.org/10.1007/s00531-018-1623-9.

García-Sansegundo, J. (1991). *Estratigrafía y estructura de la Zona Axial pirenaica en la transversal del Valle de Arán y de la Alta Ribagorça*. IGME, Madrid.

Gibson, R.L. (1991). Hercynian low-pressure-high-temperature regional metamorphism and subhorizontal foliation development in the Canigou massif, Pyrenees, France – Evidence for crustal extension. *Geology*, 19, 380–383 [Online]. Available at: https://doi.org/10.1130/0091-7613(1991)019<0380:HLPHTR>2.3.CO;2.

Gibson, R.L. and Bickle, M.J. (1994). Thermobarometric constraints on the conditions of metamorphism in the Canigou massif, Pyrenees: Implications for Hercynian geothermal gradients. *Journal of the Geological Society*, 151, 987–997 [Online]. Available at: https://doi.org/10.1144/gsjgs.151.6.0987.

Giró, E. and López, G.A. (2007). Evolución metamórfica de las kinzigitas en facies granulita del macizo del Aglí. *Geogaceta*, 95–98.

Gisbert Aguilar, J. (1981). Estudio geológico-petrológico del estefaniense-pérmico de la Sierra del Cadí (Pirineo de Lérida): Diagénesis y sedimentología. PhD Thesis, Universidad de Zaragoza.

Gleizes, G., Leblanc, D., Santana, V., Olivier, P., Bouchez, J.L. (1998a). Sigmoidal structures featuring dextral shear during emplacement of the Hercynian granite complex of Cauterets-Panticosa (Pyrenees). *Journal of Structural Geology*, 20, 1229–1245 [Online]. Available at: https://doi.org/10.1016/S0191-8141(98)00060-1.

Gleizes, G., Leblanc, D., Bouchez, J.L. (1998b). The main phase of the Hercynian orogeny in the Pyrenees is a dextral transpression. *Geological Society, London, Special Publications*, 135, 267–273 [Online]. Available at: https://doi.org/10.1144/GSL.SP.1998.135.01.17.

Gleizes, G., Crevon, G., Asrat, A., Barbey, P. (2006). Structure, age and mode of emplacement of the Hercynian Bordères-Louron pluton (Central Pyrenees, France). *Int. J. Earth Sci. (Geol. Rundsch.)*, 95, 1039–1052 [Online]. Available at: https://doi.org/10.1007/s00531-006-0088-4.

Gretter, N., Ronchi, A., López-Gómez, J., Arche, A., De la Horra, R., Barrenechea, J., Lago, M. (2015). The late Palaeozoic-early Mesozoic from the Catalan Pyrenees (Spain): 60 Myr of environmental evolution in the frame of the western peri-Tethyan palaeogeography. *Earth-Science Reviews*, 150, 679–708 [Online]. Available at: https://doi.org/10.1016/j.earscirev.2015.09.001.

Guerrot, C. (1998). Résultats de datation U-Pb par dissolution sur zircons pour le granite de Cauterets, Pyrénées. Report, SMN/PEA/ISO 146/98 CG/NB, BRGM.

Guitard, G. (1960). Linéations, schistosite et phases de plissement durant l'orogenèse hercynienne dans les terrains anciens des Pyrénées orientales, leurs relations avec le métamorphisme et la granitisation [with discussion]. *Bulletin de la Société Géologique de France*, S7-II, 862–887 [Online]. Available at: https://doi.org/10.2113/gssgfbull.S7-II.7.862.

Guitard, G. (1964). Un exemple de structure en nappe de style pennique dans la chaîne hercynienne : les gneiss stratoïdes du Canigou (Pyrénées-Orientales). *Comptes Rendus Académie des Sciences*, 258, 4597–4599.

Guitard, G. (1970). *Le métamorphisme hercynien mésozonal et les gneiss œillés du massif du Canigou (Pyrénées Orientales)*. BRGM, Orléans.

Guitard, G., Vielzeuf, D., Martinez, F. (1996). *Métamorphisme hercynien*, volume 1. BRGM-ITGE, Orléans/Madrid.

Gumiaux, C., Gapais, D., Brun, J.P., Chantraine, J., Ruffet, G. (2004). Tectonic history of the Hercynian Armorican Shear belt (Brittany, France). *Geodinamica Acta*, 17, 289–307 [Online]. Available at: https://doi.org/10.3166/ga.17.289-307.

Gutiérrez-Alonso, G., Fernández-Suárez, J., Weil, A.B. (2004). Orocline triggered lithospheric delamination. *Geological Society of America Special Papers*, 383, 121–130 [Online]. Available at: https://doi.org/10.1130/0-8137-2383-3(2004)383[121:OTLD]2.0.CO;2.

Hart, N.R., Stockli, D.F., Hayman, N.W. (2016). Provenance evolution during progressive rifting and hyperextension using bedrock and detrital zircon U-Pb geochronology, Mauléon Basin, western Pyrenees. *Geosphere*, 12, 1166–1186 [Online]. Available at: https://doi.org/10.1130/GES01273.1.

Heim, A. (1919). Geologie der Schweiz. *Chr. Herm. Tauchnitz.*

Izquierdo-Llavall, E., Casas-Sainz, A., Oliva-Urcia, B., Scholger, R. (2014). Palaeomagnetism and magnetic fabrics of the Late Palaeozoic volcanism in the Castejón-Laspaúles basin (Central Pyrenees). Implications for palaeoflow directions and basin configuration. *Geological Magazine*, 151, 777–797 [Online]. Available at: https://doi.org/10.1017/S0016756813000769.

Jammes, S., Manatschal, G., Lavier, L., Masini, E. (2009). Tectonosedimentary evolution related to extreme crustal thinning ahead of a propagating ocean: Example of the western Pyrenees. *Tectonics*, 28, TC4012 [Online]. Available at: https://doi.org/10.1029/2008TC002406.

Kilzi, M.A. (2014). Étude pétro-géochimique des formations mafiques et ultramafiques associées aux formations granulitiques, métamorphiques et anatectiques des Pyrénées hercyniennes (massif du Castillon et massif de Gavarnie). PhD Thesis, Université Toulouse III Paul Sabatier.

Kilzi, M.A., Grégoire, M., Bosse, V., Benoît, M., Driouch, Y., de Saint Blanquat, M., Debat, P. (2016). Geochemistry and zircon U-Pb geochronology of the ultramafic and mafic rocks emplaced within the anatectic series of the Variscan Pyrenees: The example of the Gavarnie-Heas dome (France). *Comptes Rendus Geoscience*, 348, 107–115 [Online]. Available at: https://doi.org/10.1016/j.crte.2015.06.014.

Kriegsman, L.M., Aerden, D.G.A.M., Bakker, R.J., den Brok, S.W.J., Schutjens, P.M.T.M. (1989). Variscan tectonometamorphic evolution of the eastern Lys-Cail-laouas massif, Central Pyrenees – Evidence for late orogenic extension prior to peak metamorphism. *Geologie en Mijnbouw*, 68, 323–333.

Kunz, B.E., Manzotti, P., von Niederhäusern, B., Engi, M., Darling, J.R., Giuntoli, F., Lanari, P. (2018). Permian high-temperature metamorphism in the Western Alps (NW Italy). *Int. J. Earth Sci. (Geol. Rundsch.)*, 107, 203–229 [Online]. Available at: https://doi.org/10.1007/s00531-017-1485-6.

Lagabrielle, Y., Asti, R., Fourcade, S., Corre, B., Poujol, M., Uzel, J., Labaume, P., Clerc, C., Lafay, R., Picazo, S., Maury, R. (2019). Mantle exhumation at magma-poor passive continental margins. Part I. 3D architecture and metasomatic evolution of a fossil exhumed mantle domain (Urdach lherzolite, north-western Pyrenees, France). *Bulletin de la Société Géologique de France*, 190 [Online]. Available at: https://doi.org/10.1051/bsgf/2019007.

Lagabrielle, Y., Asti, R., Duretz, T., Clerc, C., Fourcade, S., Teixell, A., Labaume, P., Corre, B., Saspiturry, N. (2020). A review of cretaceous smooth-slopes extensional basins along the Iberia-Eurasia plate boundary: How pre-rift salt controls the modes of continental rifting and mantle exhumation. *Earth-Science Reviews*, 201(December), 103071 [Online]. Available at: https://doi.org/10.1016/j.earscirev.2019.103071.

Lagarde, J.-L. (1978). La déformation des roches dans le domaine à schistosité subhorizontale – Applications à la nappe du Canigou-Roc de France (Pyrénées Orientales) et au complexe cristallophyllien de Champtoceaux (Massif Armoricain). Thesis, Université de Rennes.

Lago, M., Arranz, E., Pocoví, A., Galé, C., Gil-Imaz, A. (2004a). Permian magmatism and basin dynamics in the southern Pyrenees: A record of the transition from late Variscan transtension to early Alpine extension. *Geological Society, London, Special Publications*, 223, 439–464 [Online]. Available at: https://doi.org/10.1144/GSL.SP.2004.223.01.19.

Lago, M., Arranz, E., Pocoví, A., Galé, C., Gil-Imaz, A. (2004b). Lower Permian magmatism of the Iberian Chain, Central Spain, and its relationship to extensional tectonics. *Geological Society, London, Special Publications*, 223, 465–490 [Online]. Available at: https://doi.org/10.1144/GSL.SP.2004.223.01.20.

Laumonier, B. (2015). Les Pyrénées alpines sud-orientales (France, Espagne) – Essai de synthèse. *Revue de Géologie pyrénéenne.*

Laumonier, B. and Autran, A. (2001). Un chevauchement hercynien majeur dans les Pyrénées orientales : le chevauchement du Puigmal. *Comptes Rendus de l'Académie des Sciences – Series IIA – Earth and Planetary Science*, 332, 585–594 [Online]. Available at: https://doi.org/10.1016/S1251-8050(01)01569-5.

Laumonier, B., Geyssant, J., Guitard, G. (1984). Les plissements hercyniens tardifs dans le Paléozoïque inférieur du versant nord du Canigou. 2e partie : essai de restitution rétrotectonique et cartographique des mégastructures. *Géologie de la France*, 127–147.

Laurent, O., Couzinié, S., Zeh, A., Vanderhaeghe, O., Moyen, J.-F., Villaros, A., Gardien, V., Chelle-Michou, C. (2017). Protracted, coeval crust and mantle melting during Variscan late-orogenic evolution: U-Pb dating in the eastern French Massif Central. *Int. J. Earth Sci. (Geol. Rundsch.)*, 106, 421–451 [Online]. Available at: https://doi.org/10.1007/s00531-016-1434-9.

Le Bayon, B. and Cochelin, B. (2020). Anatomy of an extensional shear zone leading to the exhumation of the middle crust within the Canigou dome (Eastern Pyrenees, Axial Zone). *Journal of Structural Geology*, 141, 104200 [Online]. Available at: https://doi.org/10.1016/j.jsg.2020.104200.

Le Pichon, X. (1968). Sea-floor spreading and continental drift. *J. Geophys. Res.*, 73, 3661–3697 [Online]. Available at: https://doi.org/10.1029/JB073i012p03661.

Leblanc, D., Gleizes, G., Roux, L., Bouchez, J.L. (1996). Variscan dextral transpression in the French Pyrenees: New data from the Pic des Trois-Seigneurs granodiorite and its country rocks. *Tectonophysics*, 261, 331–345 [Online]. Available at: https://doi.org/10.1016/0040-1951(95)00174-3.

Lemirre, B. (2018). Origine et développement de la thermicité dans les Pyrénées varisques. PhD Thesis, Université Paul Sabatier Toulouse III.

Lemirre, B., Cochelin, B., Duchene, S., de Saint Blanquat, M., Poujol, M. (2019). Origin and duration of late orogenic magmatism in the foreland of the Variscan belt (Lesponne – Chiroulet – Neouvielle area, French Pyrenees). *Lithos*, 336/337, 183–201 [Online]. Available at: https://doi.org/10.1016/j.lithos.2019.03.037.

Lloret, J., Ronchi, A., López-Gómez, J., Gretter, N., De la Horra, R., Barrenechea, J.F., Arche, A. (2018). Syn-tectonic sedimentary evolution of the continental late Palaeozoic-early Mesozoic Erill Castell-Estac basin and its significance in the development of the central Pyrenees basin. *Sedimentary Geology*, 374, 134–157 [Online]. Available at: https://doi.org/10.1016/j.sedgeo.2018.07.014.

Lopez-Sanchez, M.A., García-Sansegundo, J., Martínez, F.J. (2018). The significance of early Permian and early Carboniferous U-Pb zircon ages in the Bossòst and LysCaillaouas granitoids (Pyrenean Axial Zone). *Geological Journal*, 54, 2048–2063 [Online]. Available at: https://doi.org/10.1002/gj.3283.

Lotout, C., Poujol, M., Pitra, P., Anczkiewicz, R., Van Den Driessche, J. (2020). From burial to exhumation: Emplacement and metamorphism of mafic eclogitic terranes constrained through multimethod petrochronology, case study from the Lévézou massif (French Massif Central, Variscan Belt). *J. Petrology* [Online]. Available at: https://doi.org/10.1093/petrology/egaa046.

Lucas, C. (1985). Le grès rouge du versant nord des Pyrénées : essai sur la géodynamique de dépôts continentaux du Permien et du Trias. PhD Thesis, Université Paul Sabatier de Toulouse.

Lundmark, A.M., Gabrielsen, R.H., Strand, T., Ohm, S.E. (2018). Repeated post-Caledonian intra-cratonic rifting in the central North Sea: Evidence from the volcanic record in the Embla oil field. *Marine and Petroleum Geology*, 92, 505–518 [Online]. Available at: https://doi.org/10.1016/j.marpetgeo.2017.11.018.

Martínez Catalán, J.R. (2011). Are the oroclines of the Variscan belt related to late Variscan strike-slip tectonics? *Terra Nova*, 23, 241–247 [Online]. Available at: https://doi.org/10.1111/j.1365-3121.2011.01005.x.

Martínez Catalán, J.R., Arenas, R., Abati, J., Martínez, S.S., García, F.D., Suárez, J.F., Cuadra, P.G., Castiñeiras, P., Barreiro, J.G., Montes, A.D. et al. (2009). A rootless suture and the loss of the roots of a mountain chain: The Variscan belt of NW Iberia. *Comptes Rendus Geoscience*, 341, 114–126 [Online]. Available at: https://doi.org/10.1016/j.crte.2008.11.004.

Martínez Catalán, J.R., Pascual, F.J.R., Montes, A.D., Fernández, R.D., Barreiro, J.G., Silva, Í.D.D., Clavijo, E.G., Ayarza, P., Alcock, J.E. (2014). The late Variscan HT/LP metamorphic event in NW and Central Iberia: Relationships to crustal thickening, extension, orocline development and crustal evolution. *Geological Society, London, Special Publications*, 405, 225–247 [Online]. Available at: https://doi.org/10.1144/SP405.1.

Mattauer, M. (1968). Les traits structuraux essentiels de la chaîne pyrénéenne. *Revue de Géographie Physique et de Géologie Dynamique X*, 3–12.

Matte, P. (1969). Le problème du passage de la schistosité horizontale à la schistosité verticale dans le dôme de Garonne (Paléozoïque des Pyrénées Centrales). *Comptes rendus de l'Académie des sciences*, 268, 1841–1844.

Matte, P. (2001). The Variscan collage and orogeny (480-290 Ma) and the tectonic definition of the Armorica microplate: A review. *Terra Nova*, 13, 122–128 [Online]. Available at: https://doi.org/10.1046/j.1365-3121.2001.00327.x.

Matte, P. (2002). Les plis hercyniens kilométriques couchés vers l'ouest-sud-ouest dans la région du pic du Midi d'Ossau–col du Somport (zone axiale des Pyrénées occidentales). *Comptes Rendus Geoscience*, 334, 773–779.

Matte, P. (2007). Variscan thrust nappes, detachments, and strike-slip faults in the French Massif Central: Interpretation of the lineations. *Geological Society of America Memoirs*, 200, 391–402 [Online]. Available at: https://doi.org/10.1130/2007.1200(20).

Maurel, O., Respaut, J.-P., Monié, P., Arnaud, N., Brunel, M. (2004). U-Pb emplacement and 40Ar/39Ar cooling ages of the eastern Mont-Louis granite massif (Eastern Pyrenees, France). *Comptes Rendus Geoscience*, 336, 1091–1098 [Online]. Available at: https://doi.org/10.1016/j.crte.2004.04.005.

McKenzie, D.P. and Parker, R.L. (1967). The North Pacific: An example of tectonics on a sphere. *Nature*, 216, 1276–1280.

Mencos, J., Carrera, N., Muñoz, J.A. (2015). Influence of rift basin geometry on the subsequent postrift sedimentation and basin inversion: The Organyà basin and the Bóixols thrust sheet (south central Pyrenees). *Tectonics*, 34, 2014TC003692 [Online]. Available at: https://doi.org/10.1002/2014TC003692.

Mendia, M.S. and Ibarguchi, J.I.G. (1991). High-grade metamorphic rocks and peridotites along the Leiza Fault (Western Pyrenees, Spain). *Geol. Rundsch.*, 80, 93–107 [Online]. Available at: https://doi.org/10.1007/BF01828769.

Messaoudi, H. (1990). Le complexe plutonique de Millas (Pyrénées Orientales) : pétrologie, géochimie, structurologie et cinématique de mise en place. Thesis, Université Paul Sabatier de Toulouse.

Mey, P.H.W. (1968). The geology of the upper Ribagorzana and Tor Valleys, Central Pyrenees, Spain sheet 8, 1: 50000. *Leidse Geol. Mededelingen*, 41, 229–292.

Mezger, J.E. (2005). Comparison of the western Aston-Hospitalet and the Bossòst domes: Evidence for polymetamorphism and its implications for the Variscan tectonic evolution of the Axial Zone of the Pyrenees. *Journal of the Virtual Explorer*, 19 [Online]. Available at: https://doi.org/10.3809/jvirtex.2005.00122.

Mezger, J.E. (2009). Transpressional tectonic setting during the main Variscan deformation: Evidence from four structural levels in the Bossòst and Aston-Hospitalet mantled gneiss domes, central Axial Zone, Pyrenees. *Bulletin de la Société Géologique de France*, 180, 199–207 [Online]. Available at: https://doi.org/10.2113/gssgfbull.180.3.199.

Mezger, J.E. (2010). Mimicking syntectonic growth: Cordierite overgrowth of earlier rotated staurolite porphyroblasts, strain caps and deflected foliation. *Journal of Structural Geology*, 32, 703–708 [Online]. Available at: https://doi.org/10.1016/j.jsg.2010.04.012.

Mezger, J.E. and Gerdes, A. (2016). Early Variscan (Visean) granites in the core of central Pyrenean gneiss domes: Implications from laser ablation U-Pb and Th-Pb studies. *Gondwana Research*, 29, 181–198 [Online]. Available at: https://doi.org/10.1016/j.gr.2014.11.010.

Mezger, J.E. and Passchier, C.W. (2003). Polymetamorphism and ductile deformation of staurolite-cordierite schist of the Bossòst dome: Indication for Variscan extension in the Axial Zone of the central Pyrenees. *Geological Magazine*, 140, 595–612 [Online]. Available at: https://doi.org/10.1017/S0016756803008112.

Mezger, J.E. and Régnier, J.-L. (2016). Stable staurolite-cordierite assemblages in K-poor metapelitic schists in Aston and Hospitalet gneiss domes of the central Pyrenees (France, Andorra). *J. Metamorph. Geol.*, 34, 167–190 [Online]. Available at: https://doi.org/10.1111/jmg.12177.

Mezger, J.E., Passchier, C.W., Régnier, J.-L. (2004). Metastable staurolite-cordierite assemblage of the Bossòst dome: Late Variscan decompression and polyphase metamorphism in the Axial Zone of the central Pyrenees. *Comptes Rendus Geoscience*, 336, 827–837 [Online]. Available at: https://doi.org/10.1016/j.crte.2003.12.024.

Michard, A., Soulaimani, A., Hoepffner, C., Ouanaimi, H., Baidder, L., Rjimati, E.C., Saddiqi, O. (2010). The south-western branch of the Variscan Belt: Evidence from Morocco. *Tectonophysics*, 492, 1–24 [Online]. Available at: https://doi.org/10.1016/j.tecto.2010.05.021.

Mirouse, R. (1962). Recherches géologiques dans la partie occidentale de la zone primaire axiale des Pyrénées, Tome II. Thesis, Université de Toulouse.

Mouthereau, F., Filleaudeau, P.-Y., Vacherat, A., Pik, R., Lacombe, O., Fellin, M.G., Castelltort, S., Christophoul, F., Masini, E. (2014). Placing limits to shortening evolution in the Pyrenees: Role of margin architecture and implications for the Iberia/Europe convergence. *Tectonics*, 33, 2014TC003663 [Online]. Available at: https://doi.org/10.1002/2014TC003663.

Müller, J. and Roger, P. (1977). L'évolution structurale des Pyrénées (Domaine central et occidental). Le segment hercynien, la chaîne de fond alpine. Research report, Laboratoire de Géologie de l'Université 1 de Grenoble.

Muñoz, J.A. (1992). Evolution of a continental collision belt: ECORS-Pyrenees crustal balanced cross-section. In *Thrust Tectonics*, McClay, K.R. (ed.). Springer, the Netherlands [Online]. Available at: https://doi.org/10.1007/978-94-011-3066-0_21.

Neumann, E.-R., Wilson, M., Heeremans, M., Spencer, E.A., Obst, K., Timmerman, M.J., Kirstein, L. (2004). Carboniferous-Permian rifting and magmatism in southern Scandinavia, the North Sea and northern Germany: A review. *Geological Society, London, Special Publications*, 223, 11–40.

Odlum, M.L. and Stockli, D.F. (2019). Thermotectonic evolution of the North Pyrenean Agly Massif during early cretaceous hyperextension using multi-mineral U-Pb thermochronometry. *Tectonics*, 38, 1509–1531 [Online]. Available at: https://doi.org/10.1029/2018TC005298.

Odlum, M.L. and Stockli, D.F. (2020). Geochronologic constraints on deformation and metasomatism along an exhumed mylonitic shear zone using apatite U-Pb, geochemistry, and microtextural analysis. *Earth and Planetary Science Letter*, 538, 116177.

Olivier, P., Gleizes, G., Paquette, J.L. (2004). Gneiss domes and granite emplacement in an obliquely convergent regime: New interpretation of the Variscan Agly Massif (Eastern Pyrenees, France). *Geological Society of America Special Papers*, 380, 229–242.

Olivier, P., Gleizes, G., Paquette, J.-L., Muñóz Sáez, C. (2008). Structure and U-Pb dating of the Saint-Arnac pluton and the Ansignan charnockite (Agly Massif): A cross-section from the upper to the middle crust of the Variscan Eastern Pyrenees. *Journal of the Geological Society*, 165, 141–152.

Paquette, J.-L., Gleizes, G., Leblanc, D., Bouchez, J.-L. (1997). Le granite de Bassiès (Pyrénées) : un pluton syntectonique d'âge Westphalien. Géochronologie U-Pb sur zircons. *Comptes rendus de l'Académie des sciences. Série 2. Sciences de la terre et des planètes*, 324, 387–392.

Pereira, M.F., Castro, A., Chichorro, M., Fernández, C., Díaz-Alvarado, J., Martí, J., Rodríguez, C. (2014). Chronological link between deep-seated processes in magma chambers and eruptions: Permo-Carboniferous magmatism in the core of Pangaea (Southern Pyrenees). *Gondwana Research*, 25, 290–308.

Pochat, S. and Van Den Driessche, J. (2011). Filling sequence in Late Paleozoic continental basins: A chimera of climate change? A new light shed given by the Graissessac-Lodève basin (SE France). *Palaeogeography, Palaeoclimatology, Palaeoecology*, 302, 170–186 [Online]. Available at: https://doi.org/10.1016/j.palaeo.2011.01.006.

Pohl, F., Froitzheim, N., Obermüller, G., Tomaschek, F., Schröder, O., Nagel, T.J., Sciunnach, D., Heuser, A. (2018). Kinematics and age of syn-intrusive detachment faulting in the Southern Alps: Evidence for Early Permian crustal extension and implications for the Pangea A versus B controversy. *Tectonics*, 37, 3668–3689.

Poitrenaud, T. (2018). Le gisement périgranitique à tungstène et or de Salau (Pyrénées, France), histoire polyphasée d'un système minéralisé tardivarisque. PhD Thesis, Université d'Orléans.

Poitrenaud, T., Poujol, M., Augier, R., Marcoux, E. (2020). The polyphase evolution of a late Variscan W/Au deposit (Salau, French Pyrenees): Insights from REE and U/Pb LA-ICP-MS analyses. *Miner. Deposita*, 55, 1127–1147.

Pouget, P. (1989). Évolution géodynamique hercynienne des Pyrénées Centrales. Contraintes structurales, métamorphiques, magmatiques et sédimentologiques. Thesis, Université Paul Sabatier de Toulouse.

Pouget, P. (1991). Hercynian tectonometamorphic evolution of the Bosost dome (French-Spanish central Pyrenees). *Journal of the Geological Society*, 148, 299–314.

Ramsay, J.G. (1963). Stratigraphy, structure and metamorphism in the western Alps. *Proceedings of the Geologists' Association*, 74, 357.

Roberts, M.P., Pin, C., Clemens, J.D., Paquette, J.-L. (2000). Petrogenesis of mafic to felsic plutonic rock associations: The calc-alkaline Quérigut complex, French Pyrenees. *J. Petrology*, 41, 809–844 [Online]. Available at: https://doi.org/10.1093/petrology/41.6.809.

Rodríguez-Méndez, L., Cuevas, J., Esteban, J.J., Tubía, J.M., Sergeev, S., Larionov, A. (2014). Age of the magmatism related to the inverted Stephanian-Permian basin of the Sallent area (Pyrenees). *Geological Society, London, Special Publications*, 394, 101–111.

Rodríguez-Méndez, L., Cuevas, J., Tubía, J.M. (2016). Post-Variscan basin evolution in the central Pyrenees: Insights from the Stephanian-Permian Anayet basin. *Comptes Rendus Geoscience*, 348, 333–341.

Roure, F., Choukroune, P., Berastegui, X., Munoz, J.A., Villien, A., Matheron, P., Bareyt, M., Seguret, M., Camara, P., Deramond, J. (1989). ECORS deep seismic data and balanced cross sections: Geometric constraints on the evolution of the Pyrenees. *Tectonics*, 8, 41–50.

de Saint Blanquat, M. (1993). La faille normale du massif du Saint Barthélémy. Évolution hercynienne des massifs nord-pyrénéens catazonaux considérée du point de vue de leur histoire thermique. *Geodinamica Acta*, 6, 59–77.

de Saint Blanquat, M., Lardeaux, J.M., Brunel, M. (1990). Petrological arguments for high-temperature extensional deformation in the Pyrenean Variscan crust (Saint Barthélémy Massif, Ariège, France). *Tectonophysics, International IGCP Conference Project*, 233(177), 245–262.

Saspiturry, N. (2019). Évolution sédimentaire, structurale et thermique d'un rift hyperaminci : de l'héritage post-hercynien à l'inversion alpine : exemple du bassin de Mauléon (Pyrénées). PhD Thesis, Université de Bordeaux 3.

Saspiturry, N., Cochelin, B., Razin, P., Leleu, S., Lemirre, B., Bouscary, C., Issautier, B., Serrano, O., Lasseur, E., Baudin, T. et al. (2019a). Tectono-sedimentary evolution of a rift system controlled by Permian post-orogenic extension and metamorphic core complex formation (Bidarray Basin and Ursuya dome, Western Pyrenees). *Tectonophysics*, 768, 228180 [Online]. Available at: https://doi.org/10.1016/j.tecto.2019.228180.

Saspiturry, N., Razin, P., Baudin, T., Serrano, O., Issautier, B., Lasseur, E., Allanic, C., Thinon, I., Leleu, S. (2019b). Symmetry vs. asymmetry of a hyper-thinned rift: Example of the Mauléon basin (Western Pyrenees, France). *Marine and Petroleum Geology*, 104, 86–105 [Online]. Available at: https://doi.org/10.1016/j.marpetgeo.2019.03.031.

Saspiturry, N., Lahfid, A., Baudin, T., Guillou-Frottier, L., Razin, P., Issautier, B., Le Bayon, B., Serrano, O., Lagabrielle, Y., Corre, B. (2020a). Paleogeothermal gradients across an inverted hyperextended rift system: Example of the Mauléon fossil rift (Western Pyrenees). *Tectonics*, 39, e2020TC006206 [Online]. Available at: https://doi.org/10.1029/2020TC006206.

Saspiturry, N., Razin, P., Allanic, C., Issautier, B., Baudin, T., Lasseur, E., Leleu, S. (2020b). Closure of a hyperextended system in an orogenic lithospheric pop-up, Western Pyrenees: The role of mantle buttressing and rift structural inheritance. *Terra Nova*, 32(4), 253–260 [Online]. Available at: https://doi.org.10.1111/ter.12457.

Saspiturry, N., Issautier, B., Razin, P., Andrieu, S., Lasseur, E., Allanic, C., Serrano, O., Baudin, T., Cochelin, B. (2021a). Review of the syn-rift to early post-rift depositional systems of the Cretaceous Mauléon rift: Sedimentary record of continental crust hyperextension and mantle denudation (Western Pyrenees). *Bulletin de la Société Géologique de France*, 192, 49 [Online]. Available at: https://doi.org. 10.1051/bsgf/2021044.

Saspiturry, N., Issautier, B., Razin, P., Baudin, T., Asti, R., Lagabrielle, Y., Allanic, C., Serrano, O. (2021b). Review of Iberia-Eurasia plate-boundary basins: Role of sedimentary burial on depth-dependent continental crust ductile thinning during rifting and continental breakup. *Basin Research*, 33, 1626–1661 [Online]. Available at: https://doi.org. 10.1111/bre.12529.

Saura, E. and Teixell, A. (2006). Inversion of small basins: Effects on structural variations at the leading edge of the Axial Zone antiformal stack (Southern Pyrenees, Spain). *Journal of Structural Geology*, 28.

Schnapperelle, S., Mezger, J.E., Stipp, M., Hofmann, M., Gärtner, A., Linnemann, U. (2020). Polyphase magmatic pulses along the Northern Gondwana margin: U-Pb zircon geochronology from gneiss domes of the Pyrenees. *Gondwana Research*, 81, 291–311.

Séguret, M. (1972). Étude tectonique des nappes et séries décollées de la partie centrale du versant sud des Pyrénées : caractère synsédimentaire, rôle de la compression et de la gravité. PhD Thesis, Université de Montpellier.

Séguret, M. and Proust, F. (1968). Tectonique hercynienne des Pyrénées centrales : signification des schistosités redressées, chronologie des déformations. *C.R. Acad. Sci. Ser. D*, 984–987.

Siron, G., Goncalves, P., Marquer, D., Pierre, T., Paquette, J.-L., Vanardois, J. (2020). Contribution of magmatism, partial melting buffering and localized crustal thinning on the late Variscan thermal structure of the Agly massif (French Pyrenees). *Journal of Metamorphic Geology*, 38, 799–829 [Online]. Available at: https://doi.org/10.1111/jmg.12549.

de Sitter, L.U. and Zwart, H.J. (1960). Tectonic development in supra and infrastructures of a mountain chain. *Proceedings of the 21st International Geological Congress of Cophenhagen*, 18, 248–256.

Soler, D., Teixell, A., García-Sansegundo, J. (1998). Amortissement latéral du chevauchement de Gavarnie et sa relation avec les unités sud-pyrénéennes. *Comptes Rendus de l'Académie des Sciences – Series IIA – Earth and Planetary Science*, 327, 699–704.

Soliva, J., Salel, J.F., Brunnel, M. (1989). Shear deformation and emplacement of the gneissic Canigou thrust nappe (Eastern Pyrenees). *Geologie en Mijnbouw*, 68, 357–388.

Soula, J.-C. (1982). Characteristics and mode of emplacement of gneiss domes and plutonic domes in central-eastern Pyrenees. *Journal of Structural Geology*, 4, 313–342.

Soula, J.-C., Lucas, C., Bessiere, G. (1979). Genesis and evolution of Permian and Triassic basins in the Pyrenees by regional simple shear acting on older Variscan structures: Field evidence and experimental models. *Tectonophysics*, 58, T1–T9.

Soula, J.-C., Debat, P., Déramond, J., Guchereau, J.-Y., Lamouroux, C., Pouget, P., Roux, L. (1986a). Évolution structurale des ensembles métamorphiques, des gneiss et des granitoïdes dans les Pyrénées centrales. *Bulletin de la Société Géologique de France*, 8, 79–93.

Soula, J.-C., Debat, P., Deramond, J., Pouget, P. (1986b). A dynamic model of the structural evolution of the Hercynian Pyrenees. *Tectonophysics*, 129, 29–51.

Spalla, M.I., Zanoni, D., Marotta, A.M., Rebay, G., Roda, M., Zucali, M., Gosso, G. (2014). The transition from Variscan collision to continental break-up in the Alps: Insights from the comparison between natural data and numerical model predictions. *Geological Society, London, Special Publications*, 405, 363–400.

Speksnijder, A. (1985). Anatomy of a strike-slip fault controlled sedimentary basin, Permian of the Southern Pyrenees, Spain. *Sedimentary Geology*, 44, 179–223.

Stampfli, G.M., Hochard, C., Vérard, C., Wilhem, C., von Raumer, J. (2013). The formation of Pangea. *Tectonophysics*, 593, 1–19.

Teixell, A. (1996). The Ansó transect of the southern Pyrenees: Basement and cover thrust geometries. *Journal of the Geological Society*, 153, 301–310.

Teixell, A., Labaume, P., Lagabrielle, Y. (2016). The crustal evolution of the west-central Pyrenees revisited: Inferences from a new kinematic scenario. *Comptes Rendus Geoscience*, 348, 257–267.

Teixell, A., Labaume, P., Ayarza, P., Espurt, N., de Saint Blanquat, M., Lagabrielle, Y. (2018). Crustal structure and evolution of the Pyrenean-Cantabrian belt: A review and new interpretations from recent concepts and data. *Tectonophysics*, 724/725, 146–170.

Ternet, Y., Barrère, P., Canérot, J., Majeste-Menjoulas, C. (2003). Carte géol. France (1/50 000). *Feuille Laruns-Somport*, 1069.

Teyssier, C., Tikoff, B., Markley, M. (1995). Oblique plate motion and continental tectonics. *Geology*, 23, 447–450.

Tikoff, B. and Teyssier, C. (1994). Strain modeling of displacement-field partitioning in transpressional orogens. *Journal of Structural Geology*, 16, 1575–1588.

Tournaire Guille, B. (2017). Relation entre magmatisme et métamorphisme haute-température basse-pression. Réexamen du massif de l'Agly (Pyrénées Orientales). PhD Thesis, Université Paris 6.

Tournaire Guille, B., Olivier, P., Paquette, J.-L., Bosse, V., Guillaume, D. (2019). Evolution of the middle crust of the Pyrenees during the Paleozoic: New data on the plutonic rocks from the North Pyrenean Agly massif. *Int. J. Earth Sci. (Geol. Rundsch.)*, 108, 245–265.

Triboulet, C., Guitard, G., Katona, I., Navidad, M. (2005). Évolution pression-température des amphibolites de la zone axiale au cours du métamorphisme hercynien des Pyrénées orientales. *Comptes Rendus Geoscience*, 337, 1244–1249.

Tugend, J., Manatschal, G., Kusznir, N.J., Masini, E., Mohn, G., Thinon, I. (2014). Formation and deformation of hyperextended rift systems: Insights from rift domain mapping in the Bay of Biscay-Pyrenees. *Tectonics*, 33, 2014TC003529.

Vacherat, A., Mouthereau, F., Pik, R., Bellahsen, N., Gautheron, C., Bernet, M., Daudet, M., Balansa, J., Tibari, B., Pinna Jamme, R. et al. (2016). Rift-to-collision transition recorded by tectonothermal evolution of the northern Pyrenees. *Tectonics*, 35, 2015TC004016.

Vacherat, A., Mouthereau, F., Pik, R., Huyghe, D., Paquette, J.-L., Christophoul, F., Loget, N., Tibari, B. (2017). Rift-to-collision sediment routing in the Pyrenees: A synthesis from sedimentological, geochronological and kinematic constraints. *Earth Science Reviews*, 43–74.

Van den Eeckhout, B. (1986). A case study of a mantled gneiss antiform, the Hospitalet massif, Pyrenees (Andorra, France). *Geologica Ultraiectina*, 45, 1–189.

Van den Eeckhout, B. and Zwart, H.J. (1988). Hercynian crustal-scale extensional shear zone in the Pyrenees. *Geology*, 16, 135–138.

Van Lichtervelde, M., Grand'Homme, A., de Saint Blanquat, M., Olivier, P., Gerdes, A., Paquette, J.-L., Melgarejo, J.C., Druguet, E., Alfonso, P. (2017). U-Pb geochronology on zircon and columbite-group minerals of the Cap de Creus pegmatites, NE Spain. *Miner. Petrol.*, 111, 1–21.

Vanardois, J., Trap, P., Goncalves, P., Marquer, D., Gremmel, J., Siron, G., Baudin, T. (2020). Kinematics, deformation partitioning and late Variscan magmatism in the Agly massif, Eastern Pyrenees, France. *Bulletin de la Société Géologique de France*, 191.

Vanderhaeghe, O. and Teyssier, C. (2001). Crustal-scale rheological transitions during late-orogenic collapse. *Tectonophysics*, 335, 211–228.

Vanderhaeghe, O., Burg, J.-P., Teyssier, C. (1999). Exhumation of migmatites in two collapsed orogens: Canadian Cordillera and French Variscides. *Geological Society, London, Special Publications*, 154, 181–204.

Verhoef, P.N.W., Vissers, R.L.M., Zwart, H.J. (1984). A new interpretation of the structural and metamorphic history of the western Aston massif (Central Pyrenees, France). *Geol. Mijnb*, 63, 399–410.

Vielzeuf, D. (1984). Relations de phases dans le faciès granulite et implications géodynamiques : l'exemple des granulites des Pyrénées. PhD Thesis, Université de Clermont-Ferrand.

Vilà, M., Pin, C., Liesa, M., Enrique, P. (2007). LP-HT metamorphism in a late orogenic transpressional setting, Albera massif, NE Iberia: Implications for the geodynamic evolution of the Variscan Pyrenees. *Journal of Metamorphic Geology*, 25, 321–347.

Vissers, R.L.M. (1992). Variscan extension in the Pyrenees. *Tectonics*, 11, 1369–1384.

Weil, A.B., Gutiérrez-alonso, G., Conan, J. (2010). New time constraints on lithospheric-scale oroclinal bending of the Ibero-Armorican Arc: A palaeomagnetic study of earliest Permian rocks from Iberia. *Journal of the Geological Society*, 167, 127–143.

Weil, A.B., Gutiérrez-Alonso, G., Johnston, S.T., Pastor-Galán, D. (2013). Kinematic constraints on buckling a lithospheric-scale orocline along the northern margin of Gondwana: A geologic synthesis. *Tectonophysics*, 582, 25–49.

Zwart, H.J. (1979). The geology of the Central Pyrenees. *Leidse Geol. Mededelingen*, 50, 1–74.

Zwart, H.J. (1986). The Variscan geology of the Pyrenees. *Tectonophysics*, 129, 9–27.

# 2

# Structure and Texture of Pb-Zn Mineralization: Example of a Multiscale Study in the Pyrenees to Constrain One-forming Processes and Critical Metals Mobility

**Alexandre CUGERONE[1,2], Bénédicte CENKI[1], Émilien OLIOT[1] and Manuel MUÑOZ[1]**

[1] *Géosciences Montpellier, CNRS, Université de Montpellier, France*

[2] *Department of Earth Sciences, University of Geneva, Switzerland*

## 2.1. Introduction

In orogenic basements, the genetic relationship between ore mineralization and host-rock is often difficult to unravel due to complex geometries and the general transposition of structures. Tracking key structural field relationships between mineralization and host-rock is essential to understand their mode of formation and emplacement mechanisms. Moreover, during tectono-metamorphic events, mineralized bodies may have lost their primary features, as well as having

For a color version of all the figures in this chapter, see www.iste.co.uk/saspiturry/evolution1.zip.

*Evolution of the Pyrenees during the Variscan and Alpine Cycles 1*,
coordinated by Nicolas SASPITURRY, Jessica UZEL, Alexandre ORTIZ. 

undergone substantial changes in chemistry and texture, making any interpretation based on these techniques difficult.

Figure 2.1 shows an overview of possible structural relationships between host-rocks and mineralization in folded and deformed metamorphic terranes. In metallogeny, a syngenetic mineralization is considered synchronous to the deposition of sediments. A mineralization is considered epigenetic if it has emplaced after the sedimentation history (Figures 2.1(a) and 2.1(b)). Veins cross-cutting stratification are typically interpreted as evidence of an epigenetic mineralization. However, the observation of a stratiform or stratabound (sub-concordant to stratification) mineralization in appearance is not sufficient to define an ore-body as syngenetic, due to the possible trapping of mineralization in lithological interface or the presence of transposed structures (e.g. stratification parallel to tectonic structures). Commonly, characterizing whether it is syngenetic or epigenetic is key to defining the type of ore mineralization and its mode of formation. In deformed metamorphic terranes, these geometrical relationships can be classified as pre-, syn- or post-deformation structures (Figure 2.1(c)). The structural characterization necessitates a large amount of field structural observations as well as a detailed petrographic examination of the mineral assemblages in the mineralization and the host-rocks as illustrated in Figure 2.1(d).

In the world, Pb-Zn mineralizations are mostly associated with the circulation, in the crust, of metal-rich fluids of magmatic and hydrothermal affinities (Saini-Eidukat et al. 2009; Bodnar et al. 2013; Belissont et al. 2014, 2016; Bauer et al. 2019; Burisch et al. 2019). The largest Pb-Zn world resources are located in ancient orogens. As a result, they must have experienced deformation and metamorphic events, leading to changes in the primary geometry, mineralization grain size and chemistry (Lawrence 1973; Moore et al. 1986; Marshall and Gilligan 1993; Kampunzu et al. 2009). Some of the largest known Pb-Zn resources are hosted in (meta-)sedimentary terranes and interpreted as syngenetic as sulfide is considered to have formed on the seafloor with or without associated volcanic products like volcanogenic massive sulfide (VMS) or sedimentary-exhalative (SEDEX) models, respectively (Wilkinson 2013; Piercey et al. 2015). Numerous examples of world-class Pb-Zn deposits or districts are currently classified into one of these types, such as the Red Dog (USA: Lange et al. 1985; Kelley and Jennings 2004), Mt-Isa inlier (Australia: Perkins and Bell 1998; Huston et al. 2006; Gibson et al. 2017), Neves Corvo (Portugal: Relvas et al. 2006; Moura 2008; Tornos et al. 2008), Kidd Creek (Canada: Mercier-Langevin et al. 2011; DeWolfe et al. 2018) or Xicheng belt (China: Wang et al. 2014). In many of these deposits, even if mineralized bodies are generally sub-concordant to the sedimentary stratification (i.e. these mineralizations are considered stratabound), mineralization emplacement is often controlled by faults and other structural features. In addition, the timing of mineralization is still questioned in many cases (Perkins and Bell 1998; Marignac et al. 2003; Chauvet et al. 2004; Murphy 2004; Chauvet 2019; Cave et al. 2020a).

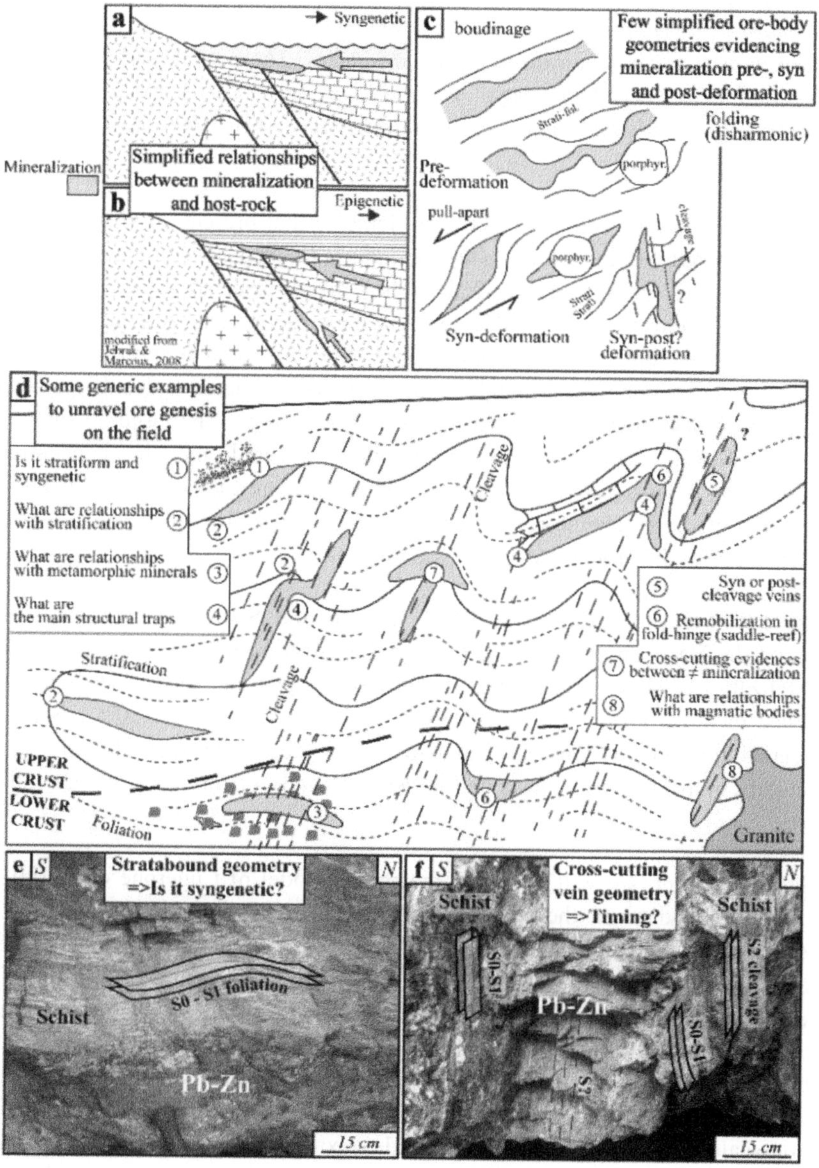


**Figure 2.1.** *Differences between syngenetic and epigenetic mineralization*

LEGEND FOR FIGURE 2.1. – *(a and b) Basic sketches showing fundamental difference between a syngenetic and an epigenetic mineralization (modified from Jébrak and Marcoux 2008). (c) Few generic examples to unravel ore-body geometry on the field. (d) Typical example of complex geometries hosting an ore-body (linked to poly-phase deformation). Some relevant examples are presented to understand relationships between mineralization and structures. These relationships are key to understanding their genetic origin and guide ore exploration. (e and f) Typical examples of stratabound and vein mineralizations (from the Liat and Anglas deposits).*

In the last decades, numerous critical elements that present an affinity with Zn such as germanium (Ge), gallium (Ga) and indium (In) have become more and more valuable for the mining supply chain. They are considered critical by the EU (European Commission 2020), due to their high supply risk and their economic importance in the optic and electronic industry. In undeformed environments, these critical metals are commonly found in ppm contents associated with sulfides such as sphalerite or chalcopyrite. Nonetheless, in orogens, despite the large abundance of zinc ore bodies, critical metals are poorly recovered and the concentration mechanisms of zinc by-products during metamorphism and deformation are still poorly understood. Therefore, understanding the mobility of base and critical elements during the orogenic cycle is essential.

In the Pyrenean Axial Zone (PAZ (France and Spain)), which has mainly recorded the Variscan orogeny, we have studied stratabound and cross-cutting vein mineralizations. These Pb-Zn ores are hosted in metasedimentary sequences that recorded metamorphism from low (<350°C) to medium grade (>550°C) as well as associated deformation phases. The PAZ is an ideal case study to unravel the structural control on ore genesis and revisit the mode of formation of mineralized veins.

In this chapter, we will complete the recent model of Cugerone et al. (2018b) with additional data on the structural framework of Pb-Zn mineralization in the PAZ. Here, we will study the stratabound Horcall and Crabioules deposits belonging to the Bossòst and Lys-Caillaouas massifs, respectively, and the vein deposits of the Arre-Anglas district in the Pierrefitte massif (Figure 2.2(b)). Together with those already described in the Bossòst massif (Cugerone et al. 2018a, 2018b, 2020), we highlight the structural control on the formation and transposition of several types of Pb-Zn mineralization in orogenic basement. In the final section, we will describe the micron-scale relationships between sulfide texture and critical metals such as germanium (Ge) in order to constrain general ore-forming processes and critical metals mobility in an orogenic context.

## 2.2. Geological overview of the Pyrenean Axial Zone

### 2.2.1. *Structural setting*

The PAZ, made up of Paleozoic basement, materialized the antiformal crustal nappe stack of the Pyrenean mountain belt that results from the collision between the Iberian and Eurasian plates during the Pyrenean-Alpine orogeny (from Late Cretaceous–Miocene, see Figure 2.2(a)). The PAZ is composed of Paleozoic metasedimentary rocks intruded by Ordovician and Carboniferous granitoids. It is limited to the north and the south by the North and South Pyrenean zones, respectively, that are mainly composed of Mesozoic and Cenozoic rocks (Figure 2.2(b)).

Nonetheless, most of the pervasive deformation and metamorphism recorded in the PAZ is not Pyrenean-Alpine in age, but Variscan (i.e. Carboniferous to Early Permian). The deeper structural level, commonly named the Infrastructure, is composed of high-grade rocks and is structurally positioned below the Suprastructure, which on the other hand represents the upper crustal level, mostly composed of low-grade and moderately deformed metamorphic rocks (Kleinsmiede 1960; Zwart 1979; Carreras and Druguet 2014; Denèle et al. 2014).

On the field throughout these two major domains, the main foliation S1 is generally E–W-trending, parallel to and transposing the S0 stratification, also being axial plane of kilometric recumbent folds that reflect an early crustal thickening event called D1 (Garcia-Sansegundo and Alonso 1989; Vilà et al. 2007; Cugerone et al. 2018b; Chapter 1 of this book). The D1 deformation phase is commonly associated with medium temperature-medium pressure (MT-MP) regional metamorphism characterized by a staurolite-garnet peak assemblage (580°C and 5.5 kbar) (Mezger and Passchier 2003; de Hoÿm de Marien et al. 2019).

A D2 deformation event related to magmatism and metamorphic dome formation is reflected by the local presence of an E–W-trending vertical S2 cleavage, which is axial planar of upright to m-scale isoclinal folds. This second deformation phase D2 occurred in relation to HT-LP (high temperature-low pressure) regional metamorphism characterized by cordierite superimposed assemblages (575–600°C and 3 kbar (Mezger and Passchier 2003)). A late D3 transcurrent deformation is reported in the PAZ and localized in regional E–W-trending shear zones and faults (Carreras 2001; Mezger 2009; Mezger et al. 2012). It is important to note that the emplacement of metamorphic domes in the PAZ belong to a continuous dynamic process during the Variscan orogenic cycle and that may overlap in time within different structural levels of the crust (Rey et al. 2011, 2017).

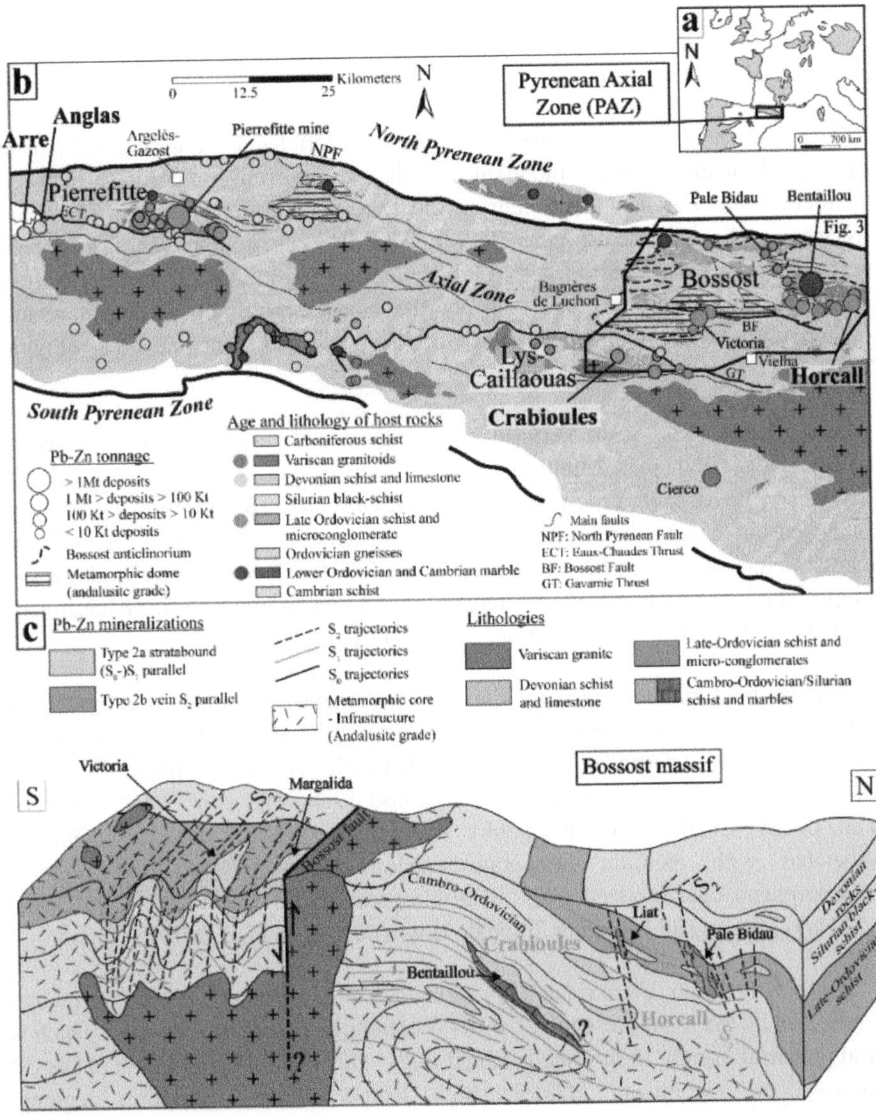


**Figure 2.2.** *(a) Location of the Pyrenean Axial Zone (PAZ) in the western part of the Variscan belt. (b) Simplified geological map of the central and western part of the PAZ. Lithology and Pb-Zn deposits are emphasized. Location of Figure 2.3 is noted. (c) 3D structural sketch of the Bossòst massif with representation of Pb-Zn deposits*

### 2.2.2. *Pyrenean Pb-Zn mineralizations*

In the Pyrenees, about 100 Pb-Zn deposits were mined during the 20th century, constituting the second largest Pb-Zn district in France with ~400,000 t Zn and ~180,000 t Pb extracted (Cugerone et al. 2018b). Most of the Pb-Zn deposits are located in the central and western part of the PAZ and hosted in the Bossòst and the Pierrefitte massif (Figure 2.2(b)). They appear as either stratabound or cross-cutting veins hosted in Cambrian-Ordovician, Devonian or Carboniferous metasedimentary rocks (Bois et al. 1976; Pouit and Bois 1986; Pesquera and Velasco 1989; Pujals 1992; Cugerone et al. 2018b). Cross-cutting veins are also reported in Carboniferous magmatic rocks (Castroviejo Bolibar and Serrano 1983; Johnson et al. 1996; Fanlo et al. 1998). Lead-zinc mineralization is essentially composed of sphalerite (ZnS), and galena (PbS) in a gangue of quartz, calcite and rare barite ($BaSO_4$).

The Bossòst massif is located in the central part of the PAZ and hosts about 20 Pb-Zn deposits. This massif is composed of Cambro-Silurians rocks and present a metamorphic core in the southern part (Figure 2.3). Located in its eastern part close to a kilometric recumbent F1 fold hinge, the Bentaillou deposit was the largest Pb-Zn mine of the Bossòst massif, made of stratabound bodies oriented sub-parallel to S1 foliation transposing S0 stratification (Garcia-Sansegundo and Alonso 1989; Cugerone et al. 2021) (Figures 2.2(b) and 2.2(c)). Similar stratabound bodies are reported in the southern part of the Bossòst massif, in the Victoria-Margalida-Solitaria district (Figures 2.2(b) and 2.2(c)). Nonetheless, in these deposits, stratabound mineralizations exhibit isoclinal E–W F2 folds and are considered syn or slightly post-metamorphism. In the northern part of the Bossòst massif, cross-cutting vein mineralization is dominant and observed in the Argut-dessus, Pale Bidau and Pale de Rase deposits. They mostly appear vertical and parallel to S2 cleavage (Figures 2.2(b) and 2.2(c)) (Cugerone et al. 2018a).

The Pierrefitte massif is located at the western part of the PAZ and hosts the Pierrefitte Pb-Zn district. At the Pierrefitte mine, Pb-Zn mineralizations are mostly composed of stratabound veins hosted in Ordovician and Devonian rocks, oriented parallel to (S0-)S1 foliation and hosted in metric D1 Variscan low-dipping thrusts. At the western edge of the Pierrefitte massif (Figure 2.2(b)), some vein systems are reported such as in the Arre-Anglas district (Reyx 1973; Cugerone et al. 2018b).

## 2.3. Structures of the studied Pb-Zn mineralizations

The Pb-Zn deposits studied in this chapter share significant similarities with those already described in the Bossòst massif in terms of structures and mineralization types. The Horcall and Crabioules are hosted in the central part of the

PAZ, at the eastern tip of the Bossòst massifs and the northern termination of the Lys-Caillaouas dome, respectively. These deposits are all located in the Suprastructure of the PAZ. In the Horcall and Crabioules deposits, S0 stratification, S1 foliation and S2 cleavage generally trend between N090E and N110E with various plunge direction and dipping angles. Moreover, in this study, additional information on the Arre-Anglas district are presented to better constrain the regional framework of the Pb-Zn vein emplacement close and structurally below the regional Eaux-Chaudes thrust.

### 2.3.1. *Horcall and Crabioules stratabound mineralizations*

The Horcall Pb-Zn deposit is located at the eastern tip of the Bossòst massif, southeast of the Bentaillou deposit (Figure 2.3(a)). The mineralization appears stratabound with a decimetric to metric thickness and is hosted in low-grade Cambro-Ordovician schist (Orla schist: Garcia-Sansegundo and Alonso 1989). At the north of the Horcall deposit (Figure 2.4(a)), a succession of kilometric recumbent F1 folds is visible, subsequently affected by F2 upright folds. At Horcall, Pb-Zn bodies are generally parallel to sub-horizontal S0-S1 foliation trending N090-100E, but some remobilizations appear in the fold hinge of N100-110E upright folds, interpreted as F2 folds and represent the thicker mineralized bodies (Figures 2.4(b) and 2.4(c)). The Crabioules Pb-Zn deposit is located in Ordovician low-grade calc-schist rocks (Figure 2.5(a)). S1 foliation is generally parallel to the stratification and trends N090-110E with a general regular dip of 30–40°S. Decimetric to metric veins are mostly parallel to S0-S1 (Figure 2.5(b)). Close to the various gallery entrances, mineralization is thicker with a pluri-metric extend. In the stratabound veins, clasts of calc-schist appear (Figure 2.5(c)). Nonetheless, stratabound bodies with constant dip are not observed everywhere. The presence of extensional pull-apart features with a sinistral movement top to the southeast is observed locally where the S0-S1 foliation is dipping steeper (40–50°S) (Figure 2.5(d)). These structures are in the direct continuity of the widespread general stratabound mineralization and are frequently surrounded by a high density of fractures sealed with quartz. In the two deposits, the mineralogy of the stratabound mineralization is mostly composed of coarse polygonal grains of sphalerite (up to few hundreds of µm; 80% volume), with minor galena, magnetite, pyrrhotite, arsenopyrite and chalcopyrite. At microscopic scale, dark-red sphalerite grains at Horcall are observed in pressure shadows of garnet parallel to S1 foliation, interpreted as syn-D1 kinematic (Figures 2.4(d) and 2.4(e)).

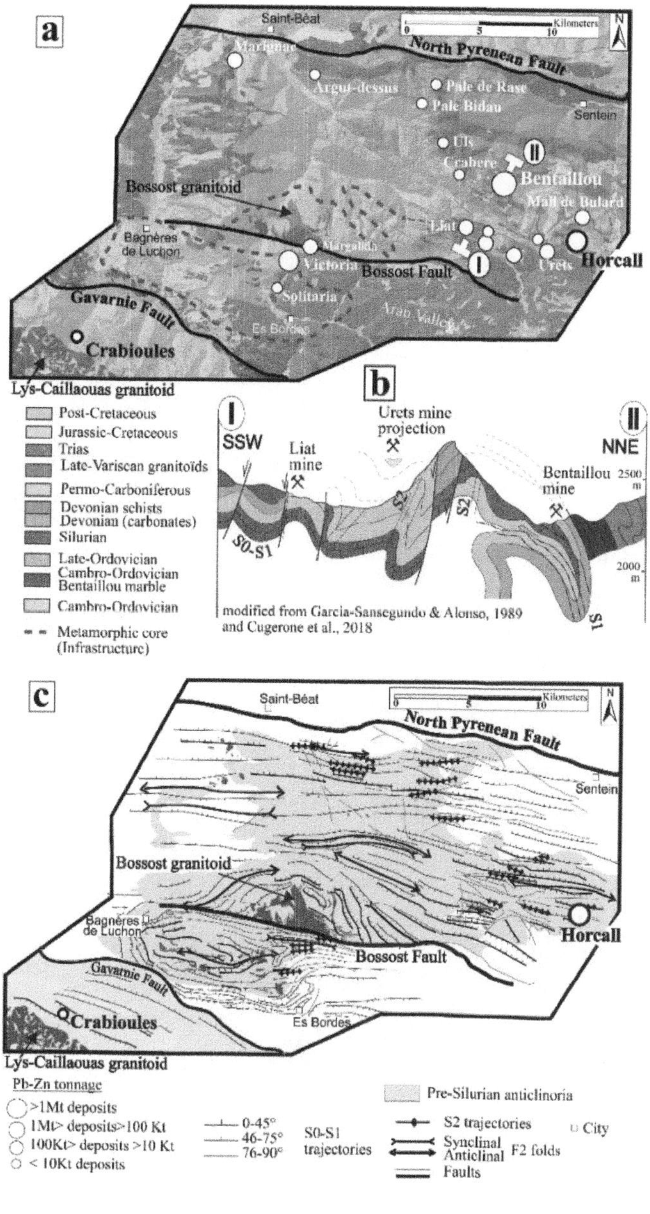


**Figure 2.3.** *The Bossòst and north of the Lys-Caillaouas massifs*

LEGEND FOR FIGURE 2.3.– *(a) Lithological map and the location of the main Pb-Zn deposits. Lithologies are based on geological map of BRGM (France: BRGM 1972;*

*Barrère et al. 1984; Clin et al. 1989) and IGME (Spain, Aran Valley; García-Sansegundo et al. 2013). (b) Cross-section at the east of the Bossòst massif. (c) Structural map with foliation trajectories of S0-S1, S2 and related F2 folds.*

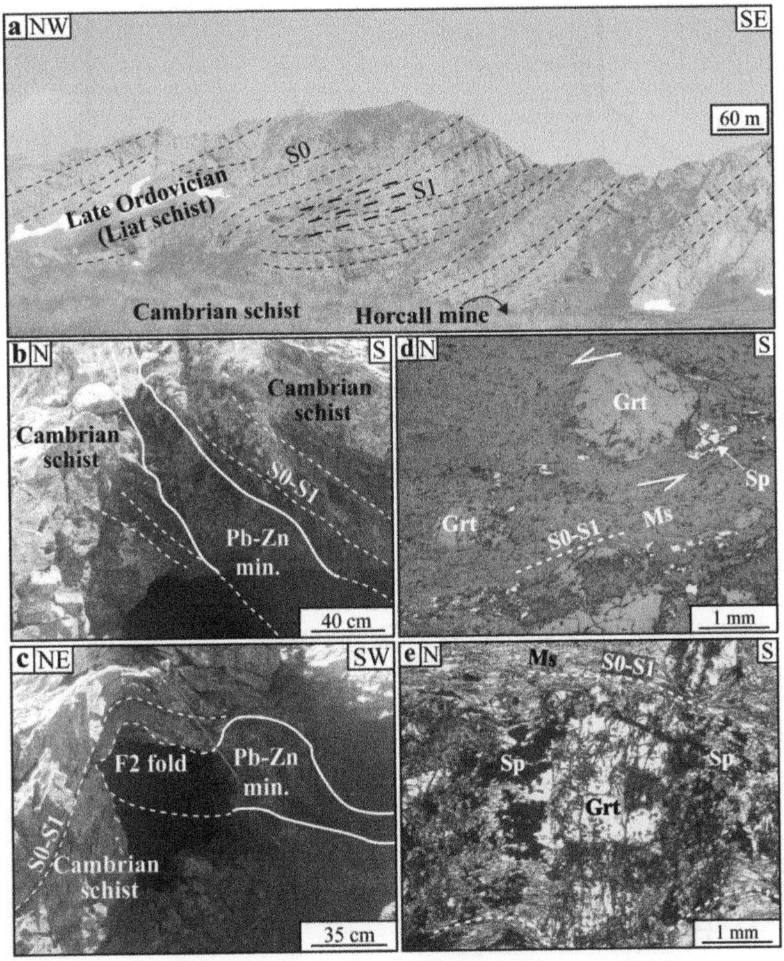


**Figure 2.4.** *Horcall mine (abbreviations: Grt = garnet; Ms = muscovite; Sp = sphalerite)*

LEGEND FOR FIGURE 2.4.– *(a) Panoramic sketch of the surrounding rocks at Horcall showing recumbent folds (F1). (b) Stratabound Pb-Zn mineralization sub-parallel to S0-S1 foliation in Cambrian schist. (c) Stratabound Pb-Zn mineralization remobilized and thicker in F2 folds. (d) Microphotograph of sphalerite in shadow pressure of garnet (reflected light). (e) Microphotograph of stratabound sphalerite replacing garnet (plane polarized light).*

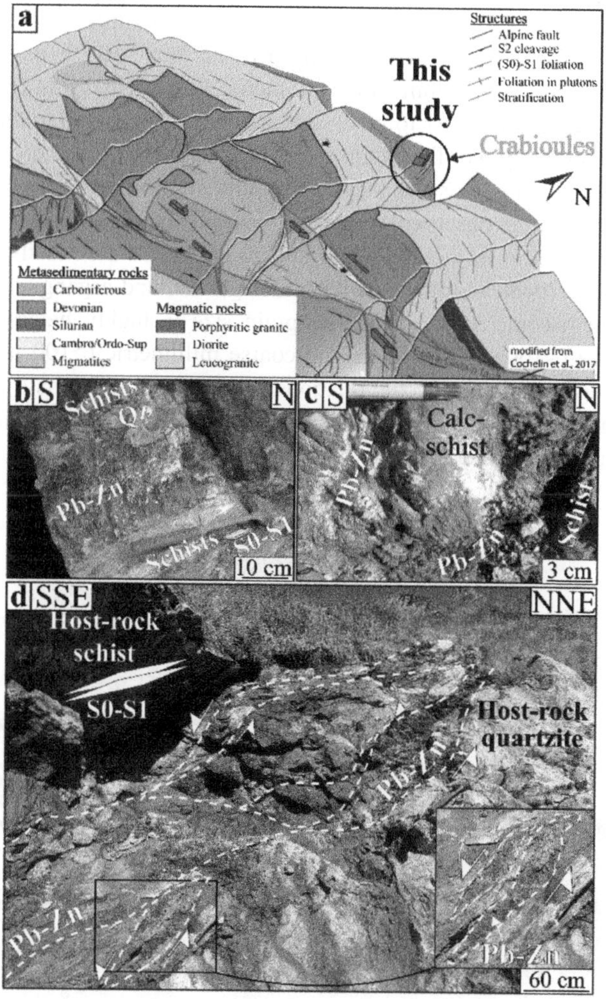


**Figure 2.5.** *Crabioules deposit (dashed white lines: S0-S1 trace)*

LEGEND FOR FIGURE 2.5. – *(a) Structural 3D model modified from Cochelin et al. (2017) and location of the Crabioules deposit. (b) Typical view of Pb-Zn stratabound mineralization. (c) Clast of schist surrounded by Pb-Zn mineralization. (d and e) Outcrop photograph of mineralized pull-apart structures related to steeper S0-S1 foliation.*

### 2.3.2. *Arre and Anglas vein mineralizations*

Arre and Anglas veins are, respectively, hosted in Devonian limestones and cordierite-bearing schists (Figure 2.2(b)). These centimetric to metric vein mineralization are N040-90°E-trending in Arre and N090-100°E in Anglas. These are located close to the fold hinge of a kilometric F2 anticlinorium (Figure 2.6(a)). These veins crosscut the S0-S1 foliation and are generally parallel to the sub-vertical S2 cleavage, marked by cordierite minerals (Figures 2.6(b) and 2.6(c)). At Anglas, centimetric Pb-Zn veins crosscut magmatic aplite and the presence of mineralized extensional features is observed (Figure 2.6(d)). Only minor veins appearing as stratabound are present in the area. Arre and Anglas veins are mostly composed of sphalerite (about 70–80% volume) with minor galena, and a gangue of quartz and calcite. At microscopic scale, sphalerite grain size is highly heterogeneous with small grain-size (<100 μm) domains and coarse millimetric (>100 μm) grains. At sample and thick section scale, the relationship between these small grain domains and coarse sphalerite grains evidence dextral simple shear with a similar trend than S2 cleavage in the host-rock.

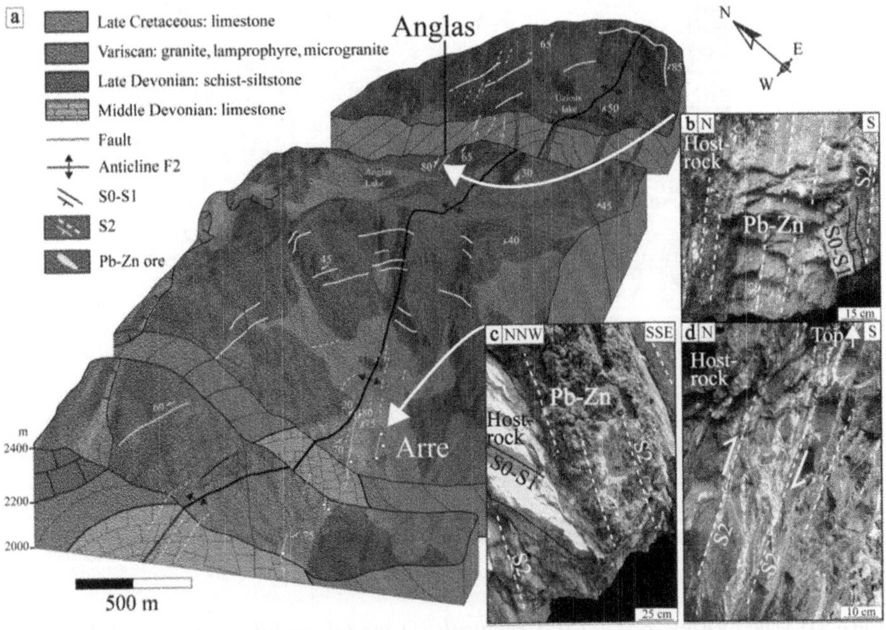


**Figure 2.6.** *Arre-Anglas mining district*

LEGEND FOR FIGURE 2.6.– *(a) Structural 3D model of the Arre-Anglas district. (b and c) Outcrop photograph of the Anglas and Arre veins, respectively. These veins are mostly composed of light-brown sphalerite and are oriented parallel to S2 cleavage. Sphalerite is also imprinted by a cleavage, probably S2. (d) Pull apart features in schist, filled with calcite, quartz and light-brown sphalerite.*

## 2.4. A novel interest in base metal mineralizations: the presence of critical metals

Once we have gained a robust understanding of the regional structural framework for mineralization emplacement, the textural relationships between mineralization and host-rock need to be examined as it is key to understanding the genesis of mineralization responsible for the concentration of valuable critical metals with Zn affinity, such as germanium (Ge), indium (In) and gallium (Ga). These critical metals can be either found in the crystallographic lattice of sulfides like sphalerite (ZnS) and chalcopyrite (CuFeS2), with contents up to a few thousands of ppm, or more rarely in highly concentrated micrometer-sized minerals, with content up to a few tens of percent (e.g. brunogeierite, $Fe_2GeO_4$ that contains about 29 wt% of Ge).

Sulfide dynamic recrystallisation is a primary driver of the mobility and concentration of critical metals (Cugerone et al. 2019, 2020). Therefore, after a detailed structural and petrological investigation, the textural analysis at the microscopic scale is essential to characterize the degree of deformation of sulfides.

In massive sphalerite (as in other sulfide phases), the geometry of grain boundaries reflecting the degree of deformation or recrystallisation is difficult to observe under the microscope (transmitted or reflected light) without etching the thin section surface (hypochlorous acid HClO for sphalerite). The use of the electron back-scattering diffraction (EBSD) technique enables the visualization of the grain boundaries and additionally allows the characterization of crystallographic orientation of grains, which is essential to unravel internal deformation mechanisms.

Coupled to EBSD textural analyses, in situ chemical mapping techniques are used to characterize the distribution of critical metals, either in the sphalerite crystal lattice or in highly enriched minerals for which the critical metals are major elements. Laser-induced breakdown spectroscopy (LIBS) mapping is an innovative technique to map a large portion of thin section, as well as detect a large spectrum of elements at a low detection limit (about 10 ppm for Ge) and a low spatial resolution (about 13 μm).

Sphalerite textures observed for the four studied deposits are shown in Figure 2.7. At Crabioules and Horcall, stratabound sphalerite crystals exhibit homogeneous coarse polygonal grains (>100 µm) with dark-red to orange colors under the optical microscope and present no pronounced color zonation (Figures 2.7(a) and (b)). Germanium and other critical metals are relatively depleted (below ~10 ppm Ge) in the sphalerite lattice and highly concentrated critical metal accessory minerals are absent.

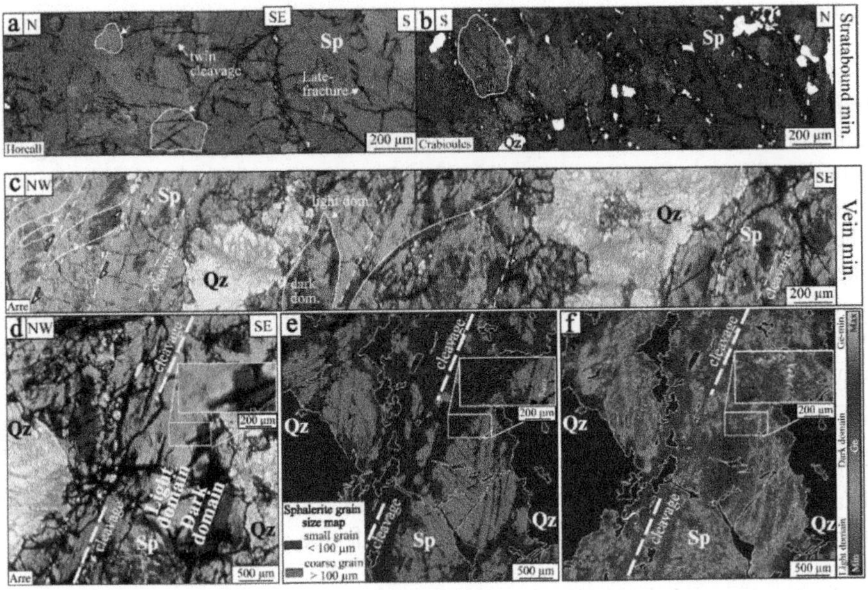


**Figure 2.7.** *Examples of textural and chemical features of the Pyrenean sphalerite (ZnS; thick section: 100 µm) (abbreviations: Qz = quartz; Sp = sphalerite) observed in plane polarized light*

LEGEND FOR FIGURE 2.7.– *(a and b) Stratabound sphalerite from the Horcall and Crabioules deposits. Grain boundaries are represented in white. (c) Thick section scale scan of Arre vein sphalerite. A cleavage (probably S2) is observed and imprint light and dark domains (color zoning) in sphalerite. (d) Zoom on Arre sphalerite with color zoning. (e) EBSD grain size map in the same area as (d). Blue and light-brown colors correspond to small and coarse sphalerite grain fraction, respectively. (f) LIBS map showing Ge concentrations in sphalerite (low and middle concentrations up to 600 ppm Ge) and in Ge-minerals (up to 25 wt.% Ge).*

In the Arre-Anglas vein mineralization, sphalerite generally appears with heterogeneous grain size (Figure 2.7) as shown on the EBSD granulometric map (Figure 2.7(e)). Coarse sphalerite crystals (>100 μm) are mantled by fine recrystallized sphalerite grains (<100 μm) following S2 cleavage (Figure 2.7(e)). This significant textural heterogeneity associated with internal ductile deformation of the coarse grain (Cugerone et al. 2019) matches the large variability observed in sphalerite chemistry (Figure 2.7(f)). In coarse "parent" grains, Ge may reach up to 600 ppm, whereas in recrystallized sphalerite, concentrations are only up to few tens of ppm Ge. Highly concentrated critical metal accessory minerals like Ge-minerals (oxides, sulfides with Ge contents up to 25 wt.%) are present, located at grain boundaries within recrystallized sphalerite domains or close to twin boundary of coarse "parent" crystals.

## 2.5. Discussion and conclusions

Macrostructural observations of Pb-Zn vein mineralizations in the PAZ at regional scale highlight two main types of ore bodies: (i) stratabound mineralizations which are mostly parallel to the (S0-)S1 foliation. These Pb-Zn mineralizations were previously interpreted as SEDEX and syngenetic with ages ranging from Ordovician to Devonian (Bois et al. 1976; Pouit and Bois 1986). However, our detailed structural study indicates that their emplacement is associated with the Variscan orogeny. The source and the exact timing of emplacement of these mineralizations are still unclear but may be related to Late-Orogenic fluid-assisted metamorphism and deformation (Cugerone et al. 2018b, 2019, 2020). (ii) Vein mineralizations are generally parallel to S2 cleavage and probably syn- to late orogenic deformation. They frequently exhibit pull-apart-type structures, and the timing may be related to Late Variscan or even Mesozoic tectonics. Similar veins are reported in various locations in the Pyrenees (Johnson et al. 1996; Fanlo et al. 1998; Subias et al. 1999; Munoz et al. 2016). The in situ mineralogical, textural and chemical characterizations of Pyrenean Pb-Zn mineralizations show that vein mineralizations are the main host for critical metals and exhibit complex sulfide dynamic recrystallization features (coarse parent grains surrounded by small recrystallized crystals).

This study highlights the importance of performing a detailed multiscale geology framework before interpretating Pb-Zn mineralizations geneses. Describing simple stratiform or stratabound veins, without an associated coherent regional structural study, is not sufficient to interpret mineralization genesis. In the Pyrenees, evidence of orogenic mineralizations are often poorly exposed due to general transposition of stratification and deformation structures. It may explain the previous controversies on their geneses. This study delivers some important keys to target specific textural locations and structures at regional or deposit-scale, which may be relevant in other terranes where the question between syngenetic and epigenetic genesis is still not resolved.

The remobilization of Ge during dynamic recrystallization of sphalerite can be explained as follows (Cugerone et al. 2019, 2020): Pb-Zn veins were naturally deformed at low metamorphic grade (<400 °C) promoting the sphalerite dynamic recrystallization and inducing a vertical late cleavage plane (parallel to S2 cleavage affecting the host-rock). During sphalerite recrystallization, the germanium hosted in undeformed millimetric crystals, is spatially remobilized, leached into the chemical system and redistributed in newly formed small Ge-minerals (oxides, sulfide), mostly hosted at grain boundaries of the recrystallized sphalerite grains. A simple mass balance calculation shows that this redistribution occurs at sample scale in a closed system (Cugerone et al. 2020).

One major consequence of this work is to reconsider exploration strategies for critical metals. The exploration of critical metals may not only focus on coarse and undeformed mineral host, but also on domains with smaller and potentially recrystallized grains. In addition, when focusing on bulk chemical analyses, this information might be lost. Germanium and other critical metals may have been remobilized and concentrated in accessory minerals during fluid-assisted tectono-metamorphic events that lead to the dynamic recrystallization of sulfides.

These ideas may have large implications when considering other metamorphosed base-metal deposits located in orogens, such as sediment or volcanic hosted massive sulfide deposit (SEDEX, VHMS) or deformed carbonate-hosted deposits (Ye et al. 2011; Belissont et al. 2019; Wei et al. 2019; Cave et al. 2020b). Further investigations on critical metal remobilization are required to understand the processes of concentration of these valuable critical metal resources, especially in orogens. The integration of various complementary methods at different scales, from the regional field scale to in situ chemical analyses in thin section, is necessary in order to fully understand how base and critical metals ore deposits form.

## 2.6. References

Barrère, P., Bouquet, C., Debroas, E.J., Pelissonnier, H., Peybernes, B., Soulé, J.C., Souquet, P., Ternet, Y. (1984). Arreau. BRGM geological map 1/50 000 with note.

Bauer, M.E., Burisch, M., Ostendorf, J., Krause, J., Frenzel, M., Seifert, T., Gutzmer, J. (2019). Trace element geochemistry of sphalerite in contrasting hydrothermal fluid systems of the Freiberg district, Germany: Insights from LA-ICP-MS analysis, near-infrared light microthermometry of sphalerite-hosted fluid inclusions, and sulfur isotope geochemistry. *Mineralium Deposita*, 54, 237–262 [Online]. Available at: https://doi.org/10.1007/s00126-018-0850-0.

Belissont, R., Boiron, M.-C., Luais, B., Cathelineau, M. (2014). LA-ICP-MS analyses of minor and trace elements and bulk Ge isotopes in zoned Ge-rich sphalerites from the Noailhac-Saint-Salvy deposit (France): Insights into incorporation mechanisms and ore deposition processes. *Geochimica et Cosmochimica Acta*, 126, 518–540 [Online]. Available at: https://doi.org/10.1016/j.gca.2013.10.052.

Belissont, R., Munoz, M., Boiron, M.-C., Luais, B., Mathon, O. (2016). Distribution and oxidation state of Ge, Cu and Fe in sphalerite by µ-XRF and K-edge µ-XANES: Insights into Ge incorporation, partitioning and isotopic fractionation. *Geochimica et Cosmochimica Acta*, 177, 298–314 [Online]. Available at: https://doi.org/10.1016/j.gca.2016.01.001.

Belissont, R., Munoz, M., Boiron, M.-C., Luais, B., Mathon, O. (2019). Germanium crystal chemistry in Cu-bearing sulfides from micro-XRF mapping and micro-XANES spectroscopy. *Minerals*, 9, 1–12 [Online]. Available at: https://doi.org/10.3390/min9040227.

Bodnar, R.J., Lecumberri-Sanchez, P., Moncada, D., Steele-MacInnis, M. (2013). *Fluid Inclusions in Hydrothermal Ore Deposits: Treatise on Geochemistry*, 2nd edition. Elsevier, Oxford.

Bois, J.P., Pouit, G., Gros, Y., Guerangue, B., Picot, P. (1976). Les minéralisations de Zn (Pb) de l'anticlinorium de Pierrefitte : un exemple de gisements hydrothermaux et sédimentaires associés au volcanisme dans le Paléozoïque des Pyrénées centrales. *Bulletin du BRGM*, 6, 543–567.

BRGM (1972). Pic de Mauberme. BRGM geological map 1/50 000 with note.

Burisch, M., Gerdes, A., Meinert, L.D., Albert, R., Seifert, T., Gutzmer, J. (2019). The essence of time – Fertile skarn formation in the Variscan Orogenic Belt. *Earth and Planetary Science Letters*, 519, 165–170.

Carreras, J. (2001). Zooming on Northern Cap de Creus shear zones. *Journal of Structural Geology*, 23, 1457–1486.

Carreras, J. and Druguet, E. (2014). Framing the tectonic regime of the NE Iberian Variscan segment. In *The Variscan Orogeny: Extent, Timescale and the Formation of the European Crust*, Schulmann, K., Catalán, J.R.M., Lardeaux, J.M., Janoušek, V., Oggiano, G. (eds). Geological Society, London.

Castroviejo Bolibar, R. and Serrano, F.M. (1983). Estructura y metalogenia del campo filoniano de Cierco (Pb-Zn-Ag), en el Pirineo de Lérida. *Boletín Geológio y Minero*, 1983, 291–320.

Cave, B., Lilly, R., Barovich, K. (2020a). Textural and geochemical analysis of chalcopyrite, galena and sphalerite across the Mount Isa Cu to Pb-Zn transition: Implications for a zoned Cu-Pb-Zn system. *Ore Geology Reviews*, 124, 103647 [Online]. Available at: https://doi.org/10.1016/j.oregeorev.2020.103647.

Cave, B., Lilly, R., Hong, W. (2020b). The effect of co-crystallising sulphides and precipitation mechanisms on sphalerite geochemistry: A case study from the Hilton Zn-Pb. *Minerals*, 10, 1–22 [Online]. Available at: https://doi.org/10.3390/min10090797.

Chauvet, A. (2019). Structural control of ore deposits: The role of pre-existing structures on the formation of mineralised vein systems. *Minerals*, 9, 1–22, [Online]. Available at: https://doi.org/10.3390/min9010056.

Chauvet, A., Onézime, J., Charvet, J., Barbanson, L., Faure, M. (2004). Syn- to late-tectonic stockwork emplacement within the Spanish section of the Iberian Pyrite Belt: Structural, textural, and mineralogical constraints in the Tharsis and La Zarza areas. *Economic Geology*, 99, 1781–1792 [Online]. Available at: https://doi.org/10.2113/gsecongeo.99.8.1781.

Clin, M., Taillefer, F., Pouchan, P., Muller, A. (1989). Bagnères de Luchon. BRGM geological map 1/50 000 with note.

Cochelin, B., Chardon, D., Denèle, Y., Gumiaux, C., Le Bayon, B. (2017). Vertical strain partitioning in hot Variscan crust: Syn-convergence escape of the Pyrenees in the Iberian-Armorican syntax. *Bulletin de la société géologique de France*, 188, 39 [Online]. Available at: https://doi.org/10.1051/bsgf/2017206.

Cugerone, A., Cenki-Tok, B., Chauvet, A., Le Goff, E., Bailly, L., Alard, O., Allard, M. (2018a). Relationships between the occurrence of accessory Ge-minerals and sphalerite in Variscan Pb-Zn deposits of the Bossòst anticlinorium, French Pyrenean Axial Zone: Chemistry, microstructures and ore-deposit setting. *Ore Geology Reviews*, 95, 1–19 [Online]. Available at: https://doi.org/10.1016/j.oregeorev.2018.02.016.

Cugerone, A., Oliot, E., Chauvet, A., Gavaldà Bordes, J., Laurent, A., Le Goff, E., Cenki-Tok, B. (2018b). Structural control on the formation of Pb-Zn deposits: An example from the Pyrenean Axial Zone. *Minerals*, 8, 1–20 [Online]. Available at: https://doi.org/10.3390/min8110489.

Cugerone, A., Cenki-tok, B., Oliot, E., Muñoz, M., Barou, F., Motto-Ros, V., Le Goff, E. (2019). Redistribution of germanium during dynamic recrystallization of sphalerite. *Geology*, 48, 236–241 [Online]. Available at: https://doi.org/10.1130/G46791.1.

Cugerone, A., Cenki-Tok, B., Muñoz, M., Kouzmanov, K., Oliot, E., Motto-Ros, V., Le Goff, E. (2020). Behavior of critical metals in metamorphosed Pb-Zn ore deposits: Example from the Pyrenean Axial Zone. *Mineralium Deposita*, 1–21 [Online]. Available at: https://doi.org/10.1007/s00126-020-01000-9.

Cugerone A., Roger F., Cenki B., Oliot E., Paquette J.L. (2021). Variscan U-Th-Pb age for stratabound Pb-Zn mineralization in the Bossòst dome (Pyrenean Axial Zone). *Ore Geology Review*, 139(A), 1–18 [Online]. Available at: doi.org/10.1016/j.oregeorev.2021.104503.

Denèle, Y., Laumonier, B., Paquette, J.-L., Olivier, P., Gleizes, G., Barbey, P. (2014). Timing of granite emplacement, crustal flow and gneiss dome formation in the Variscan segment of the Pyrenees. In *The Variscan Orogeny: Extent, Timescale and the Formation of the European Crust*, Schulmann, K., Catalán, J.R.M., Lardeaux, J.M., Janoušek, V., Oggiano, G. (eds). Geological Society, London [Online]. Available at: https://doi.org/10.1144/SP405.5.

DeWolfe, Y.M., Gibson, H.L., Richardson, D. (2018). 3D reconstruction of volcanic and ore-forming environments of a giant VMS system: A case study from the Kidd Creek Mine, Canada. *Ore Geology Reviews*, 101, 532–555 [Online]. Available at: https://doi.org/10.1016/j.oregeorev.2018.07.008.

Fanlo, I., Touray, J.C., Subías, I., Fernández-Nieto, C. (1998). Geochemical patterns of a sheared fluorite vein, Parzan, Spanish central Pyrenees. *Mineralium Deposita*, 33, 620–632 [Online]. Available at: https://doi.org/10.1007/s001260050177.

Garcia-Sansegundo, J. and Alonso, J.L. (1989). Stratigraphy and structure of the southeastern Garona dome. *Geodinamica Acta*, 3, 127–134 [Online]. Available at: https://doi.org/10.1080/09853111.1989.11105180.

García-Sansegundo, J., Merino, J.R., Santisteban, R.R., Leyva, F. (2013). Mapa Geológico de España E. 1:50.000, Hoja nº 118 bis-148 (Caneján-Vielha), 1st edition. Instituto Geológico y Minero de España, Madrid.

Gibson, G.M., Hutton, L.J., Holzschuh, J. (2017). Basin inversion and supercontinent assembly as drivers of sediment-hosted Pb-Zn mineralization in the Mount Isa region, northern Australia. *Journal of the Geological Society*, 174, 2016–105 [Online]. Available at: https://doi.org/10.1144/jgs2016-105.

de Hoÿm de Marien, L., Le Bayon, B., Pitra, P., Van Den Driessche, J., Poujol, M., Cagnard, F. (2019). Two-stage Variscan metamorphism in the Canigou massif: Evidence for crustal thickening in the Pyrenees. *Journal of Metamorphic Geology*, 37, 863–888 [Online]. Available at: https://doi.org/10.1111/jmg.12487.

Huston, D.L., Stevens, B., Southgate, P.N., Muhling, P., Wyborn, L. (2006). Australian Zn-Pb-Ag ore-forming systems: A review and analysis. *Economic Geology*, 101, 1117–1157 [Online]. Available at: https://doi.org/10.2113/gsecongeo.101.6.1117.

Jébrak, M. and Marcoux, É. (2008). Géologie des ressources minérales [Online]. Available at: http://ariane.ulaval.ca/cgi-bin/recherche.cgi?qu=a2131069.

Johnson, C.A., Cardellach, E., Tritlla, J., Hanan, B.B. (1996). Cierco Pb-Zn-Ag vein deposits: Isotopic and fluid inclusion evidence for formation during the mesozoic extension in the Pyrenees of Spain. *Economic Geology*, 91, 497–506 [Online]. Available at: https://doi.org/10.5962/bhl.title.18736.

Kampunzu, A.B., Cailteux, J.L.H., Kamona, A.F., Intiomale, M.M., Melcher, F. (2009). Sediment-hosted Zn-Pb-Cu deposits in the Central African Copperbelt. *Ore Geology Reviews*, 35, 263–297 [Online]. Available at: https://doi.org/10.1016/j.oregeorev.2009.02.003.

Kelley, K.D. and Jennings, S. (2004). A special issue devoted to barite and Zn-Pb-Ag deposits in the Red Dog district, Western Brooks Range, northern Alaska. *Economic Geology*, 99, 1267–1280 [Online]. Available at: https://doi.org/10.2113/gsecongeo.99.7.1267.

Kleinsmiede, W.F.J. (1960). Geology of the Valle de Aran (Central Pyrenees). *Leidse Geol. Mededelingen*, 25, 129–245.

Lange, I.M., Nokleberg, W.J., Plahuta, J.T., Krouse, H.R., Doe, B.R. (1985). Geologic setting, petrology, and geochemistry of stratiform sphalerite-galena-barite deposits, Red Dog Creek and Drenchwater Creek areas, northwestern Brooks Range, Alaska. *Economic Geology*, 80, 1896–1926.

Lawrence, L.J. (1973). Polymetamorphism of the sulphide ores of Broken Hill, NSW, Australia. *Mineralium Deposita*, 8, 211–236.

Marignac, C., Diagana, B., Cathelineau, M., Boiron, M.C., Banks, D., Fourcade, S., Vallance, J. (2003). Remobilisation of base metals and gold by Variscan metamorphic fluids in the south Iberian pyrite belt: Evidence from the Tharsis VMS deposit: *Chemical Geology*, 194, 143–165 [Online]. Available at: https://doi.org/10.1016/S0009-2541(02)00275-9.

Marshall, B. and Gilligan, L.B. (1993). Remobilization, syn-tectonic processes and massive sulphide deposits. *Ore Geology Reviews*, 8, 39–64 [Online]. Available at: https://doi.org/10.1016/0169-1368(93)90027-V.

Mercier-Langevin, P., Goutier, J., Ross, P., Mcnicoll, V., Monecke, T., Dion, C., Dubé, B., Thurston, P., Bécu, V., Gibson, H. et al. (2011). The Blake River Group of the Abitibi greenstone belt and its unique VMS and gold-rich VMS endowment. *Geological Survey of Canada Open File*, 6869, 1–61 [Online]. Available at: https://doi.org/10.495/288757.

Mezger, J.E. (2009). Transpressional tectonic setting during the main Variscan deformation: Evidence from four structural levels in the Bossòst and Aston-Hospitalet mantled gneiss domes, central Axial Zone, Pyrenees. *Bulletin de la société géologique de France*, 180, 199–207 [Online]. Available at: https://doi.org/10.2113/gssgfbull.180.3.199.

Mezger, J.E. and Passchier, C.W. (2003). Polymetamorphism and ductile deformation of staurolite-cordierite schist of the Bossòst dome: Indication for Variscan extension in the Axial Zone of the central Pyrenees. *Geological Magazine*, 140, 595–612 [Online]. Available at: https://doi.org/10.1017/S0016756803008112.

Mezger, J.E., Schnapperelle, S., Rölke, C. (2012). Evolution of the Central Pyrenean Mérens fault controlled by near collision of two gneiss domes. *Jahrbuch für Geowissenschaften*, 34, 11–29.

Moore, D.W., Young, L.E., Modene, J.S., Plahuta, J.T. (1986). Geologic setting and genesis of the Red Dog zinc-lead-silver deposit, western Brooks Range, Alaska. *Economic Geology*, 81, 1696–1727 [Online]. Available at: https://doi.org/doi:10.2113/gsecongeo.81.7.1696.

Moura, A. (2008). Metallogenesis at the Neves Corvo VHMS deposit (Portugal). A contribution from the study of fluid inclusions. *Ore Geology Reviews*, 34, 354–368 [Online]. Available at: https://doi.org/10.1016/j.oregeorev.2008.04.004.

Munoz, M., Baron, S., Boucher, A., Béziat, D., Salvi, S. (2016). Mesozoic vein-type Pb-Zn mineralization in the Pyrenees: Lead isotopic and fluid inclusion evidence from the Les Argentières and Lacore deposits. *Comptes rendus géoscience*, 348, 322–332.

Murphy, T.E. (2004). Structural and stratigraphic controls on mineralization at the George Fisher Zn-Pb-Ag Deposit, northwest Queensland, Australia. PhD Thesis, James Cook University, Middlesbrough.

Perkins, W.G. and Bell, T.H. (1998). Stratiform replacement lead-zinc deposits: A comparison between Mount Isa, Hilton, and McArthur River. *Economic Geology*, 93, 1190–1212.

Pesquera, A. and Velasco, F. (1989). The arditurri Pb-Zn-F-Ba deposit (Cinco Villas massif, Basque Pyrenees): A deformed and metamorphosed stratiform deposit. *Mineralium Deposita*, 24, 199–209.

Piercey, S.J., Peter, J.M., Herrington, R.J. (2015). Zn-rich volcanogenic massive sulphide (VMS) deposits. *Current Perspectives on Zinc Deposits – Irish Association for Economic Geology*, 4, 1–21.

Pouit, G. and Bois, J.P. (1986). Arrens Zn (Pb), Ba Devonian deposit, Pyrénées, France: An exhalative-sedimentary-type deposit similar to Meggen. *Mineralium Deposita*, 21, 181–189.

Pujals, I. (1992). Las mineralizaciones de sulfuros en el Cambro-Ordovicico de la Val d'Aran (Pirineo Central, Lérida). PhD Thesis, Université Autónoma, Barcelona.

Relvas, J.M.R.S., Barriga, F.J.A.S., Longstaffe, F.J. (2006). Hydrothermal alteration and mineralization in the Neves-Corvo volcanic-hosted massive sulfide deposit, Portugal. II. Oxygen, hydrogen, and carbon isotopes. *Economic Geology*, 101, 791–804.

Reyx, J. (1973). Relations entre tectonique, métamorphisme de contact et concentrations métalliques dans le secteur des anciennes mines d'Arre et Anglas (Hautes-Pyrénées – Pyrénées-Atlantiques). PhD Thesis, Université Paris VI.

Saini-Eidukat, B., Melcher, F., Lodziak, J. (2009). Zinc-germanium ores of the Tres Marias Mine, Chihuahua, Mexico. *Mineralium Deposita*, 44, 363–370 [Online]. Available at: https://doi.org/10.1007/s00126-008-0222-2.

Subias, I., Fanlo, I., Yuste, A., Fernandez-Nieto, C. (1999). The Yenefrito Pb-Zn mine (Spanish Central Pyrenees): An example of superimposed metallogenetic events. *Mineralium Deposita*, 34, 220–223.

Tornos, F., Solomon, M., Conde, C., Spiro, B.F. (2008). Formation of the tharsis massive sulfide deposit, Iberian Pyrite Belt: Geological, lithogeochemical, and stable isotope evidence for deposition in a brine pool. *Economic Geology*, 103, 185–214 [Online]. Available at: https://doi.org/10.2113/gsecongeo.103.1.185.

Vilà, M., Pin, C., Liesa, M., Enrique, P. (2007). LPHT metamorphism in a late orogenic transpressional setting, Albera Massif, NE Iberia: Implications for the geodynamic evolution of the Variscan Pyrenees. *Journal of Metamorphic Geology*, 25, 321–347.

Wang, C., Deng, J., Carranza, E.J.M., Lai, X. (2014). Nature, diversity and temporal–spatial distributions of sediment-hosted Pb-Zn deposits in China. *Ore Geology Reviews*, 56, 327–351 [Online]. Available at: https://doi.org/10.1016/j.oregeorev.2013.06.004.

Wei, C., Ye, L., Hu, Y., Danyushevskiy, L., Li, Z., Huang, Z. (2019). Distribution and occurrence of Ge and related trace elements in sphalerite from the Lehong carbonate-hosted Zn-Pb deposit, northeastern Yunnan, China: Insights from SEM and LA-ICP-MS studies. *Ore Geology Reviews*, 115, 103175 [Online]. Available at: https://doi.org/10.1016/j.oregeorev.2019.103175.

Wilkinson, J.J. (2013). Sediment-hosted zinc-lead mineralization: Processes and perspectives. In *Treatise on Geochemistry*, 2nd edition, Holland, H. and Turekian, K. (eds). Elsevier, Amsterdam.

Ye, L., Cook, N.J., Ciobanu, C.L., Yuping, L., Qian, Z., Tiegeng, L., Wei, G., Yulong, Y., Danyushevskiy, L. (2011). Trace and minor elements in sphalerite from base metal deposits in South China: A LA-ICPMS study. *Ore Geology Reviews*, 39, 188–217 [Online]. Available at: https://doi.org/10.1016/j.oregeorev.2011.03.001.

Zwart, H.J. (1979). The geology of the Central Pyrenees. *Leidse Geologische Mededelingen*, 50, 1–74.

# 3

# Meso-Cenozoic Kinematics of Western Europe Applied to the Pyrenean Domain

**Paul ANGRAND and Frédéric MOUTHEREAU**
*GET, CNRS, Université Toulouse-III-Paul-Sabatier, IRD, CNES, OMP, Toulouse, France*

## 3.1. Introduction

### 3.1.1. *Recalling plate tectonics and the Wilson cycle*

The theory of plate tectonics, established in the late 1960s, is one of the last major revolutions in the Earth Sciences. Although this chapter does not directly address the theoretical aspect of plate tectonics, it seems interesting to us to briefly recall by way of introduction, the key stages concerning the appearance and progression of this scientific theory.

Accepted by the scientific community today, this theory comes from the work of Alfred Wegener (1912), who, as early as 1912, developed the concepts of continental drift. His book, *The Origin of the Continents and Oceans* (1915), employs topographical, paleontological and climatic arguments to describe the horizontal movement of continental plates, on the surface of the Earth, spatially bounded by the oceans. Continents were then considered to be "sailing" through the oceanic crust, like icebergs drifting on water. When it was published, Wegener's

For a color version of all the figures in this chapter, see www.iste.co.uk/saspiturry/evolution1.zip.

*Evolution of the Pyrenees during the Variscan and Alpine Cycles 1*,
coordinated by Nicolas SASPITURRY, Jessica UZEL, Alexandre ORTIZ. 

theory was widely discussed and refuted by part of the community, and even forgotten, up until the 1950s.

In 1952, while working with Bruce Heezen, Mary Tharp was the first to describe the presence of valleys at the bottom of the oceans, at the axis of ocean ridges. First challenged by her own colleagues, because of the innovative aspect – and the sexism of the time – this observation gradually came to be accepted. The first map depicting the Atlantic Ocean floor was published in 1959[1], and was later followed by the publication mapping the world ocean floors in 1977[2]. These observations will play a fundamental role in understanding the evolution of the lithosphere.

In 1962, Hess (1962) hypothesized that the oceanic crust accreted at the axis of ridges, generating divergence across plates, a notion extended to the entire lithosphere by Dietz in 1962. An argument in favor of this oceanic accretion was then brought to the forefront by Vine and Matthews (1963), who proposed that the alignment of parallel magnetic oceanic domains could be interpreted as a magnetic field as recorded by basaltic rocks. The distance from the ridge axis of these bands attests to the creation of new crustal material at the ridge.

In 1968, Morgan (1968) described the Earth's surface as composed of a dozen rigid and moving plates. The relative motion of these plates at the Earth's surface is called "plate tectonics" and is accomplished by locating deformation at their boundaries. The movement of the plates on the surface of the globe is done by a rotation around a Eulerian pole[3]. The same year, Xavier Le Pichon (1968) proposed one of the first global kinematic models with six plates (America, Africa, Eurasia, India, Pacific, Antarctica) for the Cenozoic. His model is based on the opening of the South Pacific, Atlantic, Arctic, North Pacific and Indian Oceans.

Global or regional kinematic models describing the evolution in time and space of tectonic plates at the Earth's surface have subsequently evolved to take into account more and more geological, geodetic and geophysical data, increasing the number of tectonic plates accordingly. A recent model, the MORVEL model, has 25 plates (DeMets et al. 2010).

Another major concept has been developed based on the notion of structural legacy. A continental rift evolves into a passive margin (oceanic accretion), then

1 *The Floors of the Oceans. I: The North Atlantic.* Marie Tharp and Bruce C. Heezen.

2 *World Ocean Floor Panorama*, Marie Tharp and Bruce C. Heezen, a painting by Heinrich C. Berann, Library of Congress, 1977.

3 Center of rotation allowing to describe movements on the surface of a sphere. The objects on the sphere describe arcs of a circle whose center of rotation represents the Eulerian pole.

reverses into subduction, before culminating in a continental collision. The Wilson cycle, developed by Wilson (1963 1965a, 1965b) and popularized by Dewey and Burke (1974), defines the deformation of the lithosphere as a cyclic episode, associating extension and convergence phases. The spherical shape of the Earth induces that an extension (oceanic opening) must be compensated at another point of the globe by a convergence. The cycles on different places of the globe affect each other, reusing preexisting structures (plate boundaries).

### 3.1.2. *Types of plate boundaries*

The boundaries between plates can be of three types. Ridges represent a divergent boundary between two plates. The creation of space along the axis of the ridge is compensated by ocean accretion, that is, the creation of new ocean floor by upwelling of the asthenosphere. The convergence of tectonic plates is accommodated by the destruction of material, oceanic or continental, along intracontinental collision zones (between two continents, e.g., the Pyrenees) or subduction zones (between two oceanic plates, e.g., Japan, or between an oceanic plate and a continental plate, e.g., the Andes) Ocean-continent subductions are characterized by a subduction of the oceanic plate under the continental plate due to the density contrast between these two crusts (the oceanic crust being denser than the continental crust, it plunges into the lithosphere). The opposite, which happens only in a very punctual way, generates ophiolites, pieces of oceanic crust carried on a continental plate during the subduction (e.g. the Chenaillet in the Alps). Plates also move laterally in relation to each other along the fault zones. When these faults segment an oceanic ridge, they are called transform faults (Wilson 1965a). These tectonic structures, orthogonal to the axis of the ridge, are responsible for the shift in the axis of the latter, which can reach several hundred kilometers (e.g. Romanche Fault in the Atlantic Ocean).

## 3.2. Geological and scientific background

The Alpine orogenic belts of Western Europe are the result of the Late Cretaceous inversion of the Atlantic continental rift domains and the closure of the Tethys oceanic remnants between North Africa and the European cratons. In the Pyrenees-Cantabrian Basin segment (Figure 3.1(a)), Late Paleozoic to Neogene evolution was controlled by successive openings of oceanic domains in the Paleozoic and Mesozoic, by the African-European convergence in the Mesozoic–Cenozoic, and finally by the opening of the western Mediterranean in the Neogene (Figure 3.1(b)). While the kinematics of Africa in relation to Europe is well understood, the kinematics of micro-plates such as Iberia and Adria, located between Africa and Europe, is still debated.

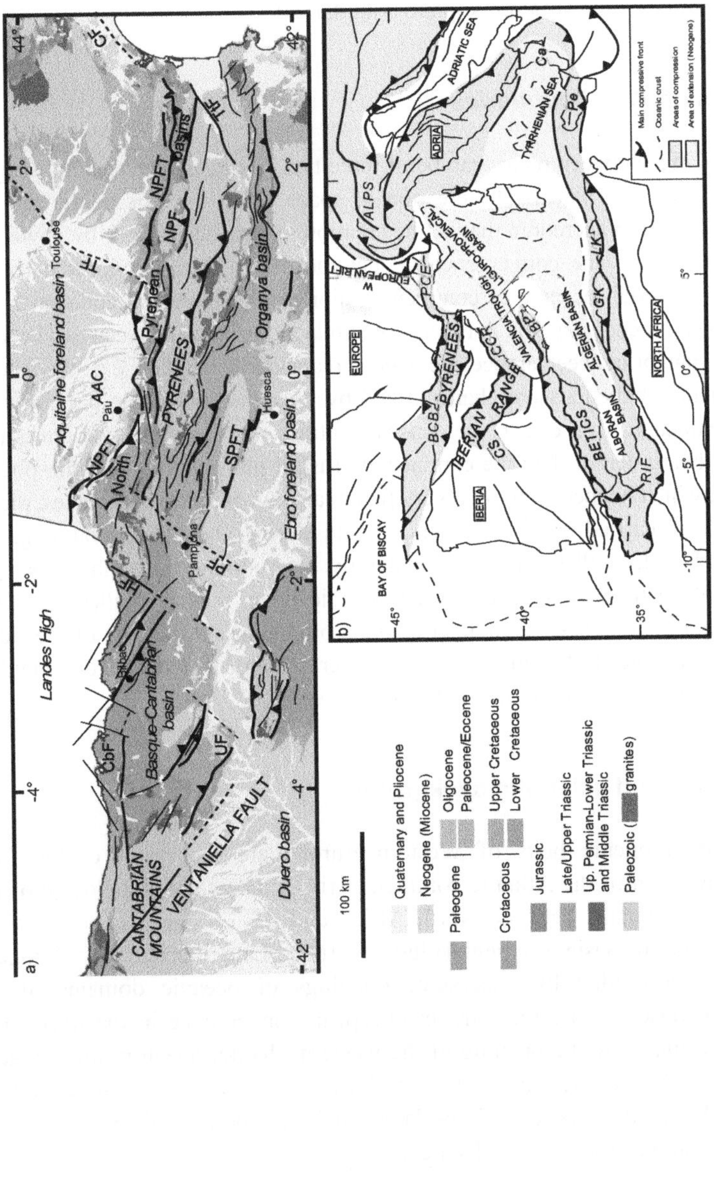


**Figure 3.1.** *Structural map of the West Mediterranean showing the main tectonic plates and deformed areas*

LEGEND FOR FIGURE 3.1.– *(a) Geological map of the Pyrenees-Cantabrian Basin domain, CbF: Carbuerniga Fault; UF: Ubernia Fault; HF: Hendaye Fault; PF: Pamplona Fault; NPFT: North Pyrenean Frontal Thrust; NPF: North Pyrenean Fault; TF: Toulouse Fault; TtF: Têt Fault; CF: Cévennes Fault; AAC: Adour-Arzacq-Comminges Basin. (b) Regional geodynamic context, gray: areas deformed during Alpine compression. Yellow: area deformed during Neogene extension (West European Rift and opening of the Liguro-Provençal oceanic basin). BCB: Basque-Cantabrian Basin; BP: Balearic Promontory; Ca: Calabria; CCR: Catalan Coastal Range; CS: Central System; GK: Greater Kabylies; LK: Lesser Kabylies; PCE: Provence; Pe: Peloritan.*

Each kinematic model has strengths and weaknesses inherent to the domain being studied and the method used. There are global-scale kinematic models (Müller et al. 2019), or regional models that focus on the evolution of a particular domain, such as the North Atlantic (Nirrengarten et al. 2018) or the Tethyan domain (van Hinsbergen et al. 2020). The study of the Pyrenean domain requires joint consideration of the evolution of the Tethyan and Atlantic oceanic domains, as well as the Late Mesozoic to Cenozoic orogenic phase associated with the Africa-Europe convergence.

Since the 2010s, the geology of the Pyrenees, in particular the northern Pyrenees, has been the subject of several national and international projects led by French institutions. The ANR PYRAMID (with: Mary Ford as its leader) and PYROPE (with: Sébastien Chevrot as its leader) projects and associated research have, respectively, contributed to understanding: (1) the evolution of the Pyrenean orogen over time and space (hinterland and foreland basins) (Mouthereau et al. 2014; Vacherat et al. 2014, 2016, 2017; Ford 2016; Rougier et al. 2016; Mouchené 2017; Grool et al. 2018); and (2) the current geometry of this orogen from geophysical imagery (Chevrot et al. 2015, 2018; Wang et al. 2016; Lehujeur et al. 2021). As for the Pyrenees project of the RGF research program, led by the *Bureau des recherches géologiques et minières* (BRGM, French geological survey), has made it possible to study a wide field of scientific domains, including Late Variscan to recent evolution of the Pyrenean belt (Cochelin et al. 2017, 2018; Taillefer et al. 2018), Cretaceous hyperextension (Corre et al. 2016, 2018), tectonic evolution of the Aquitaine foreland basin (Angrand et al. 2018) and orogen crustal geometry (Wehr et al. 2018). Finally, the OROGEN program, supported by the *Centre national de la recherche scientifique – Institut national des sciences de l'univers* (CNRS-INSU, a French national research center) in collaboration with the oil company *TotalEnergies* and the French geological institute BRGM, aimed to place the Pyrenees in the broader tectonic and kinematic context of the western Mediterranean domain, the African-Eurasian convergences and the pre-convergence tectonic history. The ambition was ultimately to apply the know-how acquired on the Pyrenees to other orogenic domains. In this case, it allowed a redefinition of the geology at the scale of the African-Iberian-European plate boundary (Angrand et al. 2020; Romagny et al. 2020).

In this chapter, we place the Pyrenees-Basque-Cantabrian system in a regional kinematic evolution, since the end of the Paleozoic. After presenting the geological context and the existing problems concerning the kinematics of the Iberian-European boundary (section 3.3), we present a kinematic model recently published by Angrand et al. (2020) and Angrand and Mouthereau (2021) (see section 3.4). Although the focus is on the evolution of the Pyrenean-Basque-Cantabrian domain, the scope of this model is intentionally broader and incorporates the evolution of the North Atlantic and the Paleo- and Neotethyan oceans, in order to have an overview of the western European domain. This model, although based on geological observations, must be confronted with other solutions presented in other studies cited in this chapter, global (Matthews et al. 2016; Müller et al. 2019) or regional (Nirrengarten et al. 2018; Romagny et al. 2020; van Hinsbergen et al. 2020), which are strongly recommended as further reading. Finally, we will discuss two remaining issues: (1) the distribution of deformation and (2) the role of tectonic legacy at the Iberian-European plate boundary.

## 3.3. State of the art: the Iberian-European plate boundary and issues

### 3.3.1. *The Pyrenees and Basque-Cantabrian system*

The Pyrenees are a 400 km long mountain range (Figure 3.1(a)) that resulted from the closure of the Lower Cretaceous Pyrenean rift at the time of the Iberian-European convergence (Late Santonian/Campanian to Middle Miocene, see Figure 3.2) (Choukroune and Mattauer 1978; Puigdefàbregas and Souquet 1986; Choukroune et al. 1989). The main structures are oriented ~N110°, parallel to the axis of the chain. Locally, NE–SW trending structures segment the orogen, in the retro-foreland basin to the north (Toulouse Fault) or in the western part of the range (Pampelona Fault). The structure of the eastern Pyrenees marks the transition with the opening of the Gulf of Lion.

The Pyrenees form a double-vergent orogen (Beaumont et al. 2000; Sinclair et al. 2005; Naylor and Sinclair 2008). The formation of the orogen retrowedge[4], on the northern slope of the Pyrenees, is characterized by the preservation of the Pyrenean rift architecture, with the exhumed mantle acting as an indenter (Wang et al. 2016; Jourdon et al. 2019; Saspiturry et al. 2020a). The inherited architecture that can be reconstructed from the northern margin of the Pyrenean system defines a highly segmented margin (Jammes et al. 2009; Masini et al. 2014; Tugend 2014a,

4 Part of the orogen located on the overriding plate of the subduction system.

2014b; Lescoutre and Manatschal 2020; Manatschal et al. 2021; Saspiturry et al. 2022). The result is a non-cylindrical orogen with a crustal root architecture that varies laterally over a few tens of kilometers and scattered occurrences of mantle material at reversed crustal depths in the North Pyrenean Zone (Daignieres et al. 1989; Casas et al. 1997; Wang et al. 2016; Chevrot et al. 2018; Wehr et al. 2018) as well as in the sediment distribution of the Aquitaine retro-foreland basin (Angrand et al. 2018). In the easternmost part of the orogen, isostatic anomalies show that the relatively high topography is not offset by a thick crustal root, and instead has a relatively attenuated root (Chevrot et al. 2018).

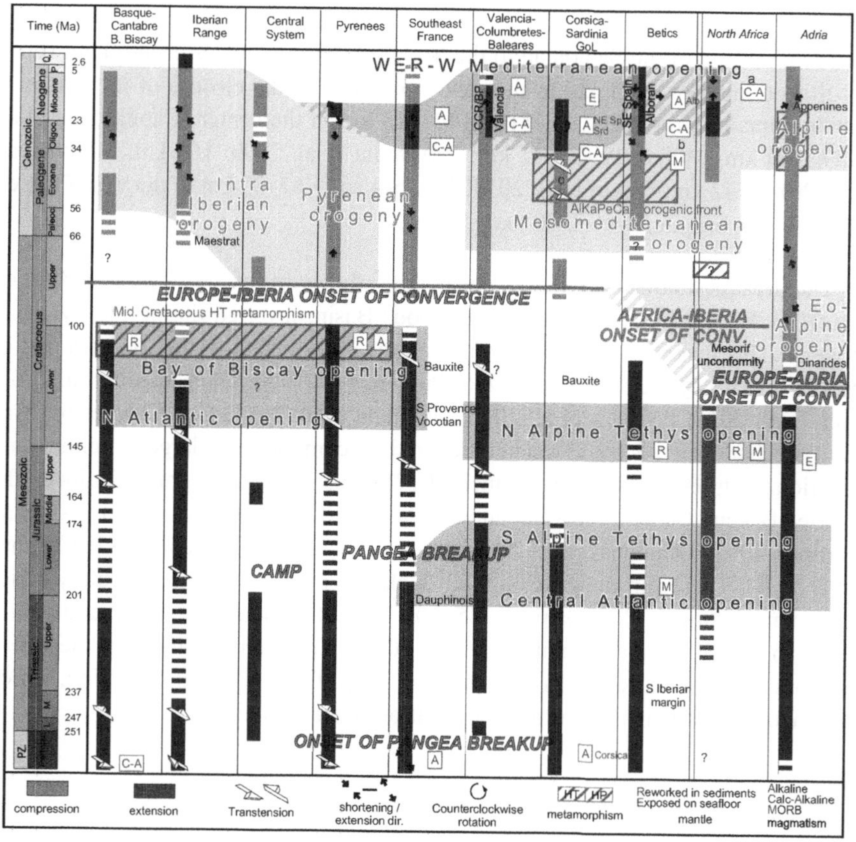


**Figure 3.2.** *Chronostratigraphic chart of main deformed areas in the western Mediterranean. According to Angrand and Mouthereau (2021). Blue: extension; orange: compression. Note that North Africa and Adria are simplified as they both represent larger domains than the other columns. BP: Balearic Promontory; CAMP: Central Atlantic Magmatic Province; CCR: Catalan Coastal Range; WER: West European Rift*

The North Pyrenean zone preserves the relics of an extensive episode, the climax of which was reached in the Lower Cretaceous (Figure 3.2), as evidenced by the remains of sub-continental mantle unearthed during the extension. These are punctually outcropped in the North Pyrenean zone or locally reworked in the extensive Albian-Cenomanian sedimentary basins (Debroas 1990; Lagabrielle and Bodinier 2008; Jammes et al. 2009; Lagabrielle et al. 2010; Saspiturry et al. 2021a) (see Chapter 5). This extensional phase is accompanied by HT-LP metamorphism affecting the pre- and synrift sedimentary cover (Albarède and Michard-Vitrac 1978; Montigny et al. 1986; Vacherat et al. 2014; Clerc et al. 2015; Ducoux et al. 2019; Saspiturry et al. 2020b, 2021b).

From the end of the Santonian to the Middle Miocene, the Pyrenees underwent a ~N-S shortening of ~111 km to the east (not counting the closure of the exhumed mantle domain; Grool et al. 2018) , ~100–165 km in the center (Choukroune et al. 1989: 100 km; Muñoz 1992: 147 km; Beaumont et al. 2000: 165 km; Mouthereau et al. 2014: 142 km; Espurt et al. 2019: 124 km) and ~67–80 km to the west (Teixell 1998: 80 km; Saspiturry et al. 2020a: 67 km).

The Basque-Cantabrian Basin, located to the west of the Pyrenees (Figure 3.1(a)), is formed by an extensive Cretaceous Basin, inverted during the Pyrenean orogeny. This inversion is synchronous with the deformation of the southern part of the Bay of Biscay (Alonso et al. 1996; Pulgar et al. 1996; Gallastegui et al. 2002; Roca et al. 2011; Pedreira et al. 2015; Quintana et al. 2015) (see Chapter 6). This basin is characterized by a weak relief and is composed of large synclines and diapiric structures (Cámara and Flinch 2017). This orogenic segment includes the inverted Basque-Cantabrian Basin and the Paleozoic Asturian massif (García-Espina Martínez 1997; Quintana et al. 2006; Roca et al. 2011; Quintana 2012; Cadenas et al. 2018). The basin preserves a total thickness of the synrift sedimentary series (Tithonian to Lower Cenomanian, 145–100 Ma) of 12 km on average and nearly 20 km in certain locales (Quintana et al. 2015). This considerable thickness was accommodated by extreme crustal extension, amplified by salt mobilization, which may have locally increased the accommodation space available.

### 3.3.2. *Kinematics at the Iberia-Europe plate boundary*

The evolution of the Iberia-Europe plate boundary results from the opening of the southern part of the North Atlantic Ocean (Figure 3.2) and the migration of Iberia eastward relative to Europe (Olivet 1996; Sibuet et al. 2004; Handy et al. 2010; Vissers and Meijer 2012; Barnett-Moore et al. 2016; Nirrengarten et al. 2018). The first model to propose a kinematics of Iberia is that by Le Pichon (1970), taking up the role of the North-Pyrenean Fault as a transforming fault linked to the opening of the Bay of Biscay. Based on this model, Choukroune and Mattauer (1978) propose that the compressive phase beginning in the Late Cretaceous is related to the eastward damping of the Iberian movement. However, this model is in direct contradiction to evidence of Late Mesozoic to Cenozoic contraction trending north–south, found in the Pyrenees and along the northern Spanish margin (Malod et al. 1982; Boillot 1986).

More recent controversies have centered on the extent and timing of Iberia's counterclockwise rotation, due to its implications for the reconstruction of the Atlantic and Tethyan domains. In particular, the significance of the "J" anomaly identified in the Atlantic Ocean and its role as a magnetic anomaly equivalent to the M0 anomaly (~120 Ma) have been debated. The J anomaly is interpreted as reflecting magmatic events in the hyper-thinned margin of western Iberia, that is, the Portuguese margin. Therefore, Bronner et al. (2011) and Nirrengarten et al. (2017) propose that it does not represent a true ocean anomaly. Two models have been proposed so far. The reconstructions that partially or do not take into account the J anomaly, all reproduce the extension in the North Pyrenean rift. However, their timing for geologic events and momentum along the boundary between the Iberian-European plates differ (Olivet 1996; Jammes et al. 2009; Tugend 2014a, 2014b; Nirrengarten et al. 2018; Angrand et al. 2020). These models diverge in particular on the amount and age of the strike-slip (W–E), compared to the pure divergence (N–S). Reconstructions that consider the J anomaly imply a convergence of 500 km and the closure of a Tethyan-related Jurassic age ocean or a hyperextended margin between the Iberian Peninsula and Europe during the Cretaceous (Srivastava et al. 2000; Sibuet et al. 2004; Vissers and Meijer 2012). Although the presence of cold material in the asthenosphere at the level of North Africa has been suggested to represent the remains of a subducted oceanic plunging panel (Barnett-Moore et al. 2016; van Hinsbergen et al. 2017), the geological implications of this scenario for the Pyrenees, such as the establishment of a large basin during the Jurassic and the development of an accretionary prism in the Cretaceous or the high pressure metamorphism, are not supported by robust geological observations.

Although all proposed reconstructions agree on a sinistral strike-slip quantity of 400–500 km between Europe and the Iberian Peninsula after the Variscan orogeny, its spatial and temporal distribution is debated. This movement has historically been considered to have been formed along the North-Pyrenean Fault delimiting the Axial Zone from the North-Pyrenean Zone (Choukroune and Mattauer 1978; Debroas 1987, 1990; Jammes et al. 2009; Lagabrielle et al. 2010). However, there is no geological constraint or robust geophysical evidence across the Pyrenees that would support strike-slip deformation of such magnitude during the Jurassic or Cretaceous (Olivet 1996; Masini et al. 2014; Canérot 2016; Chevrot et al. 2018).

An alternative scenario has recently been issued (Tugend et al. 2015; Nirrengarten et al. 2018; Tavani et al. 2018; Angrand et al. 2020; Asti et al. 2022), proposing a spatial-temporal distribution deformation in a deformation corridor wider than the Pyrenean belt alone. Transtensive deformation resulting from eastward movement of Iberia would have occurred primarily during the Late Jurassic–Early Cretaceous in northern Iberia along a 100-km-scale rift basin formed by the Iberian massifs, oriented NW–SE. Indeed, along these massifs, several extensive basins have recorded significant tectonic subsidence coupled with a strike-slip deformation between the latest Permian and the Middle Cretaceous (Alvaro et al. 1979; Salas et al. 2001; Aldega et al. 2019; Aurell et al. 2019; Soto et al. 2019; Angrand et al. 2020). However, no geological evidence attesting to strike-slip movements of lithospheric scale has yet been clearly defined in these intra-Iberian basins.

Recent reconstructions by Angrand et al. (2020) highlight the key role played by the Ebro block between the Iberian Peninsula and Europe. The basins to the south of the Ebro block separate it from the Iberian Peninsula to the south. The Ebro block is bounded to the north by the Pyrenean system. In this model, the Ebro block accommodates part of the deformation of the Atlantic domain toward the Tethyan domain, respecting for the drift of the Iberian Peninsula toward the east. This interpretation also makes it possible to minimize the strike-slip movement during the Cretaceous within the North-Pyrenean rift and to preserve a dominant orthogonal extension during the opening of the Bay of Biscay. These recent tectonic and kinematic models highlight the important role of intraplate deformation in terms of the kinematics of Iberia and deformation in the Pyrenees (Mouthereau et al. 2014; Tugend et al. 2014a, 2014b; van Hinsbergen et al. 2014; Nirrengarten et al. 2018; Angrand et al. 2020; Asti et al. 2022). Due to this, the Pyrenees more accurately reflect a collision between Eurasia and the Ebro continental block, rather than with Iberia.

## 3.4. Geodynamic evolution of the Western European domain

### 3.4.1. *Upper Permian–Triassic (270–200 Ma)*

The breakup of Pangea at the end of the Paleozoic was accompanied by the closing of the Paleotethys ocean and the opening of the Neotethys ocean (Figure 3.3(b)) (Stampfli et al. 2001; Stampfli and Borel 2002). The northward drift of the Cimmerian domain, associated with the subduction of the Paleozoic Ocean (van Hinsbergen et al. 2020), led to the opening of the Pindos Ocean and Meliata Ocean back-arc basins, respectively, during the Lower (~250 Ma, Figure 3.3(c)) and Upper (Carnian, 220 Ma, see Figure 3.3(d)) Triassic (Channell and Kozur 1997;

Stampfli et al. 2001). The westward propagation of the Neotethyan oceanic opening in the thinned lithosphere after the end of the Variscan episode favored the development of a continental extension distributed in Western Europe, North Africa as well as within the future domain of the North Atlantic Ocean (Figure 3.3) (Ziegler 1990; Leleu et al. 2016; Tavani et al. 2018; Soto et al. 2019; Angrand et al. 2020 and references therein).

From the Lower Permian to the Middle Triassic, the Arctic rift propagated southward, leading to a connection with the future North Atlantic and Neotethys Basins (Goldsmith et al. 2003). This is illustrated by the development of numerous Permian detrital basins, then Triassic salt-bearing basins in Europe, marked by the acceleration of subsidence in the regions of the Pyrenees, the Betics, Catalonia and Provence (Lucas 1985; Espurt et al. 2012; Tavani et al. 2018; Saspiturry et al. 2019a; Angrand et al. 2020). This subsidence also seems to be synchronous with the onset of the thinning of the continental crust in the future Alpine Tethys (Stampfli and Borel 2002; Schmid et al. 2008).

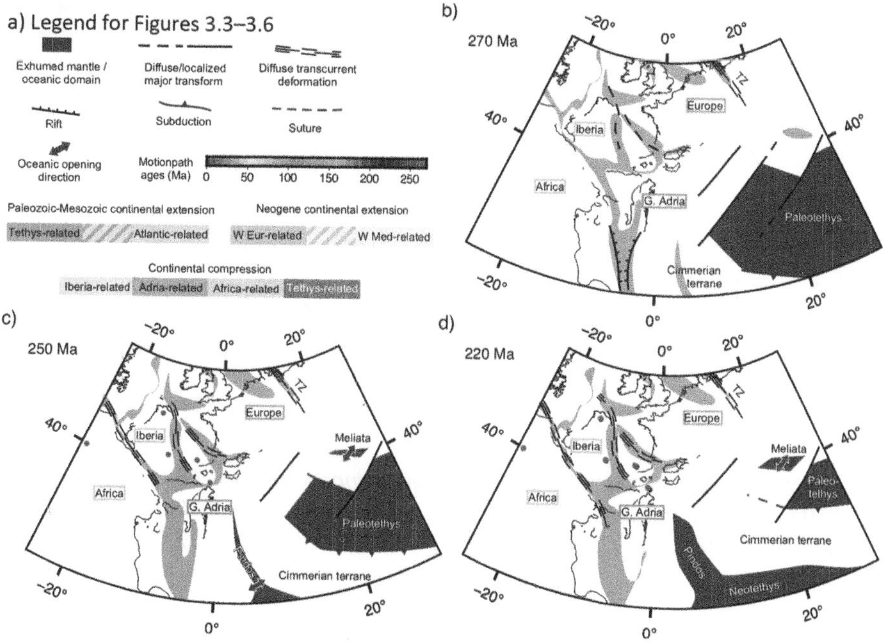


**Figure 3.3.** *(a) Legend valid for Figures 3.3–3.6; (b) reconstruction of the Atlantic-Tethys junction at 270 Ma; (c) 250 Ma; and (d) 220 Ma (modified from Angrand and Mouthereau 2021). G Adria: Greater Adria; TZ: Tornquist Zone*

### 3.4.2. *Jurassic (200–145 Ma)*

The opening of the central Atlantic Ocean at the beginning of the Jurassic propagated the extension toward the north to the southwest of the Iberian Peninsula (Murillas et al. 1990; Leleu et al. 2016). This stage was preceded by the formation of the Central Atlantic Magmatic Province (CAMP) (Olsen 1997; Marzoli et al. 1999; McHone 2000), whose magmatic intrusions extend from the Africa (Moroccan Rif – Gimeno-Vives et al. 2019) to the future Pyrenean domain (Azambre et al. 1987; Rossi et al. 2003), affecting Iberia (Cerbiá et al. 2003) before the opening of the Southern Alpine Tethys (or Maghrebine Tethys).

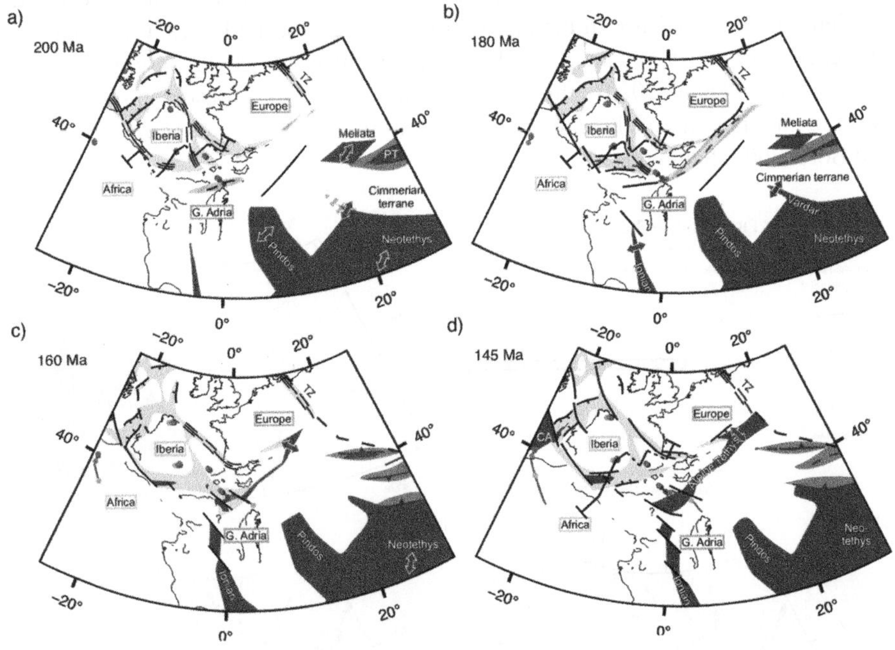


**Figure 3.4.** *Reconstruction of the Atlantic-Tethys junction at (a) 200 Ma; (b) 180 Ma; (c) 160 Ma; and (d) 145 Ma. See legend in Figure 3.3(a). Modified from Angrand and Mouthereau (2021). CA: Central Atlantic; G Adria: Greater Adria; TZ: Tornquist Zone*

Between Iberia and the Adriatic domain, the extension is located in the future Alpine Tethys (Figures 3.4 (a) and (b)) (Schmid et al. 2008; Marroni et al. 2017; Le Breton et al. 2021) due to the reactivation of extensive Triassic Neotethysian structures. This continental break-up chronology is evidenced by Early to Middle Jurassic (~190 Ma) MORB-type gabbro intrusions at the Atlantic-Tethys Junction

(Fernàndez et al. 2019) and in the southwestern termination of the Alpine Tethys, now inverted in the inner areas of Eastern Betics (~185 Ma, see Puga et al. 2011).

However, the origin of the MORB remains debated, whether it is related to the slow propagation of an oceanic rift or to magmatism associated with a hyper-stretched margin. This question has been debated in particular in the Betic region, the Moroccan Rif and the Great Kabylie (Durand-Delga et al. 2000). In our reconstruction, the Betic rift between Alboran and Iberia is assumed to propagate northward, but the latter appears to be interrupted along transfer zones (Vergés and Fernàndez 2012), forming rift segments with the Alpine Tethys, south of the Sardinia-Corsica-Peloritanian-Calabria domain, but without being directly connected to it (Figures 3.4 (c) and (d)).

The timing of MORB-type magmatism in North Africa, although slightly older than that of the eastern Betic, is consistent with the age of the onset of oceanic expansion in the northern Alpine Tethys dated to the Middle Jurassic via the postrift sedimentary series of the Alps (170–161 Ma) (Bill et al. 2001; Schaltegger et al. 2002). This chronology is confirmed by synrift subsidence associated with the opening of the southern Alpine Tethys preserved on the Calabrian paleomargin (Bouillin et al. 1988). The Jurassic Alpine Tethys thus has temporal and genetic affinities with the evolution of the Atlantic Ocean, rather than with the Neotethys Ocean.

The age of the opening of the Ionian Basin, between the Adriatic domain and Africa, is also debated. Although continental extension has been said to have started at the end of the Permian–Triassic (Stampfli et al. 2001), here we favor the tectonic model considering an onset of extension in the Lower Jurassic (~180 Ma), which is based on the timing of deformation from recent seismic data interpretation (Tugend et al. 2019). The opening of the Ionian Basin caused the Adriatic domain to shift northeastward relative to Africa, initiating extension into the future Tethys of the Southern Alps and segmentation along continental transfer faults beginning at 180 Ma. With the continued and more rapid westward drift of Adria relative to Africa, which accompanied the opening of the Alpine Tethys, a transform plate boundary was formed between Iberia and Africa at ~160 Ma (Figure 3.4(c)). The opening of the Central Atlantic caused the eastward migration and counterclockwise rotation of Africa. This movement produced subduction along the northern margin of the Neotethys (Schmid et al. 2008). During the Jurassic, tectonic reorganization was found primarily along the Alpine Tethys domain (Figures 3.4 (c) and 3.4(d)). To the west, this stage defined a tectonic quiescence along the Iberian-European boundary in the Pyrenees and Provence. From the Upper Triassic to the Middle Jurassic, deformation appeared to stop, as suggested by the stratigraphy of the Pyrenees, which indicated a slowing of tectonic subsidence, or even uplift, between

200 and 160 Ma. This contrasts with the previous stage of thinning taking place during the Permian and Triassic between the Ebro block and the rest of Iberia (Curnelle 1983; Brunet 1986, 1994; Martín-Chivelet et al. 2019).

### 3.4.3. *Early Cretaceous (145–94 Ma)*

A major change occurred from the Late Jurassic to the Early Cretaceous (145–125 Ma) with the opening of the southern North Atlantic (Figure 3.5(a)). Crustal breakup indeed occurred between 147 and 133 Ma as evidenced by the exhumation of the subcontinental mantle in the most distal parts of the southwestern Iberian hyper-stretched margin. This extensive episode reached the continental break leading to the development of an oceanic ridge between 135 and 133 Ma (Alves et al. 2009; Sallarès et al. 2013). This rifting phase seems to be generalized to the Iberian plate as evidenced by the sedimentary and tectonic record, from the Upper Jurassic to the Basal Cretaceous, in the intra-Iberian, Betics, Basque-Cantabrian and Columbretes Basins. On the contrary, the Pyrenean domain did not seem to be affected by this rifting phase before the Aptian-Albian (see Chapter 5).

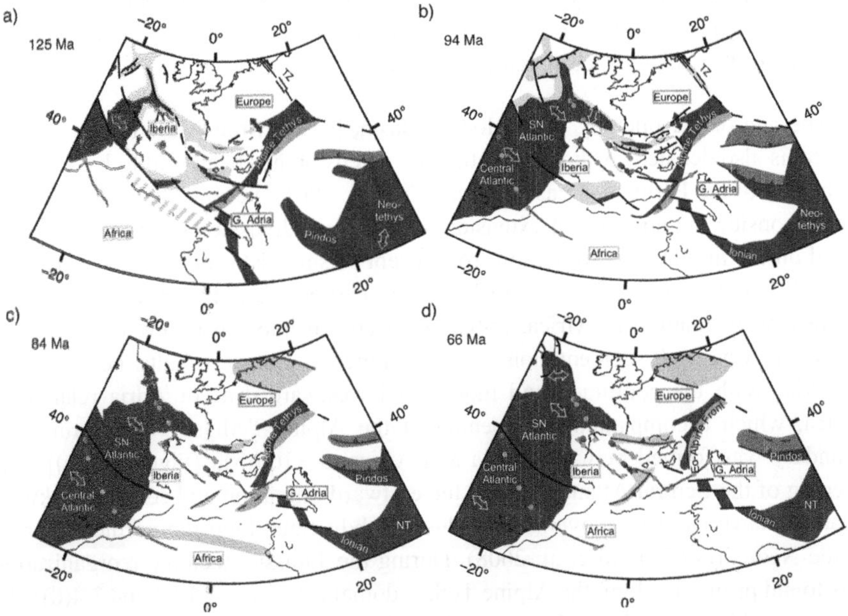


**Figure 3.5.** *Reconstruction of the Atlantic-Tethys junction at (a) 125 Ma; (b) 94 Ma; (c) 84 Ma; and (d) 66 Ma. See legend in Figure 3.3(a). Modified from Angrand and Mouthereau (2021). G Adria: Greater Adria; SN Atlantic: southern North Atlantic; NT: Neotethys*

The opening of the southern part of the North Atlantic Ocean has induced the eastward displacement of Iberia in relation to the Ebro-Sardinia-Corsica block. It also caused the relative displacement of this block with respect to Europe. Rifting then extended to the entire periphery of the Iberian plate, reactivating previous Permian–Triassic (e.g. Pyrenees, Iberian Range) (Aldega et al. 2019; Espurt et al. 2019) or Jurassic (southern Tethys) rift basins (Figures 3.5 (a) and 3.5(b)). This extensive episode appears to have been distributed in a broad northwest/southeast oriented band, materializing a broad corridor of transtensive deformation affecting the Parentis, Arzacq, North Pyrenean and Provençal domains, as well as the Basque-Cantabrian and Iberian Basins (Tugend et al. 2015; Aurell et al. 2019; Rat et al. 2019). This extensive deformation zone extended even to the Alpine Tethys located further east. Rifting in the Valencia Basin (Columbretes) suggests that extension at the southeastern end of the Iberian Basins may have spread further north into the eastern Pyrenean rift (Corbières) between the Pyrenees and Provence, reactivating an ENE–SW trending fault system along the Catalan and Languedoc margin (e.g. Nîmes Fault), inherited from Late Variscan evolution (Roure et al. 1992; Lacombe and Jolivet 2005).

After a quiescent phase, the northward propagation of the Atlantic Ocean opening resumed at 139 Ma. This resumption of rifting is supported by continental crustal thinning leading to mantle exhumation as described off the southwestern margin of Galicia (Mohn et al. 2015). This rifting episode is followed by a phase of oceanic accretion in this same area from 121 to 112 Ma (Bronner et al. 2011; Vissers and Meijer 2012). Extension then migrated and formed northward (Rat et al. 2019) leading to oceanic accretion in the Bay of Biscay (Chapter 6), hyperextension of the continental crust and exhumation of the lithospheric mantle in the Basque-Cantabrian, Parentis, and North Pyrenean Basins at ~120 Ma (Jammes et al. 2009; Lagabrielle et al. 2010; Mouthereau et al. 2014; Tugend et al. 2014a, 2014b, 2015; Saspiturry et al. 2021b).

The Pyrenean rift reached its paroxysm during the Aptian-Albian along the E–W oriented rift of Parentis (Figure 3.5(b)) (Bois and Ecors Scientific Party 1990; Jammes et al. 2009), Arzacq (Issautier et al. 2020; Ducoux et al. 2021), Organyà (Berástegui et al. 1990; Mencos et al. 2015), and in the western North Pyrenean Basins materialized by the Basque-Cantabrian Basin (Pedreira et al. 2007; Ducoux et al. 2019), the Mauléon Basin (Masini et al. 2014; Corre et al. 2016; Teixell et al. 2016; Saspiturry et al. 2019b), the Barronies and Ballongue Basin in the central Pyrenees (Debroas 1987, 1990; Espurt et al. 2019), and the Boucheville Basin in the eastern Pyrenees (Clerc et al. 2015) (Figure 3.5(b)). In the Basque-Cantabrian Basin, mantle exhumation at the base of the basin (DeFelipe et al. 2017; Ducoux et al. 2019) makes a continuity with the Pyrenean rift basin through the Pamplona Fault.

Maximum rift-related temperatures were reached at the end of extension as revealed by HT-LP metamorphism (Cuevas and Tubía 1999; DeFelipe et al. 2017) and alkaline magmatic intrusions in the Upper Albian (102 Ma) (Ubide et al. 2014; Agirrezabala 2015) (for further detail, see Chapter 5).

The Aptian-Albian synrift evolution of the Basque-Cantabrian Basin was quickly interpreted as reflecting deformation in a transtensive regime, with NW–SE structures being interpreted as left-lateral strike-slip faults (García-Mondéjar et al. 1996; Cámara 2017). In particular, the Ventaniella Fault is part of a NW–SE fault system of Triassic origins (Tavani and Granado 2015), accommodating left-lateral displacement during the Late Jurassic–Early Cretaceous, and then reversed with right-lateral kinematics during the Cenozoic (de Vicente et al. 2011; Tavani et al. 2011a; Cámara 2017). To account for the significant orthogonal extension along the NW–SE-trending Leitza-Gernika detachment and to explain the extensional segmentation and mantle exhumation along the axis of the Pyrenean-Cantabrian rift, a more recent study interprets the Basque-Cantabrian Basin as a transtensive basin linking its structural evolution to oblique extension (Pedrera et al. 2020a). The two transtensive models differ slightly depending on whether the deformation is partitioned between N–S (or NNW–SSE) extension and NW–SE strike-slip faults (García-Mondéjar et al. 1996; Cámara 2017), or whether it is not partitioned but oblique (i.e. NNE–SSW extension) toward the E–W platforms inherited from the Landes High and Duero-Ebro (Pedrera et al. 2020a). As noted by Angrand et al. (2020), the distinction between the two models can be difficult, as the sedimentary cover on which the interpretations are based is decoupled from the basement via the Triassic salt layer.

During the Late Jurassic–Early Cretaceous, the eastern extension of the European margin underwent a new rifting episode that initiated the formation of the Valaisan Basin (Schettino and Turco 2011). While it was previously considered as the western oceanic branch of the Alpine Ocean, between the Briançonnais domain and Europe, the chronology of the opening of the Valaisan Ocean suggests a close relationship with the eastward drift of the Iberian plate (Beltrando et al. 2007; Schettino and Turco 2011), or more precisely with the Ebro-Sardinia-Corsica block. The opening of the Valaisan Ocean lasted until the Middle Cretaceous (93 Ma), coinciding with the opening of the Bay of Biscay and with the intracontinental subduction and nappe stacking in the Austro-Alpine domain (e.g. Eo-Alpine orogeny; Handy et al. 2010) that resulted from the northward movement of the Adriatic domain (Müller et al. 2019). The rift clearly identified in the Pyrenees and Provence did not extend further east into the Alpine Tethys. The axis of this rift was shifted into the Vocontian Basin, which lies at the end of the Valaisan Ocean, forming another rift segment of the broader Bay of Biscay system. A major consequence is that the Briançonnais block can be placed in the continuity of Provence to the east.

From the Late Aptian to the Cenomanian, the southward movement of the Ebro resulting from orthogonal N–S extension taking place in the Bay of Biscay, Pyrenees, Aquitaine, and Parentis generated a transpressive episode between the southwestern Ebro and the eastern Western Iberian block (Sibuet et al. 2004; Barnett-Moore et al. 2016; Angrand et al. 2020). This tectonic episode induced a modest NE–SW shortening of the western Iberian Basins. This contraction is thought to have occurred in the Cameros Basin between the Upper Albian and Cenomanian, prior to the peak of HT-LP metamorphism recorded in the sedimentary series around 100 Ma (Casas-Sainz and Gil-Imaz 1998). This transpressive episode also coincides with a cessation of tectonic subsidence in the Cameros Basin (Omodeo-Salé et al. 2017). Based on thermochronological contractions and axial fold cleavages documented in the eastern part of the Cameros Basin, this episode could reflect an early contraction phase affecting a thermally unbalanced (anomalously warm) continental lithosphere. This episode would therefore be earlier than the initial collision-related cooling phase at 60 Ma imaged by low-temperature thermochronology (Rat et al. 2019). However, evidence of contraction elsewhere in the Iberian chain at this time remains to be documented.

By the end of the Cenomanian (Figure 3.5(b)), extension in the Pyrenean segment and in the Betics peaked but appears to be limited to the Iberian chain despite the high thermal gradients and metamorphism of this age (Cameros Basin). The regional HT anomaly related to the opening of the Bay of Biscay and recorded in the Pyrenean Basins (Vacherat et al. 2014; Clerc et al. 2015; Ducoux et al. 2019; Saspiturry et al. 2020a) suggests a control of the regional geodynamics by a mobile and warm asthenospheric mantle. During the Cretaceous (Figure 3.5), deformation in the Austro-Alpine domain is followed by the onset of Alpine Tethys subduction. The age of this subduction is poorly constrained due to the poor preservation of markers. Different ages have been proposed in models based on the relative movement of Adria and Iberia. It varies from 125 (our model) to 110 Ma (Schettino and Scotese 2002) or 94 Ma (van Hinsbergen et al. 2020), even 94–84 Ma (Handy et al. 2010).

### 3.4.4. *Late Cretaceous to Paleogene (84–35 Ma)*

The Santonian-Campanian transition (Figure 3.5(c)) marks the beginning of regional N–S convergence between Africa, Iberia and Europe, after the start of the northward drift of the Adriatic domain. This change induced an accelerated closure of the Alpine Tethys in the Western Alps and the inversion of most of the extensive Mesozoic Basins in Iberia and in the Ebro-Sardinia-Corsica blocks.

In the west, the convergence between Iberia and Europe is accommodated by the closure of the hyperthinned rifts staking the northern Pyrenean zone (Mouthereau et al. 2014; Ford et al. 2016; Teixell et al. 2016; Espurt et al. 2019; Saspiturry et al. 2020a) and the Basque-Cantabrian Basin (Pedreira et al. 2007; García-Senz et al. 2019). This phase is dated by the first overlaps and sedimentary inputs dated to the Upper Santonian-Campanian in the eastern and southern Pyrenees and Provence (Leleu et al. 2005; Vinciguerra 2020), the establishment of the first synorogenic sequences in the south Pyrenean slope (growth strata[5]) (Puigdefàbregas and Souquet 1986; Vergés et al. 1995), the accelerated subsidence of the Iberian and European carbonate platform (Andrieu et al. 2021; Issautier et al. 2022) and the influx of clastic series into the northern Pyrenean (Aquitaine Basin) and southern Pyrenean (Ebro Basin) foreland subsidence basins (Christophoul et al. 2003; Bilotte and Canérot 2006; Mouthereau et al. 2014; Ford et al. 2016; Rougier et al. 2016; Vacherat et al. 2016, 2017). Early synorogenic exhumation is suggested by an early cooling episode at 75–70 Ma (Mouthereau et al. 2014; Ternois et al. 2019; Waldner et al. 2021) and an initial influx of synorogenic sediments accumulated in the southern and northern Pyrenean foreland basins (Vergés et al. 1995; Sinclair et al. 2005; Grool et al. 2018).

In the Basque-Cantabrian Basin, low-temperature thermochronological analysis of the Cinco Villas massif constrains an exhumation to the early Middle Eocene (50–45 Ma) (DeFelipe et al. 2019), while slightly to the west in the Cantabrian Mountains, Fillon et al. (2016) infer an exhumation of 8–10 km between 39 and 29 Ma. The deposition of thick, well-conserved Eocene–Oligocene synorogenic turbidites in the basin reveals that the Basque-Cantabrian domain continued to deepen during the early stages of Pyrenean rift inversion. Again, this particular pattern is suggested to illustrate the impact of weak Triassic salt detachment during inversion that may have blocked the development of orogenic topography in a weakly inverted rift domain characterized by an anomalously thin crust inherited from the rifting episode (Jourdon et al. 2020).

Further east, south-verging subduction of the Alpine Tethys (Schettino and Scotese 2002; Schmid et al. 2004; Handy et al. 2010; van Hinsbergen et al. 2020) brought the eastern Alpine domain into compression during the Late Cretaceous (Kerckhove 1969; Decarlis et al. 2014). This phase is marked by the accretion of the Dent Blanche-Sesia domains north of Adria (Rubatto et al. 1999; Manzotti et al.

5 Set of asymmetrical sedimentary layers, deposited during the growth of a structure.

2014; Regis et al. 2014), associated with eoalpine high-pressure metamorphism (e.g. Villa et al. 2000) and helminthoid flysch sedimentation.

During the Early Paleogene (~66 Ma, Figure 3.5(d)), the contraction and exhumation of crustal units resulting from this compressive episode extended across the domain from Africa to Europe, as evidenced by the tectonosedimentary evolution of the Pyrenees (Choukroune et al. 1990; Muñoz 1992; Vergés et al. 1995; Mouthereau et al. 2014; Ford et al. 2016; Teixell et al. 2016; Garcés et al. 2020), the Iberian Range (Rat et al. 2019), the Basque-Cantabrian Basin (Pedreira et al. 2007; Cámara 2017), the Betics (Vergés and Fernàndez 2012; Daudet et al. 2020), and Provence (Leleu et al. 2005; Espurt et al. 2012; Bestani et al. 2016). In the Betic domain, there is evidence for the onset of exhumation of high-pressure rocks at 49–28 Ma (Monie et al. 1991; Augier et al. 2005). There is also evidence for (1) Pre–Late Oligocene deformation in the Balearic promontory and Valencia Basin (Etheve et al. 2016, 2018) and (2) Eocene (67 Ma) exhumation of the Variscan basement in southeastern Sardinia (Malusà et al. 2016). We infer that all the domains comprising Iberia, from the Betics to AlKaPeCa, and surrounding the Ebro-Sardinia-Corsica, were in contraction, uplifting and encompassed moderate exhumation in the Eocene. An orogenic domain thus extended between Europe and Africa at the location of the western Mediterranean Sea during the Eocene (Figures 3.6 and 3.7(a)). This orogenic domain included a rapidly exhumed segment (Pyrenees), an uplifted but slightly deformed domain (Sardinia-Corsica), and formed crustal zones characterized by high-pressure orogenic metamorphism such as the Betic-Rif domain. It is worth mentioning that due to the deformations that occurred at the margins of the AlKaPeCa continental domains, the basement of these domains did not necessarily undergo significant orogenic exhumation.

The closure of a remnant of the Alpine Tethys between Corsica/Briançonnais and the Adriatic led to the progressive development of the Western Alps. In the Alps, blue shale to greenschist facies units, including Jurassic–Cretaceous (Briançonnais) and European basement (Tenda in Corsica) meta-ophiolites/ metasediments, were exhumed during the Paleocene and Eocene (58–47 Ma). This first orogenic system was followed by eclogitic units of the distal European margin (Monte Rosa, Gran Paradiso, and Dora Maira) and adjacent meta-ophiolites (Viso) rapidly exhumed during the Late Eocene (44–34 Ma) (Malusà et al. 2011). Further east, deformation along the Dinarides orogen occurs in the Late Cretaceous, probably associated with subduction of the Vardar Ocean remnants (van Hinsbergen et al. 2020). The onset of convergence in the Dinarides may have begun in the Mid-Cretaceous, as evidenced by calc-alkaline magmatism (Handy et al. 2015 and references therein).

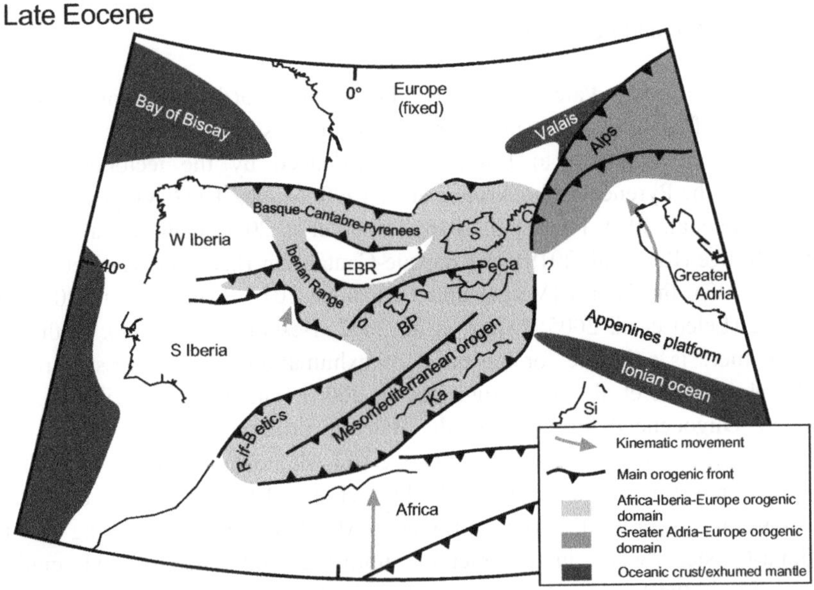


**Figure 3.6.** *Schematic reconstruction of the orogenic system between the Africa, Iberia, Adria and Europe plates in the Late Eocene. According to Angrand and Mouthereau (2021). C: Corsica; Ka: Kabylies; BP: Balearic promontory; S: Sardinia; Si: Sicily*

### 3.4.5. *Latest Paleogene to present (35-0 Ma)*

Beginning in the Late Oligocene–Early Eocene (Figure 3.7(b)), a major change occurred in the geodynamics of western Europe. As the Alpine orogenic front advanced northwestward, the West European Rift opened and propagated southward (Bois 1993; Séranne 1999; Dèzes et al. 2004). The first phase of extension is well described in Provence (35–23 Ma), in the eastern Pyrenees, and in the Valence Basin (Merle and Michon 2001; Ziegler and Dèzes 2006). At the same time, continental collision in the western Alps occurred from 35 Ma onwards, as indicated by the deposition of Alpine flysch ahead of the westward-migrating deformation front (Ford et al. 2006) and the exhumation of Paleozoic basement in the outer massifs (Bellahsen et al. 2014). The Alpine arc thus appears as a late and peculiar characteristic, unrelated to the initial stage of convergence between the Adriatic and Europe.

In the Pyrenean system, this extension is synchronous with the beginning of the extension at the Late Eocene–Early Oligocene (~30 Ma), related to the opening of the West European rift. This extension is well recorded in the Corbières and in the eastern part of the Axial Zone, which reactivates Pyrenean compressive structures into normal faults such as the NE–SW Têt Fault and further east the NE–SW Nîmes Fault (Gorini et al. 1993; Séranne et al. 1995). Low-temperature thermochronological data across the Têt Fault show that early stages of extension may have begun in the Late Oligocene (26–27 Ma) (Maurel et al. 2002). This relatively late extensional event in the Late Eocene thus predates the opening of the Gulf of Lion hinterland and the Ligurian-Provençal Basins, which occurred in a second phase around 23 Ma (Figure 3.7(d)) (Anadón et al. 1989; Bois 1993; Jolivet and Faccenna 2000). The Late Miocene reactivation of the Têt Fault (12–9 Ma) that followed the Sardinia-Corsica rotation seems to be associated with the formation of the NNW oriented oblique fault system and volcanism in Catalonia (Martí et al. 1992; Cerbiá et al. 2003). This extension may explain the current anormal topography of the eastern Pyrenees (Huyghe et al. 2020) in connection with Late Miocene alkaline magmatism, well documented in the Gulf of Lion, on the coast of Sardinia and off Corsica (Lustrino et al. 2009; Réhault et al. 2012).

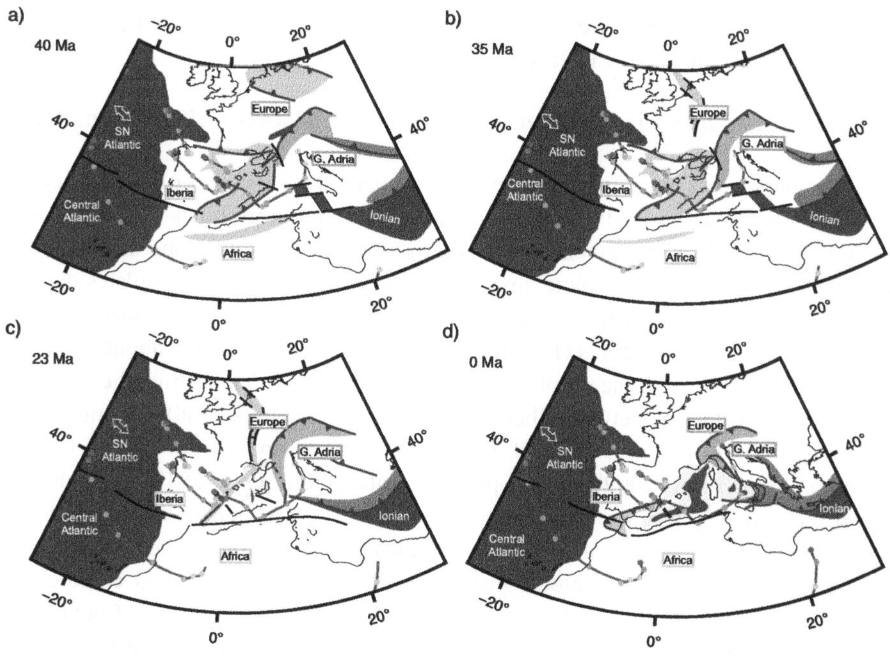


**Figure 3.7.** *Reconstruction of the Atlantic-Tethys junction at (a) 40 Ma; (b) 35 Ma; (c) 23 Ma; and (d) at the present. See legend in Figure 3.3(a). Modified from Angrand and Mouthereau (2021). G Adria: Greater Adria*

The onset of oceanization of the Ligurian-Provençal Basin between 20.5 and 15 Ma was associated with the rotation of the Sardinia-Corsica block triggered by the retreat of the Tyrrhenian subduction front. This rotation is accommodated westward by the North-Balearic transform zone. This event caused a kinematic reorganization of the westernmost part of the Mediterranean, as indicated by the propagation of volcanism toward the Algerian Mediterranean (Maillard et al. 2020). One consequence of this reorganization is the opening of the Alboran Sea by the westward retreat and delamination of the lithosphere since the Burdigalian, resulting in the dismantling of the ancient Betic orogen (Vergés and Fernàndez 2012; Daudet et al. 2020; Pedrera et al. 2020b). Beginning in the Tortonian (~11 Ma), the Betic Cordillera was rapidly exhumed and volcanism spread to the Alboran Sea and western Mediterranean. This episode is associated with mantle magmatism, the origin of which could be a change in the thermal regime of the asthenosphere, which is thought to be responsible for the uplift and anomalous elevation of the eastern Pyrenees and Iberia in the Neogene (Conway-Jones et al. 2019; Huyghe et al. 2020).

## 3.5. Discussion of the main scientific issues surrounding the kinematics of the Iberian plate

### 3.5.1. *Deformation distribution at the Iberian-European plate boundary*

The model presented in this chapter recognizes the existence of a continental domain between Iberia and Europe, corresponding to the Ebro block. The inclusion of the Ebro block in the kinematics of Iberia with respect to Europe allows for a better distribution of the absolute amount of rifting between these two plates during the Mesozoic (400–500 km). In our model, the cumulative left-lateral displacement rifting of the two rift systems (Pyrenees and intra-Iberian Basins) is about 456 km, which is consistent with the absolute 400–500 km needed to reconstruct the opening of the Atlantic Ocean between Iberia and Canada (Newfoundland). Although we consider the Ebro block to be rigid, tectonosedimentary analyses of its constituent basins (Vargas et al. 2009) and geophysical data (Wehr et al. 2018) beneath the Cenozoic foreland basin argue for tectonic subsidence related to a rifting phase during the Early Mesozoic. In addition to localized deformation on the northern and southern boundaries of this block, distributed deformation within this continental domain, between Europe and Iberia, including the Parentis and Basque-Cantabrian basins, the Iberian (Cameros, Maestrat) and Valencia (Columbretes Basin) Basins (Jammes et al. 2009; Tugend et al. 2014a, 2014b; Etheve et al. 2016; Angrand et al. 2020), must be considered in order to account for the complexity of the system.

However, in kinematic terms this distributed deformation remains poorly constrained, as no major structure can be identified within the Ebro block. At the scale of the kinematic model, and for simplicity, we have adopted here the hypothesis of a block of the Ebro behaving rigidly, so that the deformation is found only at its boundaries. This strong, purely methodological choice leads to an overestimation of the amount of deformation on the northern and southern boundaries of the Ebro block; however, there is no such structure known which could accommodate such an amount of deformation.

At the Iberia-Ebro boundary, studies have suggested that a left-lateral shear zone may be recognized along the Iberian Range and the Basque-Cantabrian rift system. Geological evidence includes the High Tagus Fault in the Iberian Range (Aldega et al. 2019; Aurell et al. 2019) and the Ventaniella Fault in the Cantabrian region (Tavani et al. 2011b). The latter is often taken into account in recent reconstructions to solely account for the left-lateral movement between Iberia and the Ebro block (Tugend et al. 2015; Nirrengarten et al. 2018). However, the estimated field-based left-lateral lateral displacement along the Ventaniella Fault is only of the order of a few kilometers (Tavani et al. 2011a) and therefore cannot be used as an equivalent for the North Pyrenean Fault delineating the axial Pyrenean zone from the Cretaceous rift basins north of it. Therefore, it seems more likely that a NW–SE fault system of Triassic origin (Tavani and Granado 2015), of which the Ventaniella structure is a part, acted as a left-lateral strike-slip zone during the Upper Jurassic–Lower Cretaceous. This was then reactivated with right-lateral kinematics in the Cenozoic (de Vicente et al. 2011; Cámara and Flinch 2017). Tectonic activity along these faults rejuvenates northeastward (Ubierna Fault: Terminal Jurassic–Lower Cretaceous; Zamanza-Oña Fault: Lower–Middle Cretaceous, salt tectonics in the center of the basin until the Middle Cretaceous) (Cámara and Flinch 2017). We suggest that the displacement of Iberia from the Ebro block may have been distributed along these structures.

The lack of markers, in the basement, to accurately quantify displacement along the tectonic structures discussed above may be explained, in part, by the significant decoupling of deformation within Triassic evaporites present at the sediment-basement interface (Ortí et al. 2017; Lagabrielle et al. 2020; Saspiturry et al. 2021b; Asti et al. 2022). The role of evaporites in mechanically decoupling the basement and its cover has been widely emphasized in the Pyrenees (Duretz et al. 2019; Grool et al. 2019; Ford and Vergés 2020; Jourdon et al. 2020; Lagabrielle et al. 2020; Saspiturry et al. 2020a), the Basque-Cantabrian Basin (Cámara and Flinch 2017), the Parentis Basin (Ferrer et al. 2012), the Cameros Basin (Rat et al. 2019), the Arzacq Basin (Issautier et al. 2020) and the Maestrat Basin (Vergés et al. 2020). Indeed, the surface expression of strike-slip motion in the basement may have been limited in the suprasalt layers. Finally, the mechanism responsible for the independent movement of the Ebro with respect to Europe and the Iberian Peninsula before the opening of the southern part of the North Atlantic Ocean remains unclear. One hypothesis is that the Ebro continental block was part of a continental unit disconnected from Iberia.

### 3.5.2. *Variscan inheritance and Late Variscan position of the Ebro, Sardinia and Corsica blocks*

Much of the Mesozoic-Cenozoic Alpine evolution is inherited from Variscan tectonic evolution. The major shear zones described in our kinematic restoration include the Pyrenean-Provençal and intra-Iberian rifts, which appear to overlap with dextral shear zones of the Variscan cycle (Figure 3.8) (Edel et al. 2015; Ballèvre et al. 2018; Molli et al. 2020; Simonetti et al. 2020). From the Late Carboniferous to the Early Permian (300–290 Ma), paleomagnetic constraints (Edel et al. 2015) suggest that Catalonia, belonging to the Ebro block as well as Sardinia, Corsica and the Maures-Esterel Massif, belonging to the Sardinia-Corsica block, were disconnected from the rest of Iberia. At this time, the westward migration of the Ebro-Sardinia-Corsica block, relative to Europe, occurred along the "eastern Variscan shear zone" (Ballèvre et al. 2018). In the western Alps, the anisotropic structure of the lithosphere related to this shear zone could also be responsible for the location of the Lower Jurassic Tethys Fault in the Alps and the subsequent oceanic expansion south of the Pennine front (Ballèvre et al. 2018). To the south, this displacement would be synchronous with the formation of the Iberian orocline during the Late Carboniferous–Lower Permian (Edel et al. 2018; Pastor-Galán et al. 2020). This work implies that the Ebro block and its Sardinia-Corsica continental continuation are tectonically distinct from Iberia. The Ebro block should not be passively reconstructed with the rest of the Iberian Peninsula (here Galicia) during the Permian, as is generally assumed in most kinematic models (Olivet 1996; Vissers and Meijer 2012; van Hinsbergen et al. 2020). This further emphasizes the key role played by the shear zone south of the Ebro during the Late Carboniferous, which has been reactivated since the Late Permian–Triassic and is currently shaping the Iberian chain. The Variscan segmentation can be further extended between Iberia and North Africa and between Iberia and North America (Fernández 2019). Despite the clear relationships between the tectonic and kinematic pattern in the Mesozoic, there are no strong constraints linking Late Variscan evolution to Alpine evolution. Therefore, more geochronological data and field studies, including new tectonic constraints from the Permian–Triassic series, are needed to fill this gap in the evolution of the western European domain during the Permian.

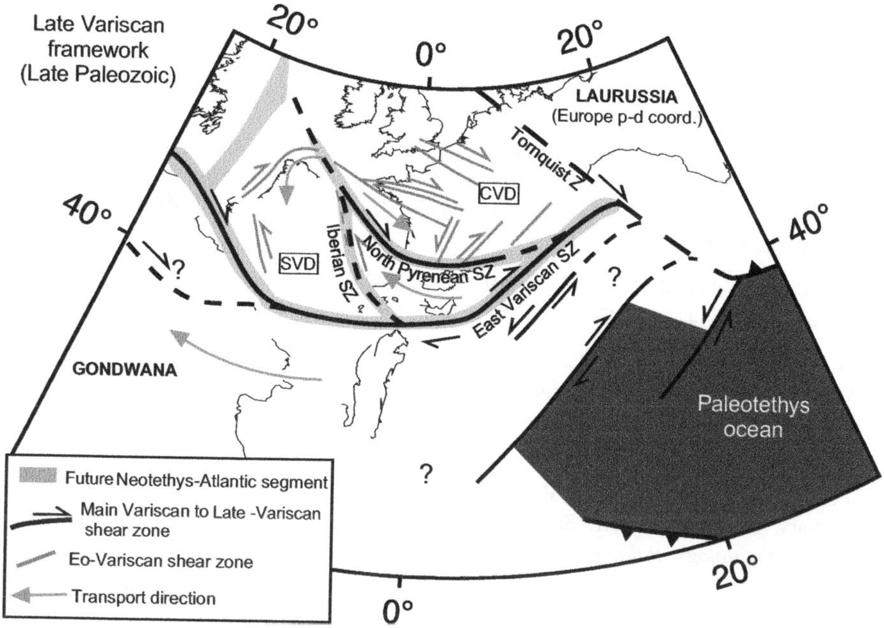


**Figure 3.8.** *Schematic reconstruction at the end of the Paleozoic. Variscan structures appear to form the later deformation and opening of the Tethyan and Alpine oceanic segments. According to Angrand and Mouthereau (2021). CVD: Central Variscan Domain; SVD: South Variscan Domain; SZ: Shear Zone*

## 3.6. Acknowledgments

We would like to thank the participants, students, PhD students, postdoctoral fellows and researchers of the ANR PYRAMID, PYROPE, RGF-Pyrenees and OROGEN projects, and more broadly all the geoscientists within the Pyrenees community.

## 3.7. References

Agirrezabala, L.M. (2015). Syndepositional forced folding and related fluid plumbing above a magmatic laccolith: Insights from outcrop (Lower Cretaceous, Basque-Can-tabrian Basin, western Pyrenees). *GSA Bulletin*, 127(7/8), 982–1000.

Albarède, F. and Michard-Vitrac, A. (1978). Age and significance of the north Pyrenean metamorphism. *Earth and Planetary Science Letters*, 40(3), 327–332 [Online]. Available at: https://doi.org/10.1016/0012-821X(78)90157-7.

Aldega, L., Viola, G., Casas-Sainz, A., Marcén, M., Román-Berdiel, T., van der Lelij, R. (2019). Unraveling multiple thermotectonic events accommodated by crustal-scale faults in northern Iberia, Spain: Insights From K-Ar Dating of Clay Gouges. *Tectonics*, 38(10), 3629–3651 [Online]. Available at: https://doi.org/10.1029/2019TC005585.

Alonso, J., Pulgar, J., García-Ramos, J., Barba, P. (1996). W5 Tertiary basins and Alpine tectonics in the Cantabrian Mountains (NW Spain). In *Tertiary Basins of Spain: The Stratigraphic Record of Crustal Kinematics*, Friend, P.F. and Dabrio, C.J. (eds). Cambridge University Press.

Alvaro, M., del Villar, R.C., Vegas, R. (1979). Un modelo de evolución geotectónica para la Cadena Celtibérica. *Acta Geológica Hispánica*, 14(1), 172–177.

Alves, T.M., Moita, C., Cunha, T., Ullnaess, M., Myklebust, R., Monteiro, J.H., Manuppella, G. (2009). Diachronous evolution of Late Jurassic-Cretaceous continental rifting in the northeast Atlantic (west Iberian margin). *Tectonics*, 28(4), 1–32 [Online]. Available at: https://doi.org/10.1029/2008TC002337.

Anadón, P., Cabrera, L., Julià, R., Roca, E., Rosell, L. (1989). Lacustrine oil-shale basins in Tertiary grabens from NE Spain (Western European rift system). *Palaeogeography, Palaeoclimatology, Palaeoecology*, 70(1/3), 7–28.

Andrieu, S., Saspiturry, N., Lartigau, M., Issautier, B., Angrand, P., Lasseur, E. (2021). Large-scale vertical movements in Cenomanian to Santonian carbonate platform in Iberia: Indicators of a Coniacian pre-orogenic compressive stress. *BSGF – Earth Sciences Bulletin*, 192, 19 [Online]. Available at: https://doi.org/10.1051/bsgf/2021011.

Angrand, P. and Mouthereau, F. (2021). Evolution of the Alpine orogenic belts in the Western Mediterranean region as resolved by the kinematics of the Europe-Africa diffuse plate boundary. *BSGF – Earth Sciences Bulletin*, 192, 42 [Online]. Available at: https://doi.org/10.1051/bsgf/2021031.

Angrand, P., Ford, M., Watts, A.B. (2018). Lateral variations in foreland flexure of a rifted continental margin: The Aquitaine Basin (SW France). *Tectonics*, 37(2), 430–449 [Online]. Available at: https://doi.org/10.1002/2017TC004670.

Angrand, P., Mouthereau, F., Masini, E., Asti, R. (2020). A reconstruction of Iberia accounting for Western Tethys-North Atlantic kinematics since the Late-Permian-Triassic. *Solid Earth*, 11(4), 1313–1332 [Online]. Available at: https://doi.org/10.5194/se-11-1313-2020.

Asti, R., Saspiturry, N., Angrand, P. (2022). The Mesozoic Iberia-Eurasia plate boundary: A wide domain of distributed transtensional deformation rather than a single wrench fault. *Earth-Science Reviews*, 230, 104040 [Online]. Available at: https://doi.org/10.1016/j.earscirev.2022.104040.

Augier, R., Agard, P., Monié, P., Jolivet, L., Robin, C., Booth-Rea, G. (2005). Exhumation, doming and slab retreat in the Betic Cordillera (SE Spain): In situ 40Ar/39Ar ages and P–T–d–t paths for the Nevado-Filabride complex. *Journal of Metamorphic Geology*, 23(5), 357–381.

Aurell, M., Fregenal-Martínez, M., Bádenas, B., Muñoz-García, M.B., Élez, J., Meléndez, N., de Santisteban, C. (2019). Middle Jurassic-Early Cretaceous tectono-sedimentary evolution of the southwestern Iberian Basin (central Spain): Major palaeogeographical changes in the geotectonic framework of the Western Tethys. *Earth-Science Reviews*, 199(July), 102983 [Online]. Available at: https://doi.org/10.1016/j.earscirev.2019.102983.

Azambre, B., Rossy, M., Lago, M. (1987). Caractéristiques pétrologiques des dolérites tholéitiques d'âge triasique (ophites) du domaine pyrénéen. *Bulletin de Minéralogie*, 110(4), 379–396.

Ballèvre, M., Manzotti, P., Dal Piaz, G.V. (2018). Pre-Alpine (Variscan) inheritance: A key for the location of the future Valaisan Basin (Western Alps). *Tectonics*, 37(3), 786–817 [Online]. Available at: https://doi.org/10.1002/2017TC004633.

Barnett-Moore, N., Hosseinpour, M., Maus, S. (2016). Assessing discrepancies between previous plate kinematic models of Mesozoic Iberia and their constraints. *Tectonics*, 35(8), 1843–1862 [Online]. Available at: https://doi.org/10.1002/2015TC004019.

Beaumont, C., Muñoz, J.A., Hamilton, J., Fullsack, P. (2000). Factors controlling the Alpine evolution of the central Pyrenees inferred from a comparison of observations and geodynamical models. *Journal of Geophysical Research: Solid Earth*, 105(B4), 8121–8145 [Online]. Available at: https://doi.org/10.1029/1999jb900390.

Bellahsen, N., Mouthereau, F., Boutoux, A., Bellanger, M., Lacombe, O., Jolivet, L., Rolland, Y. (2014). Collision kinematics in the western external Alps. *Tectonics*, 33(6), 1055–1088.

Beltrando, M., Rubatto, D., Compagnoni, R., Lister, G. (2007). Was the Valaisan basin floored by oceanic crust? Evidence of Permian magmatism in the Versoyen Unit (Valaisan domain, NW Alps). *Ofioliti*, 32(2), 85–99.

Berástegui, X., García-Senz, J.M., Losantos, M. (1990). Tecto-sedimentary evolution of the Organyà extensional basin (central south Pyrenean unit, Spain) during the Lower Cretaceous. *Bulletin de la société géologique de France*, 6(2), 251–264.

Bestani, L., Espurt, N., Lamarche, J., Bellier, O., Hollender, F. (2016). Reconstruction of the Provence Chain evolution, southeastern France. *Tectonics*, 35(6), 1506–1525 [Online]. Available at: https://doi.org/10.1002/2016TC004115.

Bill, M., O'Dogherty, L., Guex, J., Baumgartner, P.O., Masson, H. (2001). Radiolarite ages in Alpine-Mediterranean ophiolites: Constraints on the oceanic spreading and the Tethys-Atlantic connection. *Bulletin of the Geological Society of America*, 113(1), 129–143 [Online]. Available at: https://doi.org/10.1130/0016-7606(2001)113<0129:RAIAMO>2.0.CO;2.

Bilotte, M. and Canérot, J. (2006). Rôles respectifs des tectoniques fini-crétacée et éocène dans la partie orientale de la chaîne des Pyrénées. Le "Garumnien" de Cucugnan et ses relations avec le chevauchement frontal Nord-Pyrénéen (Corbières méridionales, France). *Eclogae Geologicae Helvetiae*, 99(1), 17–27 [Online]. Available at: https://doi.org/10.1007/s00015-006-1179-8.

Boillot, G. (1986). Comparison between the Galicia and Aquitaine margins. *Tectonophysics*, 129(1/4), 243–255.

Bois, C. (1993). Initiation and evolution of the Oligo-Miocene rift basins of southwestern Europe: Contribution of deep seismic reflection profiling. *Tectonophysics*, 226(1/4), 227–252.

Bois, C. and Ecors Scientific Party (1990). Major geodynamic processes studied from the ECORS deep seismic profiles in France and adjacent areas. *Tectonophysics*, 173(1/4), 397–410 [Online]. Available at: https://doi.org/10.1016/0040-1951(90)90233-X.

Bouillin, J.-P., Mouterde, R., Olivier, P., Majesté-Menjoulas, C. (1988). Le Jurassique de Longobucco (Calabre, Italie), à la jonction de la Tethys ligure et de la Tethys maghrébine. *Bulletin de la Société géologique de France*, 4(1), 93–103.

Bronner, A., Sauter, D., Manatschal, G., Péron-pinvidic, G., Munschy, M. (2011). Magmatic breakup as an explanation for magnetic anomalies at magma-poor rifted margins. *Nature Geoscience*, 4(8), 549–553 [Online]. Available at: https:// doi.org/10.1038/nphys1201.

Brunet, M.-F. (1986). The influence of the evolution of the Pyrenees on adjacent basins. *Tectonophysics*, 129(1/4), 343–354 [Online]. Available at: https://doi.org/10.1016/0040-1951(86)90260-X.

Brunet, M.-F. (1994). Subsidence in the Parentis basin (Aquitaine, France): Implications of the thermal evolution. In *Hydrocarbon and Petroleum Geology of France*, Mascle, A. (ed.). Springer, Berlin, Heidelberg [Online]. Available at: https://doi.org/10.1007/978-3-642-78849-9_14.

Cadenas, P., Fernández-Viejo, G., Pulgar, J.A., Tugend, J., Manatschal, G., Minshull, T.A. (2018). Constraints imposed by rift inheritance on the compressional reactivation of a hyperextended margin: Mapping rift domains in the North Iberian margin and in the Cantabrian mountains. *Tectonics*, 37(3), 758–785 [Online]. Available at: https://doi.org/10.1002/2016TC004454.

Cámara, P. (2017). Salt and strike-slip tectonics as main drivers in the structural evolution of the Basque-Cantabrian Basin, Spain. In *Permo-Triassic Salt Provinces of Europe, North Africa and the Atlantic Margins*, Soto, J.I., Flinch, J.F., Tari, G. (eds). Elsevier, Amsterdam.

Cámara, P. and Flinch, J.F. (2017). The Southern Pyrenees. In *Permo-Triassic Salt Provinces of Europe, North Africa and the Atlantic Margins*, Soto, J.I., Flinch, J.F., Gabor, T. (eds). Elsevier, Amsterdam [Online]. Available at: https://doi.org/10.1016/B978-0-12-809417-4.00019-7.

Canérot, J. (2016). The Iberian plate: Myth or reality? *Boletin Geologico y Minero*, 127(2/3), 563–574.

Casas, A., Kearey, P., Rivero, L., Adam, C.R. (1997). Gravity anomaly map of the Pyrenean region and a comparison of the deep geological structure of the western and eastern Pyrenees. *Earth and Planetary Science Letters*, 150(1/2), 65–78 [Online]. Available at: https://doi.org/10.1016/s0012-821x(97)00087-3.

Casas-Sainz, A.M. and Gil-Imaz, A. (1998). Extensional subsidence, contractional folding and thrust inversion of the eastern Cameros basin, northern Spain. *International Journal of Earth Sciences*, 86(4), 802–818.

Cerbiá, J.M., López-Ruiz, J., Doblas, M., Martins, L.T., Munha, J. (2003). Geochemistry of the Early Jurassic Messejana-Plasencia dyke (Portugal-Spain); implications on the origin of the Central Atlantic Magmatic Province. *Journal of Petrology*, 44(3), 547–568 [Online]. Available at: https://doi.org/10.1093/petrology/44.3.547.

Channell, J.E.T. and Kozur, H.W. (1997). How many oceans? Meliata, Vardar, Pindos oceans in Mesozoic Alpine paleogeography. *Geology*, 25(2), 183–186 [Online]. Available at: https://doi.org/10.1130/0091-7613(1997)025<0183:HMOMVA>2.3.CO;2.

Chevrot, S., Sylvander, M., Diaz, J., Ruiz, M., Paul, A., Cougoulat, G., Péquegnat, C., Wolyniec, D., Delmas, P., Grimaud, F. et al. (2015). The Pyrenean architecture as revealed by teleseismic P-to-S converted waves recorded along two dense transects. *Geophysical Journal International*, 200(1), 1096–1107 [Online]. Available at: https://doi.org/10.1093/gji/ggu400.

Chevrot, S., Sylvander, M., Diaz, J., Martin, R., Mouthereau, F., Manatschal, G., Masini, E., Calassou, S., Grimaud, F., Pauchet, H. et al. (2018). The non-cylindrical crustal architecture of the Pyrenees. *Scientific Reports*, 8(1), 1–8 [Online]. Available at: https://doi.org/10.1038/s41598-018-27889-x.

Choukroune, P. and Mattauer, M. (1978). Tectonique des plaques et Pyrénées. Sur le fonctionnement de la faille transformante nord-pyrénéenne ; comparaisons avec des modèles actuels. *Bulletin de la société géologique de France*, 7(5), 689–700.

Choukroune, P., Mediavila, F., Vauthier, A., Munoz, J.A., Servei, C.P. (1989). The ECORS Pyrenean deep seismic profile reflection data and the overall structure of and orogenic belt. *Tectonics*, 8(1), 23–39.

Choukroune, P., Roure, F., Pinet, B., Ecors Pyrenees Team (1990). Main results of the ECORS Pyrenees profile. *Tectonophysics*, 173(1/4) [Online]. Available at: https://doi.org/10.1016/0040-1951(90)90234-Y.

Christophoul, F., Soula, J.C., Brusset, S., Elibana, B., Roddaz, M., Bessiere, G., Deramond, J. (2003). Time, place and mode of propagation of foreland basin systems as recorded by the sedimentary fill: Examples of the Late Cretaceous and Eocene retro-foreland basins of the north-eastern Pyrenees. *Geological Society Special Publication*, 208(1995), 229–252 [Online]. Available at: https://doi.org/10.1144/GSL.SP.2003.208.01.11.

Clerc, C., Lahfid, A., Monié, P., Lagabrielle, Y., Chopin, C., Poujol, M., Boulvais, P., Ringenbach, J.C., Masini, E., de Saint Blanquat, M. (2015). High-temperature metamorphism during extreme thinning of the continental crust: A reappraisal of the North Pyrenean passive paleomargin. *Solid Earth*, 6(2), 643–668 [Online]. Available at: https://doi.org/10.5194/se-6-643-2015.

Cochelin, B., Lemirre, B., Denèle, Y., de Saint Blanquat, M., Lahfid, A., Duchêne, S. (2017). Structural inheritance in the central pyrenees: The variscan to Alpine tectonometamorphic evolution of the Axial Zone. *Journal of the Geological Society*, 175(2), 336–351 [Online]. Available at: https://doi.org/10.1144/jgs2017-066.

Cochelin, B., Lemirre, B., Denèle, Y., de Saint Blanquat, M., Lahfid, A., Duchêne, S. (2018). Structural inheritance in the central Pyrenees: The variscan to Alpine tectonometamorphic evolution of the Axial Zone. *Journal of the Geological Society*, 175(2), 336–351 [Online]. Available at: https://doi.org/10.1144/jgs2017-066.

Conway-Jones, B.W., Roberts, G.G., Fichtner, A., Hoggard, M. (2019). Neogene epeirogeny of Iberia. *Geochemistry, Geophysics, Geosystems*, 20(2), 1138–1163.

Corre, B., Lagabrielle, Y., Labaume, P., Fourcade, S., Clerc, C., Ballèvre, M. (2016). Deformation associated with mantle exhumation in a distal, hot passive margin environment: New constraints from the Saraillé Massif (Chaînons Béarnais, North-Pyrenean Zone). *Comptes rendus – Géoscience*, 348(/4), 279–289 [Online]. Available at: https://doi.org/10.1016/j.crte.2015.11.007.

Corre, B., Boulvais, P., Boiron, M.C., Lagabrielle, Y., Marasi, L., Clerc, C. (2018). Fluid circulations in response to mantle exhumation at the passive margin setting in the north Pyrenean zone, France. *Mineralogy and Petrology*, 112(5), 647–670 [Online]. Available at: https://doi.org/10.1007/s00710-018-0559-x.

Cuevas, J. and Tubía, J.M. (1999). The discovery of scapolite marbles in the Biscay Synclinorium (Basque-Cantabrian basin, Western Pyrenees): Geodynamic implications. *Terra Nova*, 11(6), 259–265.

Curnelle, R. (1983). Évolution structuro-sédimentaire du Trias et de l'Infra-Lias d'Aquitaine. *Bulletin des centres de recherches exploration-production Elf-Aquitaine*, 7(1), 69–99.

Daignieres, M., de Cabissole, B., Gallart, J., Hirn, A., Surinach, E., Torne, M. (1989). Geophysical constraints on the deep structure along the Ecors Pyrenees Line. *Tectonics*, 8(5), 1051–1058 [Online]. Available at: https://doi.org/10.1029/TC008i005p01051.

Daudet, M., Mouthereau, F., Brichau, S., Crespo-Blanc, A., Gautheron, C., Angrand, P. (2020). Tectono-stratigraphic and thermal evolution of the western Betic flysch: Implications for the geodynamics of South Iberian margin and Alboran Domain. *Tectonics*, 39(7), 1–29 [Online]. Available at: https://doi.org/10.1029/2020TC006093.

De Vicente, G., Cloetingh, S., Van Wees, J.D., Cunha, P.P. (2011). Tectonic classification of Cenozoic Iberian foreland basins. *Tectonophysics*, 502(1/2), 38–61 [Online]. Available at: https://doi.org/10.1016/j.tecto.2011.02.007.

Debroas, E.J. (1987). Modèle de bassin triangulaire à l'intersection de décrochements divergents pour le fossé albo-cénomanien de la Ballongue (Zone Nord-Pyrénéenne, France). *Bulletin de la société géologique de France*, 8(5), 887–898.

Debroas, E.J. (1990). Le Flysch noir albo-cénomanien témoin de la structuration albienne à sénonienne de la Zone nord-pyrénéenne en Bigorre (Hautes-Pyrénées, France). *Bulletin de la société géologique de France*, 8(2), 273–285.

Decarlis, A., Maino, M., Dallagiovanna, G., Lualdi, A., Masini, E., Seno, S., Toscani, G. (2014). Salt tectonics in the SW Alps (Italy-France): From rifting to the inversion of the European continental margin in a context of oblique convergence. *Tectonophysics*, 636, 293–314 [Online]. Available at: https://doi.org/10.1016/j.tecto.2014.09.003.

DeFelipe, I., Pedreira, D., Pulgar, J.A., Iriarte, E., Mendia, M. (2017). Mantle exhumation and metamorphism in the Basque-Cantabrian Basin (NSpain): Stable and clumped isotope analysis in carbonates and comparison with ophicalcites in the North-Pyrenean Zone (Urdach and Lherz). *Geochemistry, Geophysics, Geosystems*, 18(2), 631–652 [Online]. Available at: https://doi.org/10.1002/2016GC006690.

DeFelipe, I., Pedreira, D., Pulgar, J.A., Van der Beek, P.A., Bernet, M., Pik, R. (2019). Unraveling the Mesozoic and Cenozoic tectonothermal evolution of the eastern Basque-Cantabrian zone-western Pyrenees by low-temperature thermochronology. *Tectonics*, 38(9), 3436–3461.

DeMets, C., Gordon, R.G., Argus, D.F. (2010). Geologically current plate motions. *Geophysical Journal International*, 181(1), 1–80 [Online]. Available at: https://doi.org/10.1111/j.1365-246X.2009.04491.x.

Dewey, J.F. and Burke, K. (1974). Hot spots and continental break-up: Implications for collisional orogeny. *Geology*, 2(2), 57–60.

Dèzes, P., Schmid, S.M., Ziegler, P.A. (2004). Evolution of the European cenozoic rift system: Interaction of the Alpine and Pyrenean orogens with their foreland lithosphere. *Tectonophysics*, 389(1/2), 1–33 [Online]. Available at: https://doi.org/10.1016/j.tecto.2004.06.011.

Dietz, R.S. (1962). Ocean-basin evolution by sea-floor spreading. In *Continental Drift*, volume 3, Runcorn, S.K. (ed.). Academic Press, New York.

Ducoux, M., Jolivet, L., Callot, J.P., Aubourg, C., Masini, E., Lahfid, A., Homonnay, E., Cagnard, F., Gumiaux, C., Baudin, T. (2019). The Nappe des Marbres unit of the Basque-Cantabrian Basin: The tectono-thermal evolution of a fossil hyperextended rift basin. *Tectonics*, 38(11), 3881–3915 [Online]. Available at: https://doi.org/10.1029/2018TC005348.

Ducoux, M., Masini, E., Tugend, J., Gómez-Romeu, J., Calassou, S. (2021). Basement-decoupled hyperextension rifting: The tectono-stratigraphic record of the salt-rich Pyrenean necking zone (Arzacq Basin, SW France). *GSA Bulletin*, 34(3–4), 941–964 [Online]. Available at: https://doi.org/10.1130/B35974.1.

Durand-Delga, M., Rossi, P., Olivier, P., Puglisi, D. (2000). Situation structurale et nature ophiolitique de roches basiques jurassiques associées aux flyschs maghrébins du Rif (Maroc) et de Sicile (Italie). *Comptes rendus de l'académie de sciences – série IIa : Sciences de la terre et des planètes*, 331(1), 29–38 [Online]. Available at: https://doi.org/10.1016/S1251-8050(00)01378-1.

Duretz, T., Asti, R., Lagabrielle, Y., Brun, J., Jourdon, A., Clerc, C., Corre, B. (2019). Numerical modelling of Cretaceous Pyrenean Rifting: The interaction between mantle exhumation and syn-rift salt tectonics. *Basin Research*, 00(July), 1–16 [Online]. Available at: https://doi.org/10.1111/bre.12389.

Edel, J.-B., Schulmann, K., Lexa, O., Diraison, M., Géraud, Y. (2015). Permian clockwise rotations of the Ebro and Corso-Sardinian blocks during Iberian-Armorican oroclinal bending: Preliminary paleomagnetic data from the Catalan Coastal Range (NE Spain). *Tectonophysics*, 657, 172–186 [Online]. Available at: https://doi.org/10.1016/j.tecto.2015.07.002.

Edel, J.-B., Schulmann, K., Lexa, O., Lardeaux, J.M. (2018). Late Palaeozoic palaeomagnetic and tectonic constraints for amalgamation of Pangea supercontinent in the European Variscan belt. *Earth-Science Reviews*, 177(November), 589–612 [Online]. Available at: https://doi.org/10.1016/j.earscirev.2017.12.007.

Espurt, N., Hippolyte, J.-C., Saillard, M., Bellier, O. (2012). Geometry and kinematic evolution of a long-living foreland structure inferred from field data and cross section balancing, the Sainte-Victoire System, Provence, France. *Tectonics*, 31(4), 1–27 [Online]. Available at: https://doi.org/10.1029/2011TC002988.

Espurt, N., Angrand, P., Teixell, A., Labaume, P., Ford, M., de Saint Blanquat, M., Chevrot, S. (2019). Crustal-scale balanced cross-section and restorations of the Central Pyrenean belt (Nestes-Cinca transect): Highlighting the structural control of Variscan belt and Permian-Mesozoic rift systems on mountain building. *Tectonophysics*, 764(December), 25–45 [Online]. Available at: https://doi.org/10.1016/j.tecto.2019.04.026.

Etheve, N., de Lamotte, D.F., Mohn, G., Martos, R., Roca, E., Blanpied, C. (2016). Extensional vs contractional Cenozoic deformation in Ibiza (Balearic Promontory, Spain): Integration in the West Mediterranean back-arc setting. *Tectonophysics*, 682, 35–55.

Etheve, N., Mohn, G., Frizon de Lamotte, D., Roca, E., Tugend, J., Gómez-Romeu, J. (2018). Extreme Mesozoic crustal thinning in the Eastern Iberia margin: The example of the Columbrets Basin (Valencia Trough). *Tectonics*, 37(2), 636–662 [Online]. Available at: https://doi.org/10.1002/2017TC004613.

Fernández, O. (2019). The Jurassic evolution of the Africa-Iberia conjugate margin and its implications on the evolution of the Atlantic-Tethys triple junction. *Tectonophysics*, 750(December), 379–393 [Online]. Available at: https://doi.org/10.1016/j.tecto.2018.12.006.

Fernàndez, M., Torne, M., Vergés, J., Casciello, E., Macchiavelli, C. (2019). Evidence of segmentation in the Iberia–Africa plate boundary: A Jurassic heritage? *Geosciences (Switzerland)*, 9(8), 1–13 [Online]. Available at https://doi.org/10.3390/geosciences9080343.

Ferrer, O., Jackson, M.P.A., Roca, E., Rubinat, M. (2012). Evolution of salt structures during extension and inversion of the Offshore Parentis Basin (Eastern Bay of Biscay). *Geological Society, London, Special Publications*, 363(1), 361–380 [Online]. Available at: https://doi.org/10.1144/SP363.16.

Fillon, C., Pedreira, D., van der Beek, P., Huismans, R.S., Barbero, L., Pulgar, J.A. (2016). Alpine exhumation of the central Cantabrian mountains, northwest Spain. *Tectonics*, 35(2), 339–356.

Ford, M. and Vergés, J. (2020). Evolution of a salt-rich transtensional rifted margin, eastern North Pyrenees, France. *Journal of the Geological Society*, 178(1), 1–24.

Ford, M., Duchêne, S., Gasquet, D., Vanderhaeghe, O. (2006). Two-phase orogenic convergence in the external and internal SW Alps. *Journal of the Geological Society*, 163(5), 815–826 [Online]. Available at: https://doi.org/10.1144/0016-76492005-034.

Ford, M., Hemmer, L., Vacherat, A., Gallagher, K., Christophoul, F. (2016). Retro-wedge foreland basin evolution along the ECORS line, eastern Pyrenees, France. *Journal of the Geological Society*, 173(3), 419–437 [Online]. Available at: https://doi.org/10.1144/jgs2015-129.

Gallastegui, J., Pulgar, J.A., Gallart, J. (2002). Initiation of an active margin at the North Iberian continent-ocean transition. *Tectonics*, 21(4), 15–14 [Online]. Available at: https://doi.org/10.1029/2001TC901046.

Garcés, M., López-Blanco, M., Valero, L., Beamud, E., Muñoz, J.A., Oliva-Urcia, B., Vinyoles, A., Arbués, P., Cabello, P., Cabrera, L. (2020). Paleogeographic and sedimentary evolution of the South Pyrenean foreland basin. *Marine and Petroleum Geology*, 113, 104105 [Online]. Available at: https://doi.org/10.1016/j.marpetgeo.2019.104105.

García-Espina Martínez, R. (1997). La estructura y evolución tectonoestratigráfica del borde occidental de la cuenca vasco-cantábrica (Cordillera Cantábrica NO de España). PhD Thesis, Universidad de Oviedo.

García-Mondéjar, J., Agirrezabala, L.M., Aranburu, A., Fernández-Mendiola, P.A., Gómez-Pérez, I., López-Horgue, M., Rosales, I. (1996). Aptian-Albian tectonic pattern of the Basque-Cantabrian Basin (Northern Spain). *Geological Journal*, 31(1), 13–45 [Online]. Available at: https://doi.org/10.1002/(SICI)1099-1034(199603)31:1<13::AID-GJ689>3.0.CO;2-Y.

García-Senz, J., Ja, R.A., Pedrera, A. (2019). The North Iberian continental margin. In *The Geology of Iberia: A Geodynamic Approach: Volume 3: The Alpine Cycle*, Quesada, C. and Oliveira, J.T. (eds). Springer, Cham.

Goldsmith, P.J., Hudson, G., Van Veen, P. (2003). Triassic. In *The Millennium Atlas; Petroleum Geology of the Central and Northern North Sea*, Evans, D., Graham, C., Armour, A., Bathurst, P. (eds). Geological Society of London.

Gorini, C., Le Marrec, A., Mauffret, A. (1993). Contribution to the structural and sedimentary history of the Gulf of Lions (western Mediterranean) from the ECORS profiles, industrial seismic profiles and well data. *Bulletin de la société géologique de France*, 164(3), 353–363.

Grool, A.R., Ford, M., Vergés, J., Huismans, R.S., Christophoul, F., Dielforder, A. (2018). Insights into the crustal-scale dynamics of a doubly vergent orogen from a quantitative analysis of its forelands: A case study of the Eastern Pyrenees. *Tectonics*, 37(2), 450–476 [Online]. Available at: https://doi.org/10.1002/2017TC004731.

Grool, A.R., Huismans, R.S., Ford, M. (2019). Salt décollement and rift inheritance controls on crustal deformation in orogens. *Terra Nova*, 31(6), 562–568 [Online]. Available at: https://doi.org/10.1111/ter.12428.

Handy, M.R., Schmid, S., Bousquet, R., Kissling, E., Bernoulli, D. (2010). Reconciling plate-tectonic reconstructions of Alpine Tethys with the geological-geophysical record of spreading and subduction in the Alps. *Earth-Science Reviews*, 102(3–4), 121–158 [Online]. Available at: https://doi.org/10.1016/j.earscirev.2010.06.002.

Handy, M.R., Ustaszewski, K., Kissling, E. (2015). Reconstructing the Alps-Carpathians-Dinarides as a key to understanding switches in subduction polarity, slab gaps and surface motion. *International Journal of Earth Sciences*, 104(1), 1–26 [Online]. Available at: https://doi.org/10.1007/s00531-014-1060-3.

Hess, H.H. (1962). History of ocean basins. In *Petrologic Studies*, Engel, A.E.J., James, H.L., Leonard, B.F. (eds). Geological Society of America, Boulder, CO [Online]. Available at: https://doi.org/10.1130/Petrologic.1962.599.

van Hinsbergen, D.J.J., Vissers, R.L.M., Spakman, W. (2014). Origin and consequences of western Mediterranean subduction, rollback, and slab segmentation. *Tectonics*, 33(4), 393–419.

van Hinsbergen, D.J.J., Spakman, W., Vissers, R.L.M., van der Meer, D.G. (2017). Comment on "assessing discrepancies between previous plate kinematic models of Mesozoic Iberia and their constraints". *Tectonics*, 36(12), 3277–3285 [Online]. Available at: https://doi.org/10.1002/2016TC004418.

van Hinsbergen, D.J.J., Torsvik, T.H., Schmid, S.M., Maţenco, L.C., Maffione, M., Vissers, R.L.M., Gürer, D., Spakman, W. (2020). Orogenic architecture of the Mediterranean region and kinematic reconstruction of its tectonic evolution since the Triassic. *Gondwana Research*, 81, 79–229 [Online]. Available at: https://doi.org/10.1016/j.gr.2019.07.009.

Huyghe, D., Mouthereau, F., Segalen, L., Furió, M. (2020). Long-term dynamic topographic support during post-orogenic crustal thinning revealed by stable isotope (δ 18 O) paleo-altimetry in eastern Pyrenees. *Scientific Reports*, 10(1), 1–8.

Issautier, B., Saspiturry, N., Serrano, O. (2020). Role of structural inheritance and salt tectonics in the formation of pseudosymmetric continental rifts on the European margin of the hyperextended Mauléon basin (Early Cretaceous Arzacq and Tartas Basins). *Marine and Petroleum Geology*, 118, 104395 [Online]. Available at: https://doi.org/10.1016/j.marpetgeo.2020.104395.

Issautier, B., Lasseur, E., Saspiturry, N., Andrieu, S., Angrand, A., Serrano, O. (2022). Onset of the Iberian-European plates convergence: Late Cretaceous flexural response of a hot lithosphere. *Tectonophysics*, 843, 229504. [Online]. Available at: https://doi.org/10.1016/j.tecto.2022.229504.

Jammes, S., Manatschal, G., Lavier, L., Masini, E. (2009). Tectonosedimentary evolution related to extreme crustal thinning ahead of a propagating ocean: Example of the western Pyrenees. *Tectonics*, 28(4), 1–24.

Jolivet, L. and Faccenna, C. (2000). Mediterranean extension and the Africa-Eurasia collision. *Tectonics*, 19(6), 1095–1106.

Jourdon, A., Le Pourhiet, L., Mouthereau, F., Masini, E. (2019). Role of rift maturity on the architecture and shortening distribution in mountain belts. *Earth and Planetary Science Letters*, 512, 89–99 [Online]. Available at: https://doi.org/10.1016/j.epsl.2019.01.057.

Jourdon, A., Mouthereau, F., Le Pourhiet, L., Callot, J.P. (2020). Topographic and tectonic evolution of mountain belts controlled by salt thickness and rift architecture. *Tectonics*, 39(1), 1–14 [Online]. Available at: https://doi.org/10.1029/2019TC005903.

Kerckhove, C. (1969). La "zone du Flysch" dans les nappes de l'Embrunais-Ubaye (Alpes occidentales). *Géologie Alpine*, 45, 5–204.

Lacombe, O. and Jolivet, L. (2005). Structural and kinematic relationships between Corsica and the Pyrenees-Provence domain at the time of the Pyrenean orogeny. *Tectonics*, 24(1) [Online]. Available at: https://doi.org/10.1029/2004TC001673.

Lagabrielle, Y. and Bodinier, J.L. (2008). Submarine reworking of exhumed subcontinental mantle rocks: Field evidence from the Lherz peridotites, French Pyrenees. *Terra Nova*, 20(1), 11–21 [Online]. Available at: https://doi.org/10.1111/j.1365-3121.2007.00781.x.

Lagabrielle, Y., Labaume, P., de Saint Blanquat, M. (2010). Mantle exhumation, crustal denudation, and gravity tectonics during Cretaceous rifting in the Pyrenean realm (SW Europe): Insights from the geological setting of the lherzolite bodies. *Tectonics*, 29(4), 1–26 [Online]. Available at: https://doi.org/10.1029/2009TC002588.

Lagabrielle, Y., Asti, R., Duretz, T., Clerc, C., Fourcade, S., Teixell, A., Labaume, P., Corre, B., Saspiturry, N. (2020). A review of cretaceous smooth-slopes extensional basins along the Iberia-Eurasia plate boundary: How pre-rift salt controls the modes of continental rifting and mantle exhumation. *Earth-Science Reviews*, 201(December), 103071 [Online]. Available at: https://doi.org/10.1016/j.earscirev.2019.103071.

Le Breton, E., Brune, S., Ustaszewski, K., Zahirovic, S., Seton, M., Müller, R.D. (2021). Kinematics and extent of the Piemont-Liguria Basin – Implications for subduction processes in the Alps. *Solid Earth*, 12(4), 885–913 [Online]. Available at: https://doi.org/10.5194/se-12-885-2021.

Le Pichon, X. (1968). Sea-floor spreading and continental drift. *Journal of Geophysical Research*, 73(12), 3661–3697.

Le Pichon, X., Bonnin, J., Sibuet, J.C. (1970). La faille nord-pyrénéenne : faille transformante liée à l'ouverture du golfe de Gascogne. *C. R. Acad. Sci.*, 271(9), 1941–1944 [Online]. Available at: https://doi.org/10.1017/CBO9781107415324.004.

Lehujeur, M., Chevrot, S., Villaseñor, A., Masini, E., Saspiturry, N., Lescoutre, R., Sylvander, S. (2021). Three-dimensional shear velocity structure of the Mauleon and Arzacq basins (Pyrenees). *BSGF – Earth Sciences Bulletin*, 192, 47 [Online]. Available at: https://doi.org/10.1051/bsgf/2021039.

Leleu, S., Ghienne, J.F., Manatschal, G. (2005). Upper Cretaceous-Palaeocene basin-margin alluvial fans documenting interaction between tectonic and environmental processes (Provence, SE France). *Geological Society Special Publication*, 251, 217–239 [Online]. Available at: https://doi.org/10.1144/GSL.SP.2005.251.01.15.

Leleu, S., Hartley, A.J., van Oosterhout, C., Kennan, L., Ruckwied, K., Gerdes, K. (2016). Structural, stratigraphic and sedimentological characterisation of a wide rift system: The Triassic rift system of the Central Atlantic Domain. *Earth-Science Reviews*, 158, 89–124 [Online]. Available at: https://doi.org/10.1016/j.earscirev.2016.03.008.

Lescoutre, R. and Manatschal, G. (2020). Role of rift-inheritance and segmentation for orogenic evolution: Example from the Pyrenean-Cantabrian system. *BSGF – Earth Sciences Bulletin*, 191, 1–18 [Online]. Available at: https://doi.org/10.1051/bsgf/2020021.

Lucas, C. (1985). Le grès rouge du versant nord des Pyrénées : essai sur la géodynamique de dépôts continentaux du Permien et du Trias. Thesis, Université de Toulouse.

Lustrino, M., Morra, V., Fedele, L., Franciosi, L. (2009). Beginning of the Apennine subduction system in central western Mediterranean: Constraints from Cenozoic "orogenic" magmatic activity of Sardinia, Italy. *Tectonics*, 28(5), 1–23.

Maillard, A., Jolivet, L., Lofi, J., Thinon, I., Couëffé, R., Canva, A., Dofal, A. (2020). Transfer zones and associated volcanic province in the eastern Valencia Basin: Evidence for a hot rifted margin? *Marine and Petroleum Geology*, 119(May), 104419 [Online]. Available at: https://doi.org/10.1016/j.marpetgeo.2020.104419.

Malod, J.A., Boillot, G., Lepvier, C., Mascle, G., Taugourdeau-Lantz, J., Capdevila, R., Dupeuble, P.-A., Müller, C. (1982). Subduction and tectonics on the continental margin off northern Spain: Observations with the submersible Cyana. *Geological Society, London, Special Publications*, 10(1), 309–315.

Malusà, M.G., Faccenna, C., Garzanti, E., Polino, R. (2011). Divergence in subduction zones and exhumation of high pressure rocks (Eocene Western Alps). *Earth and Planetary Science Letters*, 310(1/2), 21–32 [Online]. Available at: https://doi.org/10.1016/j.epsl.2011.08.002.

Malusà, M.G., Danišík, M., Kuhlemann, J. (2016). Tracking the Adriatic-slab travel beneath the Tethyan margin of Corsica-Sardinia by low-temperature thermochronometry. *Gondwana Research*, 31, 135–149 [Online]. Available at: https://doi.org/10.1016/j.gr.2014.12.011.

Manatschal, G., Chenin, P., Lescoutre, R., Miró, J., Cadenas, P., Saspiturry, N., Masini, E., Chevrot, S., Ford, M., Jolivet, L. et al. (2021). The role of inheritance in forming rifts and rifted margins and building collisional orogens: A Biscay-Pyrenean perspective. *BSGF – Earth Sciences Bulletin*, 192, 55 [Online]. Available at: https://doi.org/10.1051/bsgf/2021042.

Manzotti, P., Ballèvre, M., Zucali, M., Robyr, M., Engi, M. (2014). The tectonometamorphic evolution of the Sesia-Dent Blanche nappes (internal Western Alps): Review and synthesis. *Swiss Journal of Geosciences*, 107(2/3), 309–336.

Marroni, M., Meneghini, F., Pandolfi, L. (2017). A revised subduction inception model to explain the Late Cretaceous, double-vergent orogen in the precollisional western Tethys: Evidence from the Northern Apennines. *Tectonics*, 36(10), 2227–2249 [Online]. Available at: https://doi.org/10.1002/2017TC004627.

Martí, J., Mitjavila, J., Roca, E., Aparicio, A. (1992). Cenozoic magmatism of the valencia trough (western Mediterranean): Relationship between structural evolution and volcanism. *Tectonophysics*, 203(1/4), 145–165.

Martín-Chivelet, J., López-Gómez, J., Aguado, R., Arias, C., Arribas, J., Arribas, M.E., Aurell, M., Bádenas, B., Benito, M.I., Bover-Arnal, T. et al. (2019). The Late Jurassic–Early Cretaceous rifting. In *The Geology of Iberia: A Geodynamic Approach*, Quesada, C. and Oliveira, J.T. (eds). Springer Nature, Cham.

Marzoli, A., Renne, P.R., Piccirillo, E.M., Ernesto, M., Bellieni, G., de Min, A. (1999). Extensive 200-million-year-old continental flood basalts of the Central Atlantic Magmatic Province. *Science*, 284(5414), 616–618 [Online]. Available at: https://doi.org/10.1126/science.284.5414.616.

Masini, E., Manatschal, G., Tugend, J., Mohn, G., Flament, J.M. (2014). The tectono-sedimentary evolution of a hyper-extended rift basin: The example of the Arzacq-Mauléon rift system (Western Pyrenees, SW France). *International Journal of Earth Sciences*, 103(6), 1569–1596 [Online]. Available at: https://doi.org/10.1007/s00531-014-1023-8.

Matthews, K.J., Maloney, K.T., Zahirovic, S., Williams, S.E., Seton, M., Müller, D.R. (2016). Global plate boundary evolution and kinematics since the late Paleozoic. *Global and Planetary Change*, 146, 226–250 [Online]. Available at: https://doi.org/10.1016/j.gloplacha.2016.10.002.

Maurel, O., Brunel, M., Monié, P. (2002). Exhumation cénozoïque des massifs du Canigou et de Mont-Louis (Pyrénées orientales, France). *Comptes rendus géoscience*, 334(12), 941–948 [Online]. Available at: https://doi.org/10.1016/S1631-0713(02)01834-5.

McHone, J.G. (2000). Non-plume magmatism and rifting during the opening of the central Atlantic Ocean. *Tectonophysics*, 316(3/4), 287–296 [Online]. Available at: https://doi.org/10.1016/S0040-1951(99)00260-7.

Mencos, J., Carrera, N., Muñoz, J.A. (2015). Influence of rift basin geometry on the subsequent postrift sedimentation and basin inversion: The Organyà Basin and the Bóixols thrust sheet (south central Pyrenees). *Tectonics*, 34(7), 1452–1474.

Merle, O. and Michon, L. (2001). The formation of the West European Rift; a new model as exemplified by the Massif Central area. *Bulletin de la société géologique de France*, 172(2), 213–221 [Online]. Available at: https://doi.org/10.2113/172.2.213.

Mohn, G., Karner, G.D., Manatschal, G., Johnson, C.A. (2015). Structural and stratigraphic evolution of the Iberia-Newfoundland hyper-extended rifted margin: A quantitative modelling approach. *Geological Society Special Publication*, 413(1), 53–89 [Online]. Available at: https://doi.org/10.1144/SP413.9.

Molli, G., Brogi, A., Caggianelli, A., Capezzuoli, E., Liotta, D., Spina, A., Zibra, I. (2020). Late Palaeozoic tectonics in Central Mediterranean: A reappraisal. *Swiss Journal of Geosciences*, 113(1), 1–32 [Online]. Available at: https://doi.org/10.1186/s00015-020-00375-1.

Monie, P., Galindo-Zaldivar, J., Lodeiro, F.G., Goffe, B., Jabaloy, A. (1991). 40Ar/39Ar geochronology of Alpine tectonism in the Betic Cordilleras (southern Spain). *Journal of the Geological Society*, 148(2), 289–297.

Montigny, R., Azambre, B., Rossy, M., Thuizat, R. (1986). K-Ar study of Cretaceous magmatism and metamorphism in the Pyrenees: Age and length of rotation of the Iberian Peninsula. *Tectonophysics*, 129(1/4), 257–273 [Online]. Available at: https://doi.org/10.1016/0040-1951(86)90255-6.

Morgan, W.J. (1968). Rises, trenches, great faults, and crustal blocks. *Journal of Geophysical Research*, 73(6), 1959–1982 [Online]. Available at: https://doi.org/10.1029/JB073i006p01959.

Mouchené, M., van der Beek, P., Mouthereau, F., Carcaillet, J. (2017). Controls on quaternary incision of the Northern Pyrenean foreland: Chronological and geomorphological constraints from the Lannemezan megafan, SW France. *Geomorphology*, 281, 78–93 [Online]. Available at: https://doi.org/10.1016/j.geomorph.2016.12.027.

Mouthereau, F., Filleaudeau, P.-Y., Vacherat, A., Pik, R., Lacombe, O., Fellin, M.G., Castelltort, S., Christophoul, F., Masini, E. (2014). Placing limits to shortening evolution in the Pyrenees: Role of margin architecture and implications for the Iberia/Europe convergence. *Tectonics*, 33(12), 2283–2314 [Online]. Available at: https://doi.org/10.1002/2014TC003663.

Müller, D.R., Zahirovic, S., Williams, S.E., Cannon, J., Seton, M., Bower, D.J., Tetley, M.G., Heine, C., Le Breton, E., Liu, S. et al. (2019). A global plate model including lithospheric deformation along major rifts and orogens since the Triassic. *Tectonics*, 38(6), 1884–1907 [Online]. Available at: https://doi.org/10.1029/2018TC005462.

Muñoz, J.A. (1992). Evolution of a continental collision belt: ECORS-Pyrenees crustal balanced cross-section. In *Thrust Tectonics*, McClay, K.R. (ed.). Springer, Netherlands [Online]. Available at: https://doi.org/10.1007/978-94-011-3066-0_21.

Murillas, J., Mougenot, D., Boillot, G., Comas, M.C., Banda, E., Mauffret, A. (1990). Structure and evolution of the Galicia Interior Basin (Atlantic western Iberian continental margin). *Tectonophysics*, 184, 297–319.

Naylor, M. and Sinclair, H.D. (2008). Pro- vs. retro-foreland basins. *Basin Research*, 20(3), 285–303 [Online]. Available at: https://doi.org/10.1111/j.1365-2117.2008.00366.x.

Nirrengarten, M., Manatschal, G., Tugend, J., Kusznir, N.J., Sauter, D. (2017). Nature and origin of the J-magnetic anomaly offshore Iberia-Newfoundland: Implications for plate reconstructions. *Terra Nova*, 29(1), 20–28 [Online]. Available at: https://doi.org/10.1111/ter.12240.

Nirrengarten, M., Manatschal, G., Tugend, J., Kusznir, N.J., Sauter, D. (2018). Kinematic evolution of the southern north Atlantic: Implications for the formation of hyperextended rift systems. *Tectonics*, 37(1), 89–118 [Online]. Available at: https://doi.org/10.1002/2017TC004495.

Olivet, J.L. (1996). La cinématique de la plaque ibérique. *Bulletin des centres de recherches exploration-production Elf-Aquitaine*, 20(1), 131–195.

Olsen, P.E. (1997). Stratigraphic record of the Early Mesozoic breakup of Pangea in the Laurasia-Gondwana rift system. *Annual Review of Earth and Planetary Sciences*, 25(1), 337–401 [Online]. Available at: https://doi.org/10.1146/annurev.earth.25.1.337.

Omodeo-Salé, S., Salas, R., Guimerà, J., Ondrak, R., Mas, R., Arribas, J., Suárez-Ruiz, I., Martinez, L. (2017). Subsidence and thermal history of an inverted Late Jurassic–Early Cretaceous extensional basin (Cameros, North-central Spain) affected by very low- to low-grade metamorphism. *Basin Research*, 29, 156–174 [Online]. Available at: https://doi.org/10.1111/bre.12142.

Ortí, F., Pérez-López, A., Salvany, J.M. (2017). Triassic evaporites of Iberia: Sedimentological and palaeogeographical implications for the western Neotethys evolution during the Middle Triassic–Earliest Jurassic. *Palaeogeography, Palaeoclimatology, Palaeoecology*, 471, 157–180 [Online]. Available at: https://doi.org/10.1016/j.palaeo.2017.01.025.

Pastor-Galán, D., Gutiérrez-Alonso, G., Weil, A.B. (2020). The enigmatic curvature of Central Iberia and its puzzling kinematics. *Solid Earth*, 11(4), 1247–1273 [Online]. Available at: https://doi.org/10.5194/se-11-1247-2020.

Pedreira, D., Pulgar, J.A., Gallart, J., Torné, M. (2007). Three-dimensional gravity and magnetic modeling of crustal indentation and wedging in the western Pyrenees-Cantabrian Mountains. *Journal of Geophysical Research: Solid Earth*, 112(12), 1–19 [Online]. Available at: https://doi.org/10.1029/2007JB005021.

Pedreira, D., Afonso, J.C., Pulgar, J.A., Gallastegui, J., Carballo, A., Fernàndez, M., Garcia-Castellanos, D., Jiménez-Munt, I., Semprich, J., García-Moreno, O. (2015). Geophysical-petrological modeling of the lithosphere beneath the Cantabrian Mountains and the North-Iberian margin: Geodynamic implications. *Lithos*, 230, 46–68 [Online]. Available at: https://doi.org/10.1016/j.lithos.2015.04.018.

Pedrera, A., García-Senz, J., Peropadre, C., Robador, A., López-Mir, B., Díaz-Alvarado, J., Rodríguez-Fernández, L.R. (2020a). The Getxo crustal-scale cross-section: Testing tectonic models in the Bay of Biscay-Pyrenean rift system. *Earth-Science Reviews*, 103429.

Pedrera, A., Ruiz-Constán, A., García-Senz, J., Azor, A., Marín-Lechado, C., Ayala, C., de Neira, J.A.D., Rodríguez-Fernández, L.R. (2020b). Evolution of the South-Iberian paleomargin: From hyperextension to continental subduction. *Journal of Structural Geology*, 138, 104122.

Puga, E., Fanning, M., Díaz de Federico, A., Nieto, J.M., Beccaluva, L., Bianchini, G., Díaz Puga, M.A. (2011). Petrology, geochemistry and U-Pb geochronology of the Betic Ophiolites: Inferences for Pangaea break-up and birth of the westernmost Tethys Ocean. *Lithos*, 124(3/4), 255–272 [Online]. Available at: https://doi.org/10.1016/j.lithos.2011.01.002.

Puigdefàbregas, C. and Souquet, P. (1986). Tecto-sedimentary cycles and depositional sequences of the Mesozoic and Tertiary from the Pyrenees. *Tectonophysics*, 129(1/4), 173–203 [Online]. Available at: https://doi.org/10.1016/0040-1951(86)90251-9.

Pulgar, J.A., Gallart, J., Fernández-Viejo, G., Pérez-Estaún, A., Álvarez-Marrón, J., Group, E. (1996). Seismic image of the Cantabrian Mountains in the western extension of the Pyrenees from integrated ESCIN reflection and refraction data. *Tectonophysics*, 264(1/4), 1–19.

Quintana, L. (2012). Extensión e inversión tectónica en el sector central de la Región Vasco-Cantábrica. PhD Thesis, Universidad de Oviedo.

Quintana, L., Alonso, J.L., Pulgar, J.A., Rodríguez-Fernández, L.R. (2006). Transpressional inversion in an extensional transfer zone (the Saltacaballos fault, northern Spain). *Journal of Structural Geology*, 28(11), 2038–2048 [Online]. Available at: https://doi.org/10.1016/j.jsg.2006.06.013.

Quintana, L., Pulgar, J.A., Alonso, J.L. (2015). Displacement transfer from borders to interior of a plate: A crustal transect of Iberia. *Tectonophysics*, 663, 378–398 [Online]. Available at: https://doi.org/10.1016/j.tecto.2015.08.046.

Rat, J., Mouthereau, F., Brichau, S., Crémades, A., Bernet, M., Balvay, M., Ganne, J., Lahfid, A., Gautheron, C. (2019). Tectonothermal evolution of the cameros basin: Implications for tectonics of North Iberia. *Tectonics*, 38(2), 440–469 [Online]. Available at: https://doi.org/10.1029/2018TC005294.

Regis, D., Rubatto, D., Darling, J., Cenki-Tok, B., Zucali, M., Engi, M. (2014). Multiple metamorphic stages within an eclogite-facies terrane (Sesia Zone, Western Alps) revealed by Th–U–Pb petrochronology. *Journal of Petrology*, 55(7), 1429–1456 [Online]. Available at: https://doi.org/10.1093/petrology/egu029.

Réhault, J.P., Honthaas, C., Guennoc, P., Bellon, H., Ruffet, G., Cotten, J., Sosson, M., Maury, R. (2012). Offshore Oligo-Miocene volcanic fields within the Corsica-Liguria Basin: Magmatic diversity and slab evolution in the western Mediterranean Sea. *Journal of Geodynamics*, 58, 73–95.

Roca, E., Muñoz, J.A., Ferrer, O., Ellouz, N. (2011). The role of the Bay of Biscay Mesozoic extensional structure in the configuration of the Pyrenean orogen: Constraints from the MARCONI deep seismic reflection survey. *Tectonics*, 30(2), 1–33 [Online]. Available at: https://doi.org/10.1029/2010TC002735.

Romagny, A., Jolivet, L., Menant, A., Bessière, E., Maillard, A., Canva, A., Gorini, C., Augier, R. (2020). Detailed tectonic reconstructions of the Western Mediterranean region for the last 35 Ma, insights on driving mechanisms. *BSGF – Earth Sciences Bulletin*, 191, 37 [Online]. Available at: https://doi.org/10.1051/bsgf/2020040.

Rossi, P., Cocherie, A., Fanning, C.M., Ternet, Y. (2003). U-Pb zircon SHRIMP dating of Pyrenean tholeiitic dolerites (ophites): Evidence for an age encompassing the Trias-Jurassic limit and relationships with earliest North Pyrenean "Infraliassic" tuffs. *Comptes rendus – géoscience*, 335(15), 1071–1080 [Online]. Available at: https://doi.org/10.1016/j.crte.2003.09.011.

Rougier, G., Ford, M., Christophoul, F., Bader, A.G. (2016). Stratigraphic and tectonic studies in the central Aquitaine Basin, northern Pyrenees: Constraints on the subsidence and deformation history of a retro-foreland basin. *Comptes Rendus – géoscience*, 348(3/4), 224–235 [Online]. Available at: https://doi.org/10.1016/j.crte. 2015.12.005.

Roure, F., Brun, J.-P., Colletta, B., Van Den Driessche, J. (1992). Geometry and kinematics of extensional structures in the Alpine foreland basin of southeastern France. *Journal of Structural Geology*, 14(5), 503–519.

Rubatto, D., Gebauer, D., Compagnoni, R. (1999). Dating of eclogite-facies zircons: The age of Alpine metamorphism in the Sesia-Lanzo Zone (Western Alps). *Earth and Planetary Science Letters*, 167(3/4), 141–158.

Salas, R., Guimerà, J., Mas, R., Martin-Closas, C., Melendez, A., Alonso, A. (2001). Evolution of the Mesozoic central Iberian Rift System and its Cainozoic inversion (Iberian chain). *Mémoires du Muséum national d'histoire naturelle*, 1993(186), 145–186.

Sallarès, V., Martínez-Loriente, S., Prada, M., Gràcia, E., Ranero, C., Gutscher, M.A., Bartolome, R., Gailler, A., Dañobeitia, J.J., Zitellini, N. (2013). Seismic evidence of exhumed mantle rock basement at the Gorringe Bank and the adjacent Horseshoe and Tagus abyssal plains (SW Iberia). *Earth and Planetary Science Letters*, 365, 120–131 [Online]. Available at: https://doi.org/10.1016/j.epsl.2013.01.021.

Saspiturry, N., Cochelin, B., Razin, P., Leleu, S., Lemirre, B., Bouscary, C., Issautier, B., Serrano, O., Lasseur, E., Baudin, T. et al. (2019a). Tectono-sedimentary evolution of a rift system controlled by Permian post-orogenic extension and metamorphic core complex formation (Bidarray Basin and Ursuya dome, Western Pyrenees). *Tectonophysics*, 768, 228180 [Online]. Available at: https://doi.org/10.1016/j.tecto.2019.228180.

Saspiturry, N., Razin, P., Baudin, T., Serrano, O., Issautier, B., Lasseur, E., Leleu, S. (2019b). Symmetry vs. asymmetry of a hyper-thinned rift: Example of the Mauléon Basin (Western Pyrenees, France). *Mar. Pet. Geol.*, 104, 86–105 [Online]. Available at: https://doi.org/10.1016/j.marpetgeo.2019.03.031.

Saspiturry, N., Allanic, C., Razin, P., Issautier, B., Baudin, T., Lasseur, E., Serrano, O., Leleu, S. (2020a). Closure of a hyperextended system in an orogenic lithospheric pop-up, Western Pyrenees: The role of mantle buttressing and rift structural inheritance. *Terra Nova*, 32(4), 253–260 [Online]. Available at: https://doi.org/10.1111/ter.12457.

Saspiturry, N., Lahfid, A., Baudin, T., Guillou-Frottier, L., Razin, P., Issautier, B., Le Bayon, B., Serrano, O., Lagabrielle, Y., Corre, B. (2020b). Paleogeothermal gradients across an inverted hyperextended rift system: Example of the Mauléon fossil rift (Western Pyrenees). *Tectonics*, 39(10), 1–36 [Online]. Available at: https://doi.org/10.1029/2020TC006206.

Saspiturry, N., Issautier, B., Razin, P., Andrieu, S., Lasseur, E., Allanic, C., Serrano, O., Baudin, T., Cochelin, B. (2021a). Review of the syn-rift to early post-rift depositional systems of the Cretaceous Mauléon Rift: Sedimentary record of continental crust hyperextension and mantle denudation (Western Pyrenees). *BSGF-Earth Sciences Bulletin*, 192(1), 49.

Saspiturry, N., Issautier, B., Razin, P., Baudin, T., Asti, R., Lagabrielle, Y., Allanic, C., Serrano, O. (2021b). Review of Iberia-Eurasia plate-boundary basins: Role of sedimentary burial and salt tectonic during rifting and continental breakup. *Basin Research*, 33(2), 1626–1661. [Online]. Available at: https://doi.org/10.1111/bre.12529.

Saspiturry, N., Allanic, C., Courrioux, G., Serrano, O., Lahfid, A., Le Bayon, B., Baudin, T., Razin, P., Issautier, B. (2022). Lithospheric transfer zones driving the non-cylindrical shape of the Pyrenean orogen. *Journal of Structural Geology*, 156, 104535 [Online]. Available at: https://doi.org/10.1016/j.jsg.2022.104535.

Schaltegger, U., Desmurs, L., Manatschal, G., Müntener, O., Meier, M., Frank, M., Bernoulli, D. (2002). The transition from rifting to sea-floor spreading within a magma-poor rifted margin: Field and isotopic constraints. *Terra Nova*, 14(3), 156–162.

Schettino, A. and Scotese, C. (2002). Global kinematic constraints to the tectonic history of the Mediterranean region and surrounding areas during the Jurassic and Cretaceous. *Journal of the Virtual Explorer*, 8, 149–168.

Schettino, A. and Turco, E. (2011). Tectonic history of the Western Tethys since the Late Triassic. *Bulletin of the Geological Society of America*, 123(1/2), 89–105 [Online]. Available at: https://doi.org/10.1130/B30064.1.

Schmid, S.M., Fügenschuh, B., Kissling, E., Schuster, R. (2004). Tectonic map and overall architecture of the Alpine orogen. *Eclogae Geologicae Helvetiae*, 97(1), 93–117.

Schmid, S.M., Bernoulli, D., Fügenschuh, B., Matenco, L., Schefer, S., Schuster, R., Tischler, M., Ustaszewski, K. (2008). The Alpine-Carpathian-Dinaridic orogenic system: Correlation and evolution of tectonic units. *Swiss Journal of Geosciences*, 101(1), 139–183 [Online]. Available at: https://doi.org/10.1007/s00015-008-1247-3.

Séranne, M. (1999). The Gulf of Lion continental margin (NW Mediterranean) revisited by IBS: An overview. *Geological Society, London, Special Publications*, 156(1), 15–36 [Online]. Available at: https://doi.org/10.1144/GSL.SP.1999.156.01.03.

Séranne, M., Benedicto, A., Labaum, P., Truffert, C., Pascal, G. (1995). Structural style and evolution of the Gulf of Lion Oligo-Miocene rifting: Role of the Pyrenean orogeny. *Marine and Petroleum Geology*, 12(8), 809–820.

Sibuet, J.-C., Srivastava, S.P., Spakman, W. (2004). Pyrenean orogeny and plate kinematics. *Journal of Geophysical Research: Solid Earth*, 109(B8), 1–18 [Online]. Available at: https://doi.org/10.1029/2003JB002514.

Simonetti, M., Carosi, R., Montomoli, C., Cottle, J.M., Law, R.D. (2020). Transpressive deformation in the southern European Variscan Belt: New insights from the Aiguilles Rouges massif (Western Alps). *Tectonics*, 39(6), 1–24 [Online]. Available at: https://doi.org/10.1029/2020tc006153.

Sinclair, H.D., Gibson, M., Naylor, M., Morris, R.G. (2005). Asymmetric growth of the Pyrenees revealed through measurement and modeling of orogenic fluxes. *American Journal of Science*, 305(5), 369–406 [Online]. Available at: https://doi.org/10.2475/ajs.305.5.369.

Soto, R., Casas-Sainz, A.M., Oliva-Urcia, B., García-Lasanta, C., Izquierdo-Llavall, E., Moussaid, B., Kullberg, J.C., Román-Berdiel, T., Sánchez-Moya, Y., Sopeña, A. et al. (2019). Triassic stretching directions in Iberia and North Africa inferred from magnetic fabrics. *Terra Nova*, 31(5), 465–478 [Online]. Available at: https://doi.org/10.1111/ter.12416.

Srivastava, S.P., Sibuet, J.-C., Cande, S., Roest, W.R., Reid, I.D. (2000). Magnetic evidence for slow seafloor spreading during the formation of the Newfoundland and Iberian margins. *Earth and Planetary Science Letters*, 182(1), 61–76 [Online]. Available at: https://doi.org/10.1016/S0012-821X(00)00231-4.

Stampfli, G.M. and Borel, G.D. (2002). A plate tectonic model for the Paleozoic and Mesozoic constrained by dynamic plate boundaries and restored synthetic oceanic isochrons. *Earth and Planetary Science Letters*, 196(1–2), 17–33 [Online]. Available at: https://doi.org/10.1016/S0012-821X(01)00588-X.

Stampfli, G.M., Mosar, J., Favre, P., Pillevuit, A., Vannay, J.-C. (2001). Permo-Mesozoic evolution of the western Tethyan realm: The Neotethys/East-Mediterranean connection. In *Peri-Tethys Memoir 6: Peri-Tethyan Rift/Wrench Basins and Passive Margins*, Ziegler, P.A., Cavazza, W., Robertson, A.H.F., Crasquin-Soleau, S. (eds). Mémoires du Muséum d'Histoire Naturelle, Paris.

Taillefer, A., Guillou-Frottier, L., Soliva, R., Magri, F., Lopez, S., Courrioux, G., Millot, R., Ladouche, B., Le Goff, E. (2018). Topographic and faults control of hydrothermal circulation along dormant faults in an orogen. *Geochemistry, Geophysics, Geosystems*, 19(12), 4972–4995 [Online]. Available at: https://doi.org/10.1029/2018GC007965.

Tavani, S. and Granado, P. (2015). Along-strike evolution of folding, stretching and breaching of supra-salt strata in the Plataforma Burgalesa extensional forced fold system (northern Spain). *Basin Research*, 27(4), 573–585 [Online]. Available at: https://doi.org/10.1111/bre.12089.

Tavani, S., Mencos, J., Bausà, J., Muñoz, J.A. (2011a). The fracture pattern of the Sant Corneli Bóixols oblique inversion anticline (Spanish Pyrenees). *Journal of Structural Geology*, 33(11), 1662–1680 [Online]. Available at: https://doi.org/10.1016/j.jsg.2011.08.007.

Tavani, S., Quintà, A., Granado, P. (2011b). Cenozoic right-lateral wrench tectonics in the Western Pyrenees (Spain): The Ubierna fault system. *Tectonophysics*, 509(3/4), 238–253 [Online]. Available at: https://doi.org/10.1016/j.tecto.2011.06.013.

Tavani, S., Bertok, C., Granado, P., Piana, F., Salas, R., Vigna, B., Muñoz, J.A. (2018). The Iberia-Eurasia plate boundary east of the Pyrenees. *Earth-Science Reviews*, 187, 314–337 [Online]. Available at: https://doi.org/10.1016/j.earscirev.2018.10.008.

Teixell, A. (1998). Crustal structure and orogenic material budget in the west central Pyrenees. *Tectonics*, 17(3), 395–406.

Teixell, A., Labaume, P., Lagabrielle, Y. (2016). The crustal evolution of the west-central Pyrenees revisited: Inferences from a new kinematic scenario. *Comptes rendus – géoscience*, 348(3/4), 257–267 [Online]. Available at: https://doi.org/10.1016/j.crte.2015.10.010.

Ternois, S., Odlum, M., Ford, M., Pik, R., Stockli, D., Tibari, B., Vacherat, A., Bernard, V. (2019). Thermochronological evidence of early orogenesis, eastern Pyrenees, France. *Tectonics*, 38(4), 1308–1336 [Online]. Available at: https://doi.org/10.1029/2018TC005254.

Tugend, J., Manatschal, G., Kusznir, N.J., Masini, E. (2014a). Characterizing and identifying structural domains at rifted continental margins: Application to the Bay of Biscay margins and its Western Pyrenean fossil remnants. *Geological Society Special Publication*, 413(1), 171–203 [Online]. Available at: https://doi.org/10.1144/SP413.3.

Tugend, J., Manatschal, G., Kusznir, N.J., Masini, E., Mohn, G., Thinon, I. (2014b). Formation and deformation of hyperextended rift systems: Insights from rift domain mapping in the Bay of Biscay-Pyrenees. *Tectonics*, 33(7), 1239–1276 [Online]. Available at: https://doi.org/10.1002/2014TC003529.

Tugend, J., Manatschal, G., Kusznir, N.J. (2015). Spatial and temporal evolution of hyperextended rift systems: Implication for the nature, kinematics, and timing of the Iberian-European plate boundary. *Geology*, 43(1), 15–18.

Tugend, J., Chamot-Rooke, N., Arsenikos, S., Blanpied, C., Frizon de Lamotte, D. (2019). Geology of the Ionian basin and margins: A key to the east Mediterranean geodynamics. *Tectonics*, 38(8), 2668–2702 [Online]. Available at: https://doi.org/10.1029/2018TC005472.

Ubide, T., Wijbrans, J.R., Galé, C., Arranz, E., Lago, M., Larrea, P. (2014). Age of the Cretaceous alkaline magmatism in northeast Iberia: Implications for the Alpine cycle in the Pyrenees. *Tectonics*, 33(7), 1444–1460.

Vacherat, A., Mouthereau, F., Pik, R., Bernet, M., Gautheron, C., Masini, E., Le Pourhiet, L., Tibari, B., Lahfid, A. (2014). Thermal imprint of rift-related processes in orogens as recorded in the Pyrenees. *Earth and Planetary Science Letters*, 408, 296–306.

Vacherat, A., Mouthereau, F., Pik, R., Bellahsen, N., Gautheron, C., Bernet, M., Daudet, M., Balansa, J., Tibari, B., Pinna Jamme, R. et al. (2016). Rift-to-collision transition recorded by tectonothermal evolution of the northern Pyrenees. *Tectonics*, 35(4), 907–933.

Vacherat, A., Mouthereau, F., Pik, R., Huyghe, D., Paquette, J.L., Christophoul, F., Loget, N., Tibari, B. (2017). Rift-to-collision sediment routing in the Pyrenees: A synthesis from sedimentological, geochronological and kinematic constraints. *Earth-Science Reviews*, 172, 43–74.

Vargas, H., Gaspar-Escribano, J.M., López-Gómez, J., Van Wees, J.D., Cloetingh, S., de La Horra, R., Arche, A. (2009). A comparison of the Iberian and Ebro Basins during the Permian and Triassic, eastern Spain: A quantitative subsidence modelling approach. *Tectonophysics*, 474(1/2), 160–183 [Online]. Available at: https://doi.org/10.1016/j.tecto.2008.06.005.

Vergés, J. and Fernàndez, M. (2012). Tethys-Atlantic interaction along the Iberia-Africa plate boundary: The Betic-Rif orogenic system. *Tectonophysics*, 579, 144–172 [Online]. Available at: https://doi.org/10.1016/j.tecto.2012.08.032.

Vergés, J., Millán, H., Roca, E., Muñoz, J.-A., Marzo, M., Cirés, J., Bezemer, T.D., Zoetemeijer, R., Cloetingh, S.A.P.L. (1995). Eastern Pyrenees and related foreland basins: Pre-, syn- and post-collisional crustal-scale cross-sections. *Marine and Petroleum Geology*, 12(8), 903–915 [Online]. Available at: https://doi.org/10.1016/0264-8172(95)98854-X.

Vergés, J., Poprawski, Y., Almar, Y., Drzewiecki, P.A., Moragas, M., Bover-Arnal, T., Macchiavelli, C., Wright, W., Messager, G., Embry, J., Hunt, D. (2020). Tectono-sedimentary evolution of Jurassic–Cretaceous diapiric structures: Miravete anticline, Maestrat Basin, Spain. *Basin Research*, 32(6), 1653–1684.

Villa, I.M., Hermann, J., Müntener, O., Trommsdorff, V. (2000). $^{39}$Ar–$^{40}$Ar dating of multiply zoned amphibole generations (Malenco, Italian Alps). *Contributions to Mineralogy and Petrology*, 140(3), 363–381.

Vinciguerra, C. (2020). Reconstruction des paléo-drainages des bassins précoces périorogéniques (Crétacé terminal-Paléocène) à partir des dépôts fluviatiles dans le système pyrénéen oriental. Université Michel de Montaigne–Bordeaux III.

Vine, F.J. and Matthews, D.H. (1963). Magnetic anomalies over oceanic ridges. *Nature*, 199(4897), 947–949.

Vissers, R.L.M. and Meijer, P.T. (2012). Iberian plate kinematics and Alpine collision in the Pyrenees. *Earth-Science Reviews*, 114(1/2), 61–83.

Waldner, M., Bellahsen, N., Mouthereau, F., Bernet, M., Pik, R., Rosenberg, C.L., Balvay, M. (2021). Central Pyrenees mountain building: Constraints from new LT thermochronological data from the axial zone. *Tectonics*, 40(3), e2020TC006614.

Wang, Y., Chevrot, S., Monteiller, V., Komatitsch, D., Mouthereau, F., Manatschal, G., Sylvander, M., Diaz, J., Ruiz, M., Grimaud, F. et al. (2016). The deep roots of the western Pyrenees revealed by full waveform inversion of teleseismic P waves. *Geology*, 44(6), 475–478.

Wegener, A. (1912). Die entstehung der kontinente. *Geologische Rundschau*, 3(4), 276–292.

Wegener, A. (1915). *Die Entstehung der Kontinente und Ozeane*. Vieweg, Braunschweig.

Wehr, H., Chevrot, S., Courrioux, G., Guillen, A. (2018). A three-dimensional model of the Pyrenees and their foreland basins from geological and gravimetric data. *Tectonophysics*, 734/735(April), 16–32.

Wilson, J.T. (1963). Evidence from islands on the spreading of ocean floors. *Nature*, 197(4867), 536–538.

Wilson, J.T. (1965a). A new class of faults and their bearing on continental drift. *Nature*, 207(4995), 343–347.

Wilson, J.T. (1965b). Evidence from ocean islands suggesting movement in the Earth. *Philosophical Transactions of the Royal Society of London. Series A, Mathematical and Physical Sciences*, 258(1088), 145–167.

Ziegler, P.A. (1990). *Geological Atlas of Western and Central Europe*. Shell International Petroleum Mij. B.V. and Geological Society, London.

Ziegler, P.A. and Dèzes, P. (2006). Crustal evolution of Western and Central Europe. *Geological Society, London, Memoirs*, 32(1), 43–56.

# 4

# Sedimentary Breccias Synextension in the North-Eastern Pyrenees

**Tarik KERNIF[1], Thierry NALPAS[1], Romain BOUSQUET[1,2] and Roman CHELALOU[1]**

[1] *Géosciences Rennes, CNRS, Université de Rennes 1, France*

[2] *Institute of Geosciences, Christian-Albrechts-Universität zu Kiel, Germany*

## 4.1. Introduction

Breccias correspond to an agglomeration of angular rock clasts. There are several types of breccias depending on their formation process (Figure 4.1). For each type of breccia, their characteristics are representative of the formation process. Therefore, breccias can inform us about the history of the geological system at the origin of their formation. We can already differentiate between the breccias that are formed deep in the crust (hydraulic, magmatic and tectonic) and those that are formed at the surface of the crust (volcanic, karstic, sedimentary, eluvial and impactites).

These breccias can have common characteristics, therefore confusions can sometimes be made as to the processes behind the genesis of these rocks. However, some criteria are specific to the formation process. These criteria allow for discrimination when identifying the process at the origin of the formation of the breccias.

---

For a color version of all the figures in this chapter, see www.iste.co.uk/saspiturry/evolution1.zip.

*Evolution of the Pyrenees during the Variscan and Alpine Cycles 1*,
coordinated by Nicolas SASPITURRY, Jessica UZEL, Alexandre ORTIZ. 

Many breccias have been described in the Pyrenees, and for some of them several formation processes have been proposed. These uncertainties concerning the identification of breccias can result from two reasons: (1) either the data are not sufficiently discriminating so as to be able to make a decision (which is rarely the case), or (2) the data collected are oriented toward an already preestablished model.

In this chapter, we focus on the Late Jurassic breccias known as “brèche limite”, which are preserved in a large part of the Pyrenees. Their age coincides with the first phases of extensional deformation between the Iberian and Eurasian plates (Sibuet 1974; Jammes et al. 2009) (see section 4.3 in Angrand et al. 2022). The origin of these breccias, sedimentary, tectonic or karstic, has been widely debated for several decades.

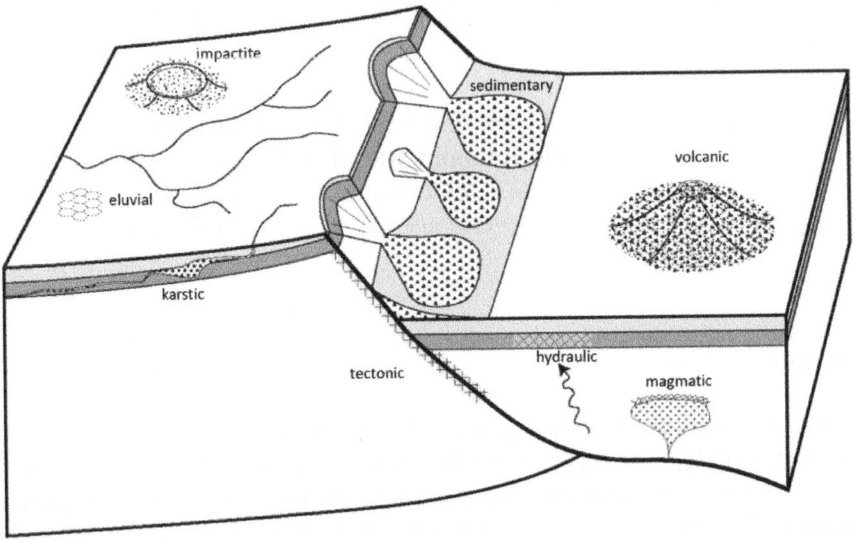


**Figure 4.1.** *The different processes at the origin of the formation of breccias*

The breccias presented in this work can reach a thickness of nearly 300 m. They are generally polygenic and poorly sorted. The carbonate clasts composing these rocks are multi-millimeter to multi-millimeter in size (blocky granules) and very angular. These breccias are located near a normal paleofault. Large pebble-granule sized clasts are found close to the fault, whereas block-sized (metric) clasts are far from the normal fault. We will show that this organization results from talus type landslides and slumps, where the clasts originate from the block lifted by the play of the normal fault (*hangingwall*). In general, the clasts become coarser as they move away from the fault, contrary to what is found in alluvial fan systems.

Our goal here is to show how, from a field study and a classical laboratory analysis, we can remove ambiguities about the processes of breccia formation in the northeastern Pyrenees and propose a coherent explanation for breccia emplacement.

## 4.2. The breccias of the north-eastern Pyrenees

### 4.2.1. *Geological background*

The Pyrenees correspond to a chain that currently extends over 400 km from the Atlantic to the Mediterranean. This double vergence chain is the result of the collision of the Iberian and Eurasian plates from the Upper Cretaceous, following an extension from the Upper Jurassic to the Lower Cretaceous (Chapter 5).

The present chain is divided into five structural zones bounded by four major structures (Bertrand 1940; Mattauer 1968; Choukroune et al. 1973) (Figure 4.2(a)): (1) the Ebro foreland basin bounded on the north by the South Pyrenean frontal thrust, (2) the South Pyrenean zone bounded on the north by the South Pyrenean fault, (3) the axial zone bounded on the north by the North Pyrenean fault, (4) the North Pyrenean zone bounded on the north by the North Pyrenean frontal thrust and finally (5) the Aquitaine foreland basin.

The study area located in the North Pyrenean zone at the extreme northeast of the Pyrenees concerns the Boucheville, Bas-Agly and Saint-Paul-de-Fenouillet synclines (Figure 4.2). These three domains represent different depocenters of Cretaceous rift basins. Their present synclinal shape is the result of the Pyrenean collision. It can be noted that a rifting phase is widely recorded in the North Atlantic Ocean as well as in the north of the Iberian Peninsula, at the end of the Jurassic, and marks the beginning of the extension of the Lower Cretaceous between the Iberian Peninsula and Europe (Sibuet 1974; Jammes et al. 2009; Asti et al. 2022), in concomitance with a Jurassic rift event in the North Atlantic. It is at this time that the Upper Jurassic deposits of the northeastern Pyrenees all show a particular geological formation called “brèche limite” (Peybernès 1976).

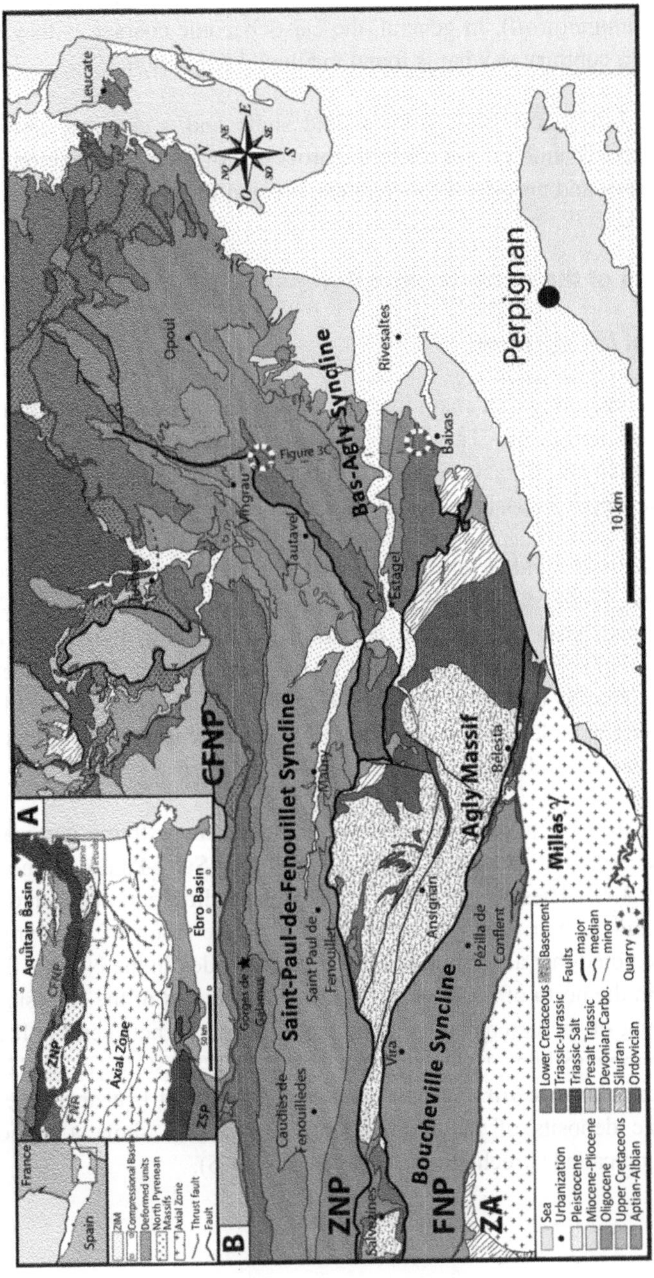


**Figure 4.2.** *Location of the study area. Modified from Kernif et al. (2022)*

LEGEND FOR FIGURE 4.2.– *(a) Structural map of part of the Pyrenean chain (from BRGM geological map 1:1,000,000; ZIM: internal metamorphic zone from Clerc (2012). (b) Simplified geological map of the study area modified from BRGM geological maps 1:50,000 of Tuchan (Berger et al. 1997), Perpignan (Berger et al. 1988), Leucate (Bles and Berger 1982), Quillan (Crochet et al. 1989) and Rivesaltes (Fonteilles et al. 1993).*

Many outcrops show breccias preserved over varying thicknesses ranging from a few meters to several hundred meters in the three synclines of Mesozoic terrain, but also wedged into the Agly massif (Figure 4.2). Most of these breccias are classified as Upper Jurassic. However, some of them are suspected to belong to the Eocene (Rivesaltes, Saint-Paul-de-Fenouillet, Tuchan maps). We are mainly interested in the Upper Jurassic breccias of the Saint-Paul-de-Fenouillet, Boucheville and Bas-Agly synclines in order to determine their emplacement process and their geodynamic interest. Their formation is attributed to a synsedimentary activity of tectonic accidents (Souquet and Debroas 1980).

The Bas-Agly syncline has the most remarkable breccias outcrops due to the presence of quarries on the southern flank (Baixas) and northern flank (Vingrau) of the syncline (Figure 4.2).

### 4.2.2. Description and organization of the breccias

Along the entire southern and northern flank of the Bas-Agly syncline, breccias are found in the Upper Jurassic just before the base of the Lower Cretaceous. These breccias could be studied in the field in the quarries of the Lafarge-Holcim group in Baixas and in the quarries of the Omya group in Vingrau. The breccias are organized differently in the two quarries in relation to the surrounding layers (number, thickness, intercalation), but the layers they form in the south and in the north have their own characteristics.

The southern flank of the Bas-Agly syncline is in fault contact with the Paleozoic basement and is composed of terrains ranging from the Upper Triassic to the Lower Cretaceous. The *substratum* on which the breccia layers are deposited shows, at the base, a set of more or less dolomitic limestone layers that show a boudinage deformation for a few meters before contact with the breccias (Figure 4.3, A1). The boundary between this *substratum* and the first breccia deposits is marked by a subvertical fault (Figure 4.3(a)).

The breccia layers are organized from base to top as follows (Figure 4.3(a)): polygenic breccias with angular to very angular clasts with very large blocks at the top (Figure 4.3, I, A2); less polygenic breccias (Figure 4.3, II); polygenic breccias with a dominant yellow matrix (Figure 4.3, III, A3); polygenic breccias (Figure 4.3, IV, A4); limestone interval; very polygenic breccias with large blocks (Figure 4.3, V; some clasts are from the Paleozoic basement), with internal slip surfaces.

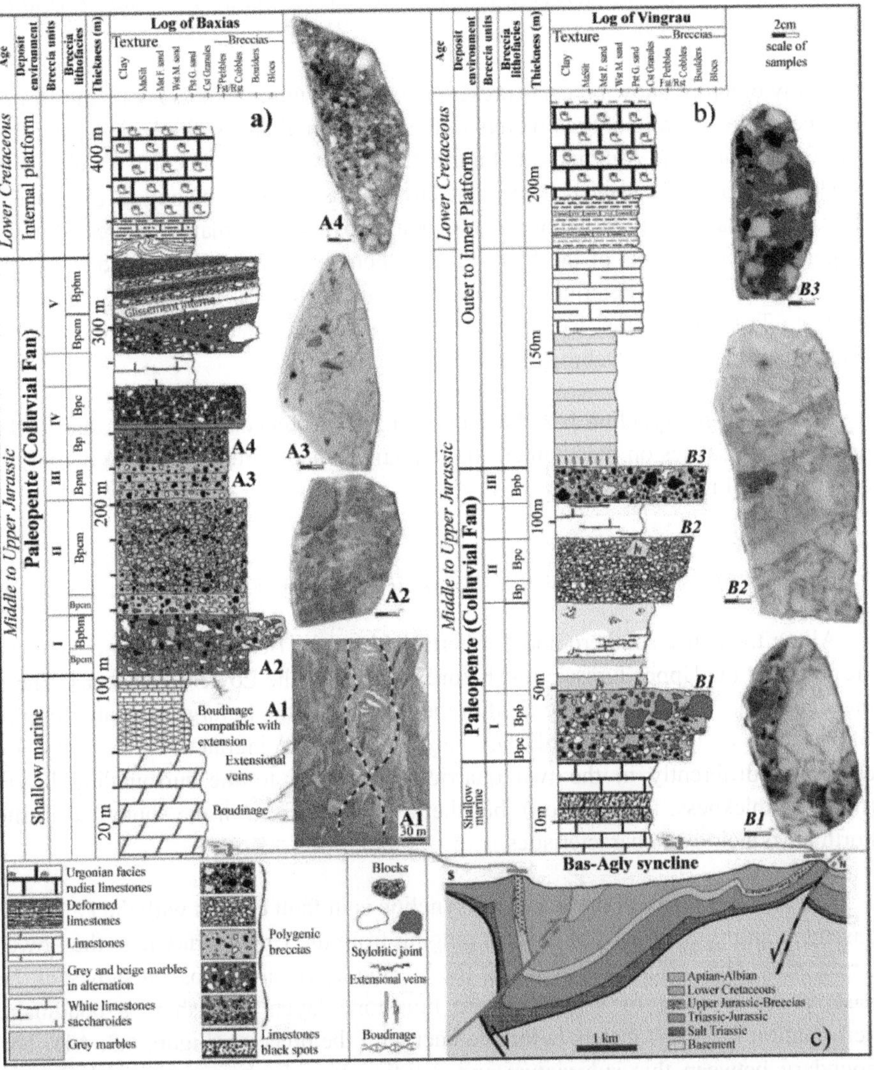


**Figure 4.3.** *Sedimentological log and geological section of the Bas-Agly syncline. Modified from Kernif et al. (2022)*

REMARKS ON FIGURE 4.3.– *(a) Sedimentological log from the Baixas quarry (Baixas, southern flank of the Bas-Agly syncline); (b) sedimentological log from the Coume Roujou quarry (Vingrau, northern flank of the Bas-Agly syncline); (c) geological cross-section through the Bas-Agly syncline illustrating the location of the two sedimentological logs.*

Layers of nonbrecciated limestone, followed by rudist limestone, rest stratigraphically on the breccia layers.

The northern flank of the Bas-Agly syncline is in fault contact with the Saint-Paul-de-Fenouillet syncline and is composed of terrain ranging from the Upper Triassic to the Lower Cretaceous. At the base of the breccia layers, we find a set of calcareous layers that are grayish in color.

| ***Facies code*** | ***Facies descriptions*** | ***Binding phase matrix*** | ***Deposition process*** |
|---|---|---|---|
| ***Massive breccias*** | | | |
| ***Bp*** | Moderately sorted breccias with very angular to angular clasts from small to large pebbles, pseudomonogenes, calcisiltite matrix | <5% | **Rock fall**<br>Generated by the collapse of the block exhumed by the normal fault |
| ***Bpc*** | Poorly sorted breccias with small to large angular clasts, occasionally boulders, pseudomonogenes to polygenics, calcerenite matrix | <10% | |
| ***Bpcm*** | Very poorly sorted breccias with very angular clasts from small to large stones, occasionally boulders, polygenic, matrix phase of granules or/and micrite | <15% | |
| ***Massive megabreccias*** | | | |
| ***Bpb*** | Unsorted breccias with very angular to subangular clasts of the size of small stones to large stones and sometimes boulders, polygenic, alignment of blocks parallel to the direction of gravity flow, supported clasts, matrix phase of granules or calcarenite | 5 - 10% | **Rock fall landslides**<br>Generated by the collapse of the block exhumed by the normal fault |
| ***Bpbm*** | Unsorted breccias with very angular to subangular clasts of the size of small to large pebbles and frequently boulders, polygenic, internal slip, supported matrices rich in encalcarenite or/and micrite | 15 - 25% | |
| ***Microbreccias with stratified breccias*** | | | |
| ***Bpm*** | Well-sorted stratified breccias with angular to subangular clasts, granule-sized clasts, some small pebbles, rarely subrounded clasts of large pebbles, monogenic to polygenic, muddy matrix | >80% | **Cohesive debris flow**<br>Generated by intense precipitation |

**Table 4.1.** *Facies of the carbonate sedimentary breccias of the syncline of the Bas-Agly syncline, and interpretations of the depositional processes. Modified from Kernif et al. (2022)*

The breccia layers are organized from the base to the top as follows (Figure 4.3(b)): chaotic polygenic breccias with angular to very angular clasts, with very large blocks at the top (Figure 4.3, I, B1); limestone interval; weakly polygenic coarsening upward breccias (Figure 4.3, II, B2); limestone interval; highly polygenic breccias with large subrounded blocks (Figure 4.3, III, B3).

The same layers of nonbrecciated limestone and rudist limestone stratigraphically cover the breccia layers. A meticulous fieldwork (cartography, sedimentological log, geometrical organization, thickness measurements, etc.) was coupled with laboratory analysis of samples collected in the field (thin sections, polished surfaces and petrography, etc.). This multi-scale approach allows (1) to propose a precise classification of these breccias and (2) to trace their genesis and preservation process (Table 4.1). The lateral organization of the facies is presented in the interpretation section (Figure 4.5). The process that may generate these very poorly sorted and angular sedimentary breccia facies is related to the process of rockfall in slope depositional systems (Kernif et al. 2022).

It should be noted that all the layers were straightened during the Pyrenean compression with the creation of faults that cut all of the layers. On the other hand, in the breccia levels there is no evidence of synsedimentary deformation (fault brecciation type), such as striations on clasts, clast fragmentations, tension cracks or styloliths.

The Saint-Paul-de-Fenouillet syncline is used as a reference area in the region for dating sedimentary terrains. The dating of certain sedimentary deposits in the region (e.g. Boucheville Basin), which are too metamorphosed to have preserved biostratigraphic markers, is done by analogy with the deposits of the Saint-Paul-de-Fenouillet syncline. The metamorphism that affected this basin is low-temperature, averaging 240°C (Chelalou et al. 2016). The northern flank of the Saint-Paul-de-Fenouillet syncline is limited in its northern part by the North-Pyrenean frontal thrust fault, which is located on the Triassic detachment level and sometimes shows tectonic breccia passes. In the Jurassic, monogenic breccias can be found which have been interpreted as tectonic. In the Upper Jurassic, along the northern flank of the Saint-Paul-de-Fenouillet syncline, we find several tens of meters of polygenic breccias with some rounded clasts. As in the case of the Bas-Agly syncline, we find an interstratification between more or less polygenic breccias and layers of limestone, as in the Gorges de Galamus to the north of Saint-Paul-de-Fenouillet (Figure 4.4(a)). These breccias of sedimentary origin are associated with the "brèche limite" formation (Peybernès 1976; Chelalou 2015). We will integrate these breccias into an overall schema, but the absence of quarries does not allow for as detailed a study as in the Bas-Agly syncline.

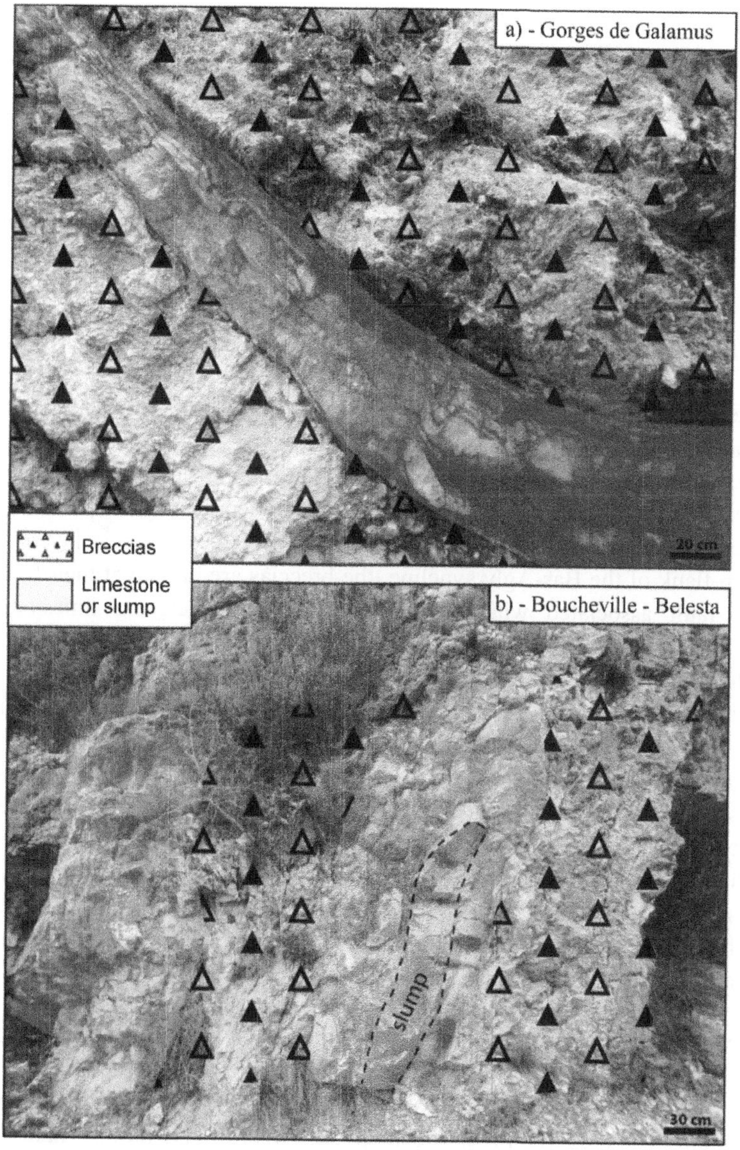


**Figure 4.4.** *Alternating breccias with limestone layers: (a) in the gorges of Galamus; (b) with slumps in the syncline of Boucheville toward Bélesta*

In the Boucheville Syncline, breccia outcrops are preserved on the northern flank in areas where terrain that is mapped as Upper Jurassic occurs. These polygenic breccias consist primarily of angular carbonate clastics and may form alternations between breccias and slumps (Figure 4.4(b)). The presence of slumps is characteristic of a syndepositional slope. This northern part of the Boucheville syncline has been interpreted as the major boundary of an extensional Lower Cretaceous basin (Chelalou et al. 2016). The Boucheville syncline was heavily affected by HT-LP metamorphism during the Lower Cretaceous extension and is part of the metamorphic inner zone (Figure 4.2) (Chapter 5). We will include these breccias in an overall scheme; however, the absence of quarries does not allow for a study as detailed as that carried out for the Bas-Agly syncline.

### 4.2.3. *Interpretation*

The breccias form different layers, whose clasts are polygenic, with very important variations of granulometry. These layers are limited by limestone deposits. Moreover, these breccias do not show any trace of deformation (fault breccias). The only observable deformations are slides on a syn-deposit slope. We can therefore only consider a sedimentary origin for these breccia layers. On the southern flank of the Bas-Agly syncline, the breccias are limited at their base by an ancient fault associated with a boudinage in the underlying limestone. This concomitance is found along the southern flank of the Bas-Agly syncline and is consistent with extensional deformation (Clerc 2012). The verticalized fault is thus an ancient normal fault bordering a synrift sedimentary basin. The organization of breccias in successive layers more or less intercalated with limestones is characteristic of successive arrivals of material following landslides. The progressive appearance of increasingly polygenic clastics with the presence of basement clasts in the last breccias is the result of a progressive exhumation of the formations that were then slumped. These observations are consistent with a rockfall process associated with the formation of a major synsedimentary normal fault at the edge of the basin, with uplift of the lower block of the fault (hangingwall uplift), and exhumation of the terrain (Figure 4.5).

The important parameters for the formation of these sedimentary breccias are (1) the creation of a relief linked to the movement of the normal fault and the uplift of the lower block; (2) the successive collapse of the exhumed lands and their sedimentation at the bottom of the slope in front of the normal fault; (3) the subsidence of the upper block of the normal fault allowing for the preservation of the breccias and their covering by new deposits, and avoiding the reorganization of the clasts by a system of longitudinal drainage; and (4) the slope of the system allows the largest blocks to acquire more energy during the fall, and to go the farthest, thus the layers of breccias present an evolution of the facies from the finest to the coarsest.

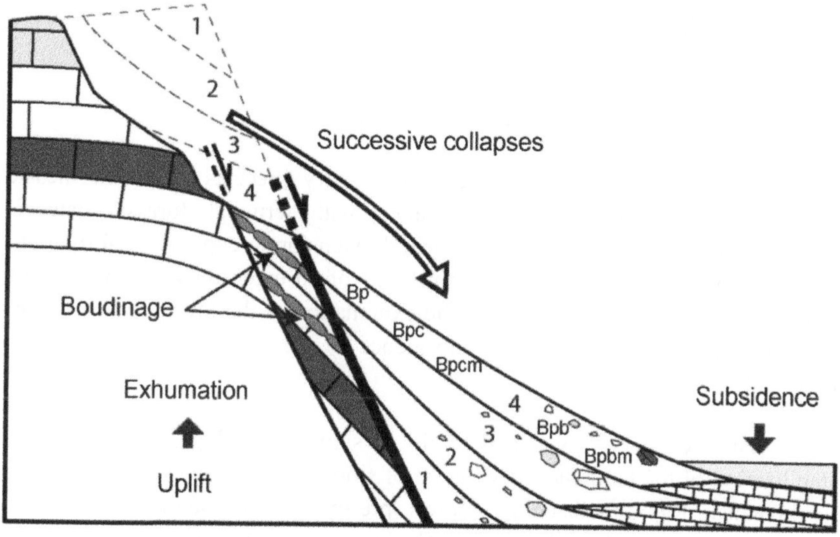


**Figure 4.5.** *Synextensional breccias formation processes in the Lower Agly Basin during the Upper Jurassic. Modified from Kernif et al. (2022)*

## 4.3. Discussion

### 4.3.1. *Implications of the interpretation that the Lower Agly breccias had on the evolution of the Pyrenees*

These breccias are interbedded with limestone levels throughout the northern flank of the Lower Agly syncline, consistent with the formation of a sedimentary basin during the slumping process that originated the breccias. This sedimentary basin is bounded by normal faults and is therefore an extensional end-Jurassic basin that precedes the extensional phase of the Lower Cretaceous (Figure 4.6), which is consistent with geodynamic events in this area (Jammes et al. 2009; Asti et al. 2022). Normal faults created during this early phase of Upper Jurassic extension form prominent landforms due to the brittle rheology of the crust. These normal faults then serve as points of weakness and promote localization of deformation during the Lower Cretaceous rifting. The Lower Cretaceous episode is described as being associated with high temperature, more ductile deformation and lithospheric boudinage (Clerc et al. 2016; Lagabrielle et al. 2016, 2020; Saspiturry et al. 2020, 2021), which are not conducive to the formation of significant landforms and the creation of talus breccias.

### 4.3.2. *Sedimentary breccias: markers of extension?*

A depositional model of preserved sedimentary breccias in the northeastern Pyrenean basins shows a clear association with terminal Jurassic extension. Important parameters during early continental extension to explain the formation of sedimentary breccias associated with rim faults of an extensional basin are (1) the steepness (±60°) and relief created by the exhumation of the lower compartment of the normal fault and (2) the rapid subsidence of the upper compartment of the normal fault.

The relief created, associated with a significant slope, allows for the destabilization of the land and the creation of slope failures downslope, which form talus breccias. Rapid subsidence allows preservation of sedimentary breccias without reworking of clasts by runoff. During early extension, rivers flow in the opposite direction to the extensional basin due to exhumation of the lower compartment of the normal fault and basin-oriented catchments are very small or nonexistent. Therefore, along the edges of the extensional basins, mass transfers are related to collapse processes such as rockfall and slump toward the basin. The major river drainage runs in the opposite direction to the extensional basin-margins and therefore does not allow for clast reworking (Figure 4.7). The major drainage system oriented away from the extensional basin-margins is characteristic of this geologic system, in contrast to compressional systems, where the major drainage goes in the direction of the compressional basin and carries rounded clasts that form very large alluvial fan deposits (Barrier et al. 2010). Therefore, the preservation of sedimentary breccias can be considered a marker of early extensional phases by being associated with major normal faults.

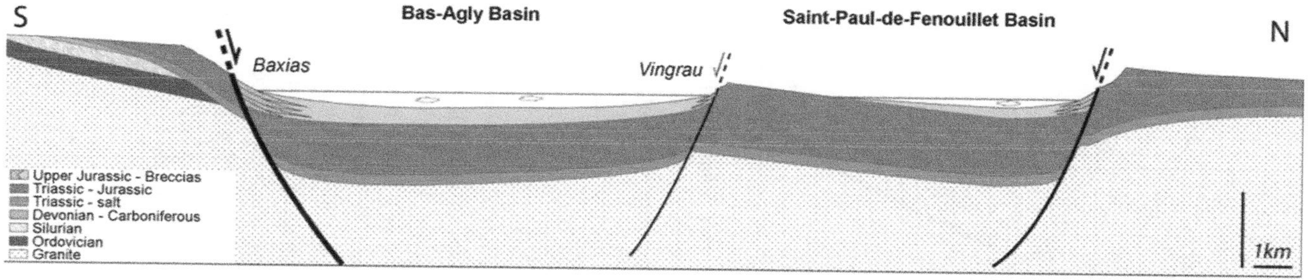


**Figure 4.6.** *Organization of the Bas-Agly and Saint-Paul-de-Fenouillet basins with the location of breccias during the extension to the Upper Jurassic*

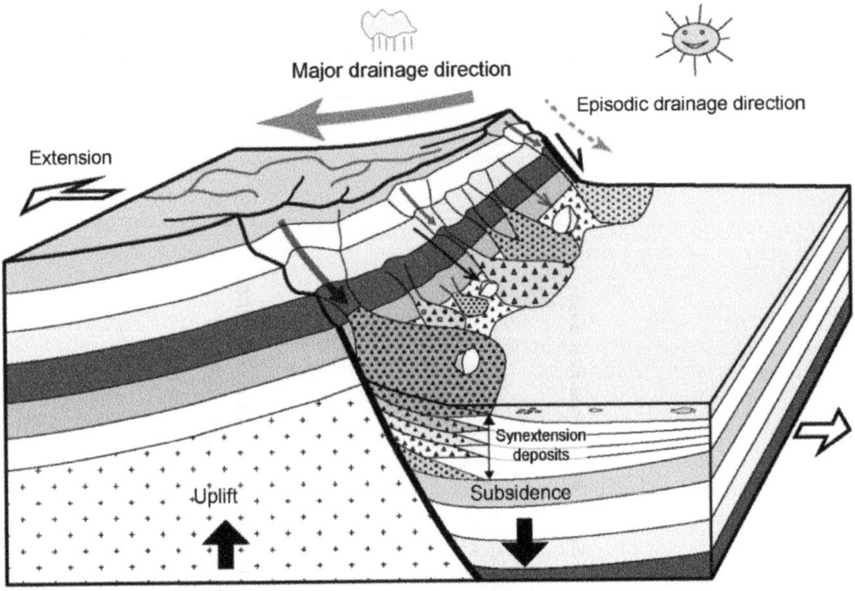


**Figure 4.7.** *Model of sedimentary breccia formation and preservation during continental extension on an extensional basin boundary normal fault. Modified from Kernif et al (2022)*

## 4.4. Conclusion

1) The sedimentary breccias present common criteria: (i) angular to very angular clasts of various sizes; (ii) a diverse origin of the clasts; and (iii) breccias organized in massive coarsening upward units. (2) These are characteristic of slope failure processes, such as talus breccias. (3) These are found on the boundary faults of extensional Upper Jurassic basins. (4) They are associated with the movement of a normal fault with uplift of the lower block of the normal fault during extension and creation of an important local topography, then its destabilization and landslide in the form of debris flow downslope. (5) Subsidence of the upper block of the normal fault during extension allows rapid burial of the slid lands and preservation of a large quantity of sedimentary breccia. (6) Formation of normal faults during extension promotes the production and preservation of sedimentary breccias that become a marker of continental extension.

## 4.5. References

Asti, R., Saspiturry, N., Angrand, P. (2022). The Mesozoic Iberia-Eurasia plate boundary: A wide domain of distributed transtensional deformation rather than a single wrench fault. *Earth-Science Reviews*, 230, 104040 [Online]. Available at: https://doi.org/10.1016/j.earscirev.2022.104040.

Barrier, L., Proust, J.N., Nalpas, T., Robin, C., Guillocheau, F. (2010). Control of alluvial sedimentation at foreland-basin active margins: A case study from the northeastern Ebro Basin (southeastern Pyrenees, Spain). *J. Sed. Res.*, 80, 728–749.

Berger, G.M., Fonteilles, M., Leblanc, D., Clauzon, G., Marshal, J.P., Vautrelle, C. (1993). Notice explicative carte 1/50 000 de Rivesaltes. BRGM, Orléans.

Berger, G.M., Bessière, G., Bilotte, M., Viallard, P. (1997). Carte géol. France (1/50000), feuille Tuchan (1078). BRGM, Orléans.

Bertrand, L. (1940). Sur la structure géologique du versant nord des Pyrénées. *Bulletin des services de la carte géologique de la France*, 204, 205–282.

Chelalou, R., Nalpas, T., Bousquet, R., Prevost, M., Lahfid, A., Poujol, M., Ringenbach, J.C., Ballard, J.F. (2016). New sedimentological, structural and paleo-thermicity data in the Boucheville Basin (eastern North Pyrenean Zone, France). *Comptes rendus géoscience. From Rifting to Mountain Building: The Pyrenean Belt*, 348, 312–321.

Choukroune, P., Seguret, M., Galdeano, A. (1973). Caractéristiques et évolution structurale des Pyrénées ; un modèle de relations entre zone orogénique et mouvement des plaques. *Bulletin de la société géologique de France*, 5/6, 600–611.

Clerc, C., Lagabrielle, Y., Neumaier, M., Reynaud, J.Y., de Saint Blanquat, M. (2012). Exhumation of subcontinental mantle rocks: Evidence from ultramafic-bearing clastic deposits nearby the Lherz peridotite body, French Pyrenees. *Bulletin de la société géologique de France*, 183, 443–459.

Jammes, S., Manatschal, G., Lavier, L., Masini, E. (2009). Tectonosedimentary evolution related to extreme crustal thinning ahead of a propagating ocean: Example of the western Pyrenees. *Tectonics*, 28, 4.

Kernif, T., Nalpas, T., Gautier, P., Bourquin, S., Poujol, M., Fournier, F. (2022). Formation and preservation of colluvial sedimentary breccias during early extension: Processes and facies organization. *Comptes rendus géoscience*, 354, 205–231.

Lagabrielle, Y., Clerc, C., Vauchez, A, Lahfid, A., Labaume, P., Azambre, B., Fourcade, F., Dautria, J.M. (2016). Very high geothermal gradient during mantle exhumation recorded in mylonitic marbles and carbonate breccias from a Mesozoic Pyrenean paleo-margin (Lherz area, North Pyrenean Zone). *Comptes rendus géoscience*, 348 (3–4), 290–300.

Lagabrielle, Y., Asti, R., Duretz, T., Clerc, C., Fourcade, S., Teixell, A., Labaume, P., Corre, B., Saspiturry, N. (2020). A review of Cretaceous smooth-slopes extensional basins along the Iberia-Eurasia plate boundary: How pre-rift salt controls the modes of continental rifting and mantle exhumation. *Earth-Science Reviews*, 201, 103071.

Peybernès, B. (1976). Le Jurassique et le Crétacé inférieur des Pyrénées franco-espagnoles entre la Garonne et la Méditerranée. PhD Thesis, Université de Toulouse 3.

Saspiturry, N., Lahfid, A., Baudin, T., Guillou-Frottier, L., Razin, P., Issautier, B., Le Bayon, B., Serrano, O., Lagabrielle, Y., Corre, B. (2020). Paleogeothermal gradients across an inverted hyperextended rift system: Example of the Mauléon fossil rift (Western Pyrenees). *Tectonics*, 39, e2020TC006206 [Online]. Available at: https://doi.org/10.1029/2020TC006206.

Saspiturry, N., Issautier, B., Razin, P., Baudin, T., Asti, R., Lagabrielle, Y., Allanic, C., Serrano, O. (2021). Review of Iberia-Eurasia plate-boundary basins: Role of sedimentary burial on depth-dependent continental crust ductile thinning during rifting and continental breakup. *Basin Research*, 33(2), 1626–1661 [Online]. Available at: https://doi.org/10.1111/bre.12529.

Sibuet, J.C. (1974). Histoire tectonique du Golfe de Gascogne, contribution du département scientifique. *Centre océanologique de Bretagne*, 137(155), 297–312.

# 5

# The North Pyrenean Zone: Hyper-thinned Rift Crust and Exhumed Mantle

**Yves LAGABRIELLE**
*Géosciences Rennes, CNRS, Université de Rennes 1, France*

## 5.1. The North Pyrenean Zone: introduction

In the Pyrenean range, north of the axial primary zone, a long continuous zone of sedimentary terrain of Mesozoic and Paleozoic age extends, forming the North Pyrenean Zone (NPZ). The NPZ itself is thrust over the foreland of the chain, the sub-Pyrenean Zone, which is in turn thrust onto the Aquitaine Basin via a major north-verging reverse fault: the North Pyrenean Frontal Thrust (NPFT). To the south, the NPZ is separated from the axial zone by the major reverse North Pyrenean fault (NPF) (Figures 5.1 and 5.2).

---

For a color version of all the figures in this chapter, see www.iste.co.uk/saspiturry/evolution1.zip.

This chapter is dedicated in memory of Bernard Azambre, who was able to pass on his knowledge and passion for the metamorphism and alkaline magmatism of the Pyrenean province and for the Pyrenees in general.

*Evolution of the Pyrenees during the Variscan and Alpine Cycles 1*,
coordinated by Nicolas SASPITURRY, Jessica UZEL, Alexandre ORTIZ. 

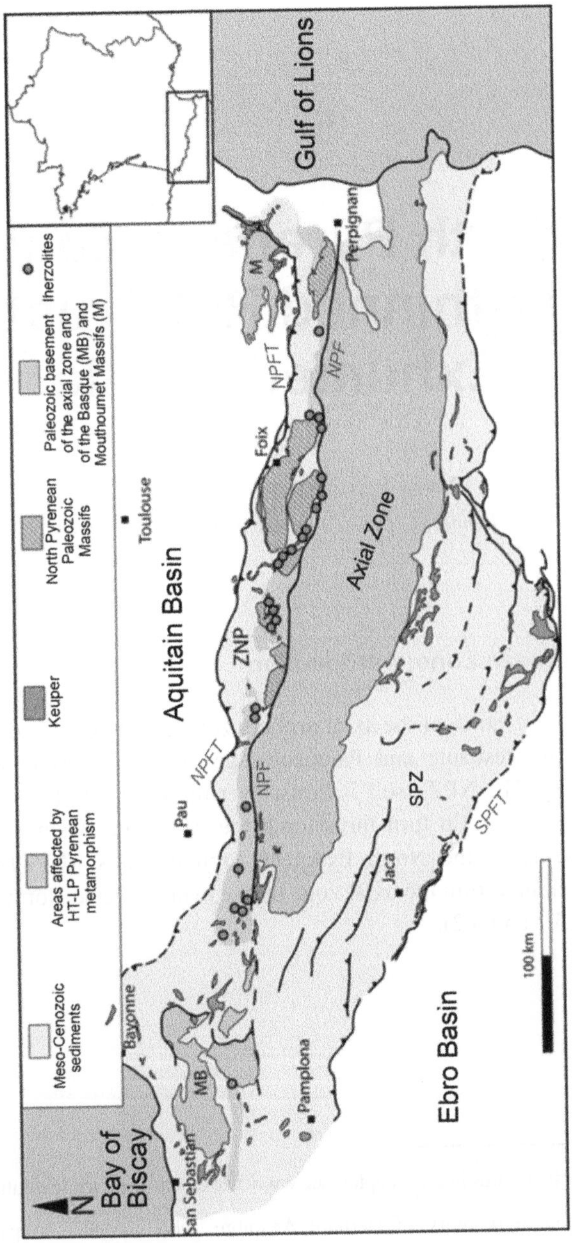


**Figure 5.1.** *Structural map of the Pyrenean range and location of the North Pyrenean Zone (NPZ). NPFT: North Pyrenean Frontal Thrust. SPFT: South Pyrenean Frontal Thrust. NPF: North Pyrenean fault. SPZ: South Pyrenean Zone*

The stratas forming the NPZ display fairly classic characteristics for folded sediments of a piedmont thrust at the front of a mountain range. The Germanic-type Triassic layers including thick Keuper evaporites is followed by layers deposited in shallow marine environments during the Jurassic. After evidence of partial emersion, the Lower Cretaceous deposits correspond to a reef platform.

Finally, fine turbidite sediments of Albian and Upper Cretaceous age are deposited in discrete basins opened within the previous carbonate platform. These latter deposits are the witness of a period of Cretaceous extension and subsidence that characterizes the NPZ. This extensional Cretaceous event has been recognized for a very long time by early geologists; in particular, the term "flysch" has been used to designate the infill of the corresponding shallow basins, undergoing active subsidence where carbonate and sandstone turbidites accumulated over few-kilometre thicknesses (Lefebvre 1928; Casteras 1933; Debroas and Souquet 1976; Peybernès 1976; Souquet et al. 1977, 1985; Debroas 1990; Souquet et al. 1990; Deramond et al. 1993).

Let us discuss here an ambiguity that often confuses geologists when approaching the geological history of the Pyrenees. As in the case of the Western Alps, flysch is usually seen as a synorogenic formation representing deposits emplaced during the early stages of convergence at the front of the growing orogens, that is, in a compressional setting. Here, in the northern Pyrenees, the term flysch is applied primarily to turbidite deposits emplaced from the Albian to the Turonian times, during a period of syn-rift pre-orogenic extension (but formerly seen as contractional). Therefore, the term flysch is used here only in its sedimentological sense. This is a first particularity of the Pyrenees. The second particularity relates to the post-Turonian–Coniacian period, when the Iberian-Eurasian convergence initiated, the Pyrenean flysch trench became a subsiding foreland area without its paleogeography changing fundamentally. Thus, we are faced with a rare case of turbiditic formations recording in continuity rifting-related subsidence followed by contractional-related subsidence in a foreland position.

**Box 5.1.** *Synrift and synorogenic flyschs*

If we consider only their sedimentary facies, the platform and basin sediments of the NPZ are finally quite similar to what can be encountered in Mediterranean alpine chains. However, in the NPZ, these sediments are surprisingly associated with subcontinental mantle rocks of the peridotite group, mostly lherzolites (olivine + clinopyroxene + orthopyroxene + spinel) (Figure 5.2). In addition, by place, the sediments of the NPZ underwent deformation and metamorphism of high temperature and low pressure (thermometamorphism) of regional extent (Figure 5.2). This type of metamorphic paragenesis is common in rocks transformed at the contact of magmatic intrusions. It has been known since the first monographs on the NPZ that this phase of deformation and metamorphism occurred prior to the Cenomanian, and thus early in the history of the chain, and that it has no obvious relationship with compression, the climax of which is Cenozoic in age (Ravier 1959; Mattauer and Proust 1965; Boillot et al. 1973; Choukroune 1976; Durand-Delga 1980, 1990).

In this chapter, we show how the careful acquisition of geological, mineralogical, geochemical and chronological information over more than 200 years has allowed us to progressively reveal the geodynamic significance of the lithological associations that characterize the NPZ. Finally, our aim is to propose a consistent explanation for the enigmas that they still pose today.

## 5.2. The NPZ: a surprising association of varied lithologies

The various characters of the lithologies of the NPZ make it an original and unique geological unit in the world. In this section, we review the main properties of the stratigraphy, magmatism, thermometamorphism of the NPZ, as well as results relative to fluid-rock relationships and lherzolites setting. Modern views on the significance of the NPZ lithologies will be discussed in section 5.3.

### 5.2.1. *Stratigraphy*

The Mesozoic sediments of the NPZ are associated with the lherzolites, as we have seen, but also with massifs of Paleozoic rocks, known as the North Pyrenean massifs. From east to west, the main massifs are the Agly, Trois-Seigneurs, Saint-Barthélémy, Castillon, Arize, Milhas and Barousse massifs. They consist of a thinned crust only 10 km thick and show at their base granulitic rocks (Fonteilles 1970; Passchier 1982; Vielzeuf 1984; Vielzeuf and Kornprobst 1984; de Saint Blanquat et al. 1990; Paquet and Mansy 1991; de Saint Blanquat 1993; Olivier et al. 2004, 2008; Cochelin et al. 2017, 2018; Tournaire-Guille et al. 2019). A significant portion of the crustal thinning of the Paleozoic massifs of the NPZ relates to the Cretaceous events discussed in this chapter (Monié et al. 1994; Asti et al. 2019; Odlum and Stöckli 2019; Aumar et al. 2022).

Since the beginning of the 20th century, stratigraphic and sedimentologic works have shown that the Cretaceous Basins of the NPZ opened in the Upper Albian times after the partial collapse of a rudist reef platform (Aptian carbonates with Urgonian facies) (Souquet et al. 1990). This led to the local reworking of bauxites established on the surrounding continental landforms and the collapse of fault scarp in the outer parts of the basins, yielding breccias reworking both Paleozoic basement clasts and Aptian limestones (Debroas 1978, 1990; Combes and Peybernès 1987, 1989) (see summary in Canérot (2017b)).

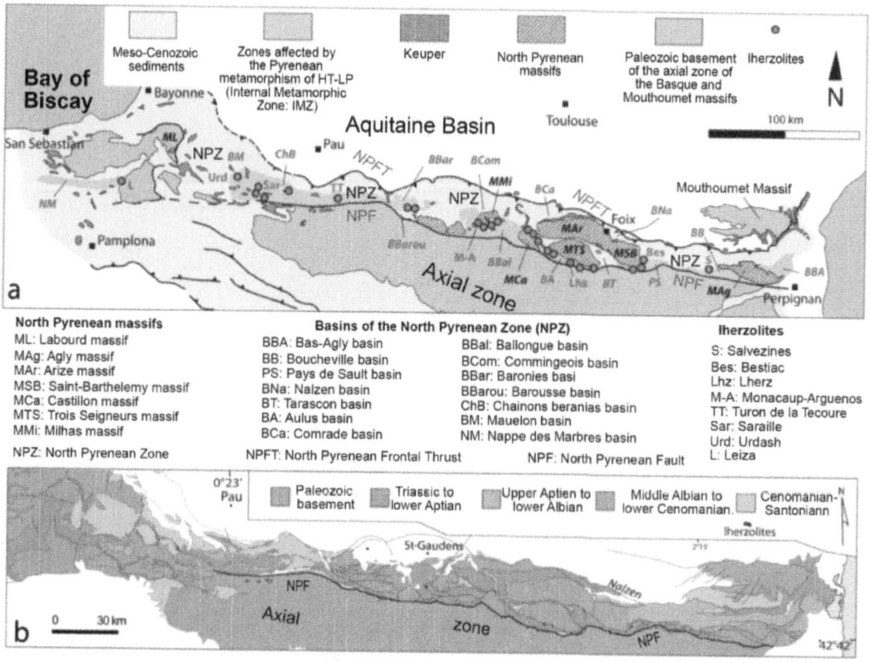


**Figure 5.2.** *The North Pyrenean Zone. (a) Structural map with main geographic labels and location of the Internal Metamorphic Zone (IMZ); (b) simplified geological map (modified from Clerc et al. 2016)*

Keuper facies, with evaporites, clays and ophite intrusions (Azambre et al. 1987) are well represented in some areas where anhydrite was mined for plaster. This is particularly the case at Tarascon-sur-Ariège and Salies-du-Salat, where the Triassic rocks are expelled in the North Pyrenean thrust (Thiébault et al. 1988, 1992). The North Pyrenean Mesozoic series comprises two distinct superposed megasequences. The early megasequence corresponds to the Jurassic-Aptian carbonate platform series (mainly dolomites and limestones) underlain by Triassic rocks that are often incomplete due to frequent tectonic truncations and polyphase diapirism (Figures 5.3 and 5.4). The Triassic strata are known to be unstable during the Jurassic and Lower Cretaceous periods prior to the Cenozoic tectonic inversion, when many diapirs were formed (Canérot 1988; James and Canérot 1988). These different characteristics of the NPZ are particularly well represented in the Bearnais Ranges showing complete and various sedimentary series (Figure 5.3).

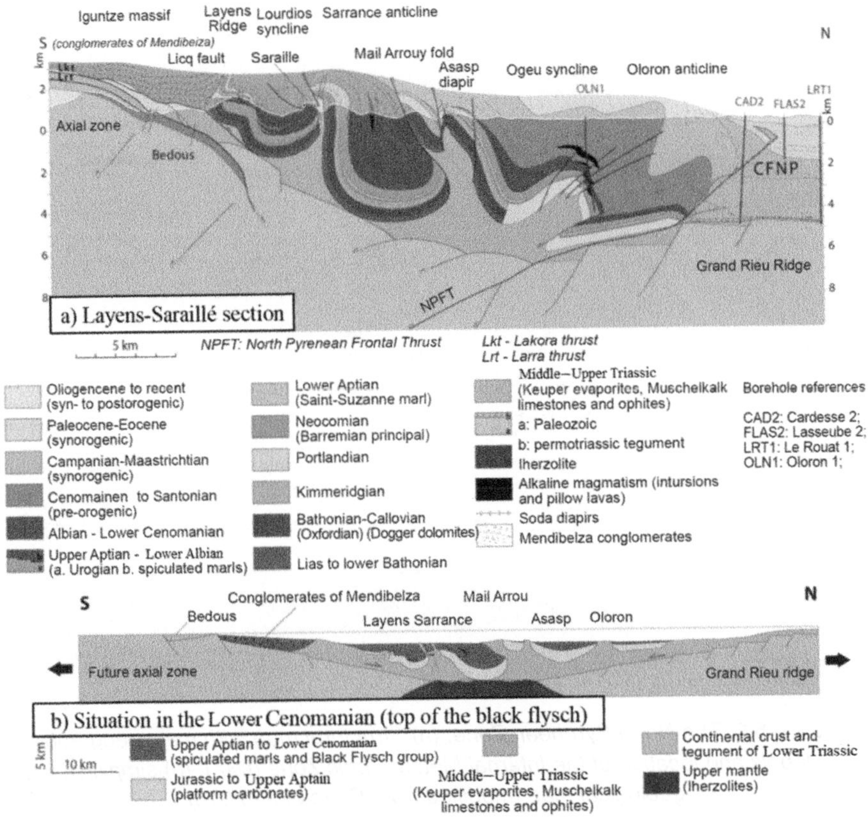


**Figure 5.3.** *The NPZ in the Bearnais Ranges. (a) N–S section of the Bearnais Ranges along the Layens–Saraillé transverse line; (b) reconstruction of the corresponding Albian Basin. Modified from Labaume and Teixell (2020)*

The synrift megasequence is a thick deposit of fine-grained turbidites, often calci-turbidites, and fine-grained sandstones of Albian to Upper Cretaceous age, referred to generically as the Black Flysch Group (Souquet et al. 1985; Debroas 1990) (Figure 5.5). These sediments are considered as symptomatic of the filling of subsiding basins opened during the North Pyrenean rifting period (Debroas and Souquet 1976; Debroas 1978, 1990; Souquet et al. 1985). In detail, the Black Flysch Group consists of three turbidite sequences, each composed of conglomerates, sandstones, siltstones and/or marls for an average thickness of 4 km (Debroas and

Souquet 1976; Debroas 1990). These deposits were emplaced in triangular-shaped basins largely controlled by transtensional faults. During the Cenomanian, subsidence increased and these basins enlarged, and then merged into a single larger basin (Peybernès and Souquet 1984; Debroas 1990). Note that the subsidence of the Aquitaine Basin and the NPZ is strongly asymmetric and decreases toward the east (Dauteuil and Ricou 1989; Brunet 1991).

**Figure 5.4.** *Some landscapes from the NPZ*

LEGEND FOR FIGURE 5.4.– *(a) The Trois-Seigneurs massif seen from the Freychinède lherzolites; (b) the Mail Arrouy range (Béarn); (c) the Saleix valley (Ariège), on the right (south) the axial zone, on the left (north) the metamorphic series of the Aulus Basin, at the bottom of the valley, the granulites; (d) the prerift Jurassic cover at the Trône du Roy summit (Béarn) (photos by Lagabrielle).*

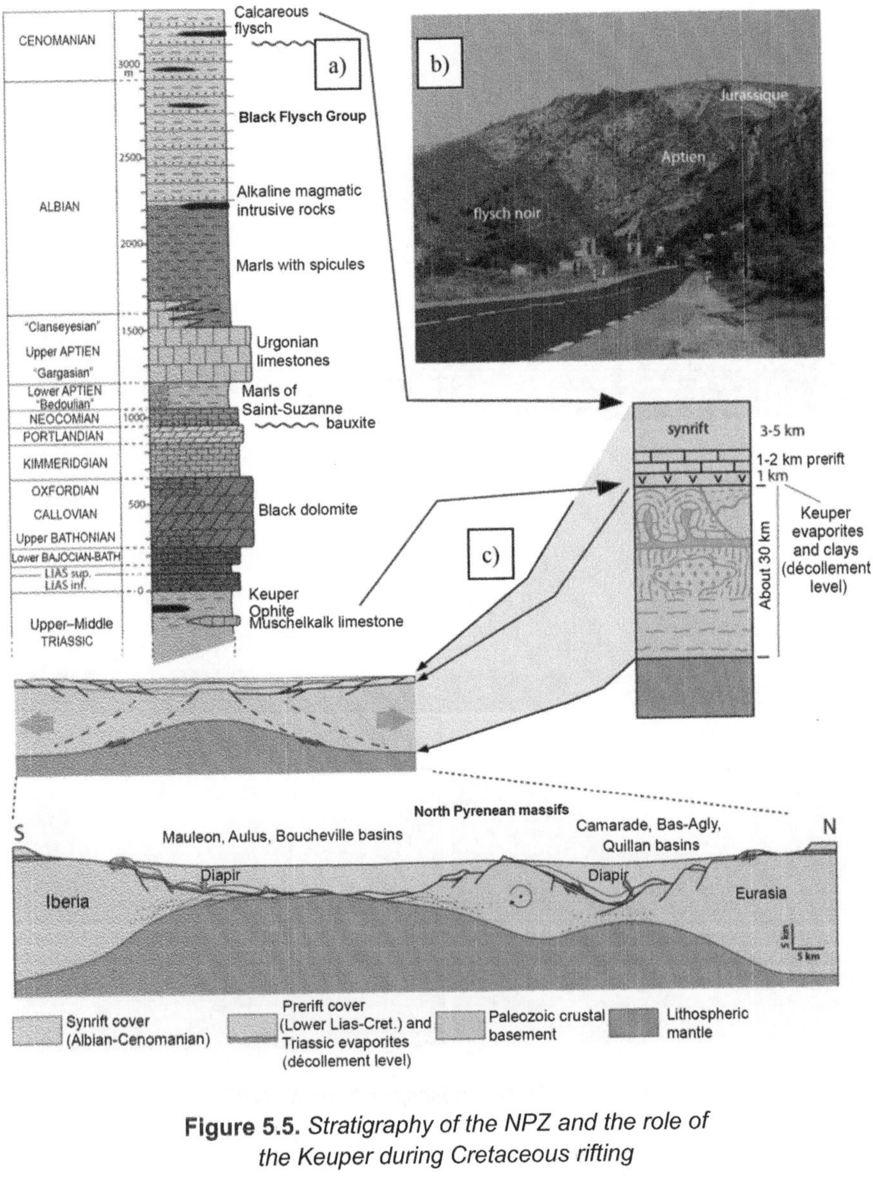


**Figure 5.5.** *Stratigraphy of the NPZ and the role of the Keuper during Cretaceous rifting*

LEGEND FOR FIGURE 5.5.– *(a) Log of the sedimentary series in the Béarn Ranges (modified from Labaume and Teixell 2020); (b) view of the NPZ cover at the reverse flank of the Saint-Paul-de-Fenouillet syncline (eastern NPZ); (c) log of the continental crust of the Pyrenean domain showing the heterogeneous Variscan basement and the Keuper décollement level (see also Figure 5.32).*

The Iberian edge of the Cretaceous Basins is locally well preserved along the future axial zone. The Mendibelza conglomerates containing decimetric pebbles of Paleozoic basement and Albian platform are found west of the Pyrenees; they pass laterally to the Albian flysch beds (Viers 1953; Castéras and Souquet 1964; Boirie 1981; Boirie and Souquet 1982) (Figure 5.3). They mark the edge of a continent in surrection. The slope on which the conglomerates are deposited dips gently northward, which is contrary to a classic passive margin tilted block geometry, where each block should have a southward dip (Teixell et al. 2016). Elsewhere, basin margins are marked by chaotic breccias with angular elements and a matrix of black flysch: the breccias of Castel Nérou in the Ballongue Basin (e.g. Debroas and Souquet 1972) (Figure 5.2).

The first detailed reconstructions of the evolution of the NPZ were published in the 1980s. Based on a structural and stratigraphic study of the western Pyrenees, Ducasse et al. (1986) and Ducasse and Velasque (1988) first proposed a model with symmetrical rifting, as commonly envisioned at that time, then they modernized their model by introducing a shallow dipping fault separating Iberia from Eurasia. This pattern was also adopted by Claude (1990) for the cases of the Ursuya massif and the Mauléon Basin. This was consistent with the ideas of the period when the enthusiasm for detachment faults started with applications to the Atlantic margins and the Jurassic Alpine Ocean (Boillot et al. 1987; Lemoine et al. 1987). Ducasse and Velasque (1988) also reported on the gliding and syn-sedimentary folding of the Cretaceous prerift Arbailles sedimentary units using Triassic salt.

### 5.2.2. *Cretaceous magmatism*

During the evolution of the Cretaceous Basins, local significant magmatic activity occurs in the form of intrusions and flows. The chemistry of the magmatic rocks is always of alkaline affinity. This is consistent with a moderate crustal thinning during narrow rift opening, without oceanization. Since its extension is over nearly 400 km (from the eastern Corbières to the Spanish Basque Country) and its emplacement occurs within a relatively narrow time interval (Albian–Turonian), the Cretaceous magmatism allows us to define the Cretaceous alkaline magmatic province of the Pyrenees (Azambre 1967; Azambre and Rossy 1976; Béziat et al. 1991; Castanares et al. 2001). The largest volumes of magmatic rocks outcrop in Spain, within the Biscay synclinorium, in the provinces of Guipuzcoa and Biscay. Magma was emplaced as submarine volcanic formations interbedded in the Cenomanian-Turonian flysch and a set of hydrothermalized intrusive rocks. The submarine lavas outcrop over nearly 50 km in length, sometimes exceeding 1,000 m in thickness. These include alkaline basalts and more differentiated lavas such as trachytes. The basaltic lavas form *pillow-lava* flows associated with volcanoclastites (pillow breccias, hyaloclastites, pyroclastites).

The chronology of alkaline flows in the NPZ and the Spanish Basque Region is well constrained by field data. The sediments interbedded with the lavas have been dated by microfossils to the Albian-Cenomanian boundary in the Oloron area (Béarn) and between the Latest Cenomanian to Turonian in the Saint-Gaudens boreholes and in the Spanish Basque Region (Azambre and Rossi 1976; Azambre et al. 1981). On the other hand, the youngest intrusive rocks cut the Albian marl in the Corbières, the Albo-Cenomanian flysch in the western Pyrenees and the volcanic series in the Spanish Basque Country. The K/Ar and Ar/Ar dating show that magmatic activity is not continuous: it seems to initiate in the central Pyrenees around 105 ± 1 Ma and to develop between 95 and 90 Ma in the Corbières and the western Pyrenees with final manifestations around 85 ± 1 Ma in the Spanish Basque Region (Montigny et al. 1986a, 1986b). These data must be viewed with caution, as the K/Ar method is no longer considered to be totally reliable. However, they do confirm that magmatism is clearly controlled by the active transtensional lineament at the Iberian-European plate boundary between 110 and 85 Ma. We will see below that these ages are also those of high temperature metamorphism and associated fluids.

### 5.2.3. *Thermometamorphism*

One of the original features of the NPZ is the presence of a narrow belt of high-temperature low-pressure (HT-LP) metamorphism affecting the Mesozoic pre- and synrift sedimentary series at various temperatures. This metamorphic belt (also called the Internal Metamorphic Zone (IMZ)) forms a 300 km long and 5 km wide belt, generally flanking the North Pyrenean fault (Figures 5.2 and 5.6).

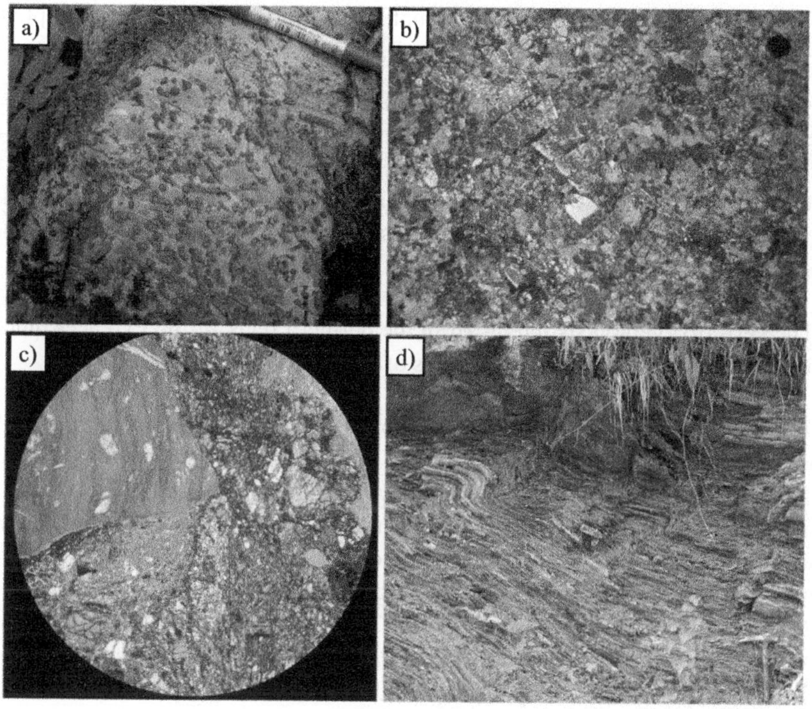


**Figure 5.6.** *Pyrenean metamorphism. (a) Scapolites in the Triassic of the Plagnaux valley (Ariège); (b) scapolite seen in thin section (polarized light analyzed); (c) left foliated marble clast in the Lherz microbreccia (polarized light analyzed) (see Figure 5.12); (d) folded schistosity in the Béarn black flysch (photos (a), (b) and (c) by Lagabrielle and (d) by Corre)*

Studies from the 1970s to the 1990s clearly show that at the scale of the NPZ, a zonation appears in the intensity of thermometamorphism. Four areas can be distinguished (Choukroune and Séguret 1973; Choukroune 1976; Bernus-Maury 1984; Golberg 1987; Golberg and Leyreloup 1990):

– in the east, the Boucheville Basins, the Sault country and the Aulus rift show the highest metamorphic grades (up to 600°C and beyond);

– in the center, the region of the Ballongue and Baronnies troughs show high metamorphic grades (400–450°C, locally higher than 550°C) and decreasing toward the west;

– to the west, in the French Basque Region and in the Béarn ranges, the lowest metamorphic grades are observed with temperatures of the order of 250–350°C maximum. The marbles are devoid of symptomatic minerals (scapolites) except very locally (Très Crouts);

– further east, in the Spanish Basque Region Basin, the thermal anomaly is concentrated in the Nappe des Marbres, on the southern slope of the belt, against the Paleozoic massif of Cinco Villas. The latter is seen as a relay zone between the western tip of the Mauléon Basin and the eastern tip of the Spanish Basque Basin located further south (Ducoux 2017; Saspiturry et al. 2019a; Lescoutre and Manatschal 2020; Lescoutre et al. 2021).

The very first geologists interested in the northern flank of the Pyrenees quickly pointed out the mineralogical peculiarities of the IMZ.

The Pyrenean metamorphism was studied in great detail by Lacroix in the 19th century (Lacroix 1894, 1895a, 1895b, 1897), and then in the 20th century, it became the subject of a preliminary analysis by Castéras (1933) in the context of multidisciplinary geological studies of the northern slopes of the Pyrenees. This was followed by a monograph by Ravier (Thèse d'État in 1959), completed by more specific studies few decades later (Golberg and Leyreloup 1985; Golberg et al. 1986; Dauteuil et al. 1987; Golberg 1987; Dauteuil and Ricou 1989). Precursors such as Casteras (1933) and Ravier (1959) assumed the Middle Cretaceous age of the Pyrenean metamorphism by noting that it affected the Mesozoic sediments, but was lacking in the top of the Upper Cretaceous sequence and within the entire Cenozoic sequence.

In the early days, a pre-Cenomanian phase was evidenced based on geological observations and was considered as the earliest tectonic event of the NPZ (Castéras 1933; Durand-Delga 1965). This phase corresponds in fact to a major paleogeographic change evidenced in the entire Pyrenean-Basco-Cantabrian domain. However, initially, this event was wrongly considered as a compressional episode giving rise to a cordillera running along the floor of the flysch trench. For Ravier (1959), the thermometamorphism of the NPZ relates to a distending pre-Cenomanian crustal fracturing, which was a visionary conception. In 1976, in his Thèse d'État, Choukroune recognized the extensional origin of the pre-Cenomanian phase and emphasized, following a series of reconstructions of the NPZ Basins, that this phase did in fact correspond to a generalized NS crustal extension initiating in the Albian times.

During the thermometamorphism, the silico-carbonated sediments of Liassic age are transformed into black schists with thin layers of quartzite, and the less pure limestones locally give marbles with symptomatic minerals (scapolites, muscovites, actinotes, hornblendes, albite, orthoclase) (Figure 5.6). The presence of scapolites underlines the influence of fluids expelled from the Triassic evaporites that are rich in chemical elements such as Na, Cl and B (allochemical metamorphism or metasomatism). Indeed, a common Jurassic limestone does not have a sufficient concentration of chlorine and sodium to allow the formation of scapolites and it is necessary to consider the contribution of these elements by active fluid circulation. Noncalcareous Triassic sediments, such as interbedded clays within evaporite sequences, are transformed into the so-called "micaceous hornfels" mainly composed of scapolite, albite and phlogopite associated with talc-chlorite schists (Ravier 1959; Ravier and Thiébault 1982). The dolomites give cargneules and dolomitic breccias. Ophite intrusions are locally transformed into green hornblende scapolitites in the case of severe metamorphism (Azambre et al. 1971). Jurassic and Lower Cretaceous limestones and dolomites are changed into marbles and dolomitic marbles valued for their properties in construction and decoration. These marbles are still actively exploited in several quarries and have been exported to different continents, even to North America where they adorn the exterior and interior of many buildings.

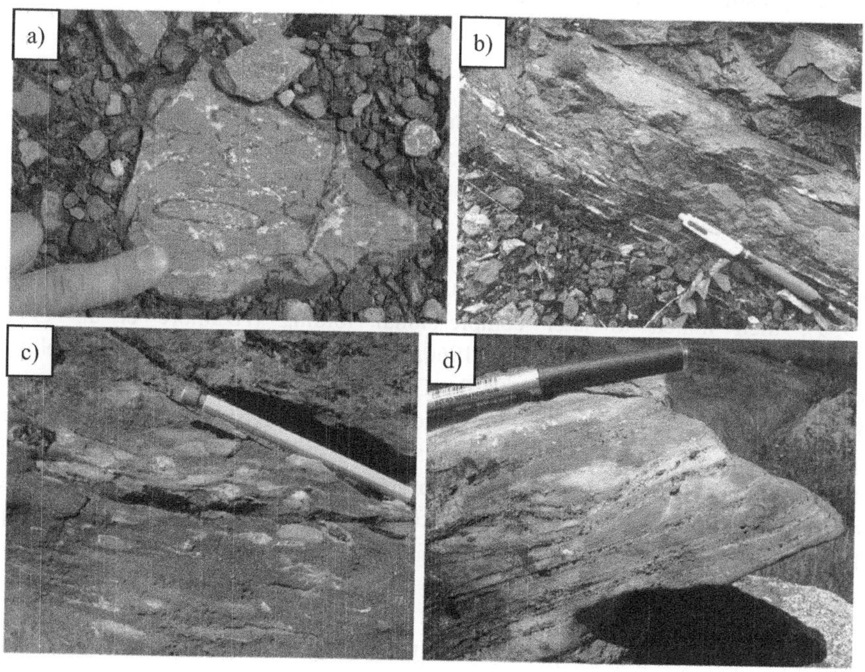


**Figure 5.7.** *Marbles of the NPZ. (a) Flattened Rudists in the Urgonian marble of Saraillé (Béarn); (b) foliated Aptian marble at Sarrance (Béarn) (the white bands are ancient Rudists); (c and d) foliated marbles of the Saleix valley (Ariège). Note the small flattened lamellibranches in c (images by Corre (a), Lagabrielle (b, d) and Clerc (c))*

In the eastern part of the IMZ, exhibiting high metamorphic grade, the layers of the black flysch group acquire a slaty cleavage or take on the appearance of black hornfels where the deformation is weak. In the Bas-Agly syncline, in the Boucheville syncline as well as in the Pays de Sault, the flysch of lower Albian-Cenomanian age (Debroas 1987) shows neocrystallizations of Ca-plagioclase orthose, albite, biotite, muscovite, scapolite and chlorite (Ravier 1959). Aligned in the foliation of the carbonate schists are numerous small crystals of metamorphic Ca-clinopyroxenes (diopsides) with pressure shadows, the presence of which is symptomatic of high-grade thermometamorphism (Golberg and Leyreloup 1990) (Figures 5.6 and 5.7).

Ravier (1959) was correct in linking the thermometamorphism of the NPZ to an early crustal extension period, even though he focused on its isochemical character. We shall see later that this view, which is opposed to that of an allochemical metamorphism where fluids actively transport metamorphic reactants, was erroneous. For Choukroune (1976) and Souquet et al. (1977), the Pyrenean metamorphism is not related to extensional tectonics, but is supposed to occur later in relation, with the convergence that begins in the Upper Cretaceous.

Choukroune and Mattauer (1978) then reinterpreted metamorphism as a consequence of crustal thinning, but with a strong thermal inertia, shifting its effects to the compressional period. This ambiguity in the pre- or synorogenic character of the thermometamorphism is due to the fact that some symptomatic minerals such as scapolites appear with ante-, syn- or post-tectonic *habitus* depending on their location in the IMZ (Choukroune 1976). This led several authors to the concept of a polyphase metamorphism, initially a burial metamorphism as claimed by Ravier (1959), then linked to later compressional folding (Azambre et al. 1971; Choukroune 1976). Therefore, in the 1960s to 1980s, the Pyrenean metamorphism was not universally associated with the Cretaceous rifting episode. Like Ravier, Bernus-Maury (1984) was able to demonstrate the existence of non-oriented metamorphic minerals (pre-deformational), proving that the Cretaceous metamorphic episode had taken place before compression. The definitive postulate will come with Vielzeuf and Kornprobst (1984), then Golberg and Leyreloup (1990), who firmly associated crustal thinning with the thermal metamorphism of the NPZ.

At the end of the 1970s, radiometric dating of minerals confirmed the Mid-Cretaceous age of the Pyrenean metamorphism.The first radiometric data obtained with K/Ar, Ar/Ar and Rb/Sr methods on mineralogical assemblages with scapolite, hornblende, biotite, phlogopite, diopside, Ca-plagioclase and orthoclase show that the micas and amphiboles in the NPZ metasediments crystallized between 87 and 98 Ma (Azambre and Rossy 1976; Albarède and Michard-Vitrac 1978; Golberg et al. 1986; Montigny et al. 1986a, 1986b; Golberg and Maluski 1988).

> The study of metamorphic equilibria has led to considerable advances in the knowledge of the pressure-temperature conditions of Pyrenean thermometamorphism. In the Boucheville syncline, east of the Pyrenees, a thermometer based on calcite-dolomite equilibria gives maximum temperatures of the order of 600–620°C (Golberg and Leyreloup 1990). Pressure calibration is more difficult to obtain, but there is considerable evidence that Pyrenean metamorphism is of the low-pressure type, requiring a burial depth of only a few kilometers. In the same region, Golberg and Leyreloup (1990) drew up the first temperature map of the metamorphism in the eastern part of the IMZ. The map shows very large variations, inconsistent with a regional metamorphic event in a chain uplifted under compressional conditions. For the first time, these authors clearly propose a link between thermometamorphism, local geothermal uplift and continental rifting involving very strong crustal thinning. An analogy with the Salton Sea situation at the junction of the Gulf of California oceanic opening and the San Andreas fault is proposed. Therefore, by the end of the 1990s, it was clearly established that high local thermicity in the IMZ is indicative of an anomalous geothermal gradient, only consistent with isothermal mantle uplift in the context of strong crustal thinning.

**Box 5.2.** *Calibrations of the NPZ thermometamorphism and decisive progress (Golberg and Leyreloup 1990)*

### 5.2.4. *Fluids and metamorphism*

Due to the large heterogeneity of isograds and isothermal zones along the NPZ, a link was quickly made between the strong thermal gradients and fluid circulation in an extensional setting (Golberg and Leyreloup 1985; Golberg et al. 1986; Dauteuil et al. 1987; Golberg 1987; Dauteuil and Ricou 1989). For Dauteuil and Ricou (1989), models of heat transfer by simple conduction in a rift do not account for a metamorphism restricted to such a hot and narrow zone. They invoke a circulation of hot fluids actively transferring heat from the thinnest crustal zone to the surface and circulating along the future North Pyrenean fault, which is regarded as a major transform zone between the Eurasian and Iberian plates.

In addition to the strong metasomatic imprint within the NPZ Mesozoic sediments, we must also consider the metamorphism and the fluid circulations that affect the Hercynian material of the North Pyrenean massifs and the edge of the axial zone during the Cretaceous extension. In the North Pyrenean massifs, regional-scale fluid flows are responsible for the metasomatic alteration of considerable volumes of rock, a phenomenon highlighted since the 2000s (Boulvais et al. 2007; Poujol et al. 2010). One of the major international references for this type of fluid-rock interaction is found in the Saint-Barthélémy massif. The Trimouns talc-chlorite deposit (Chapter 7), mined in the largest quarry in the world, results from hydrothermal alteration along a shear zone separating migmatites, micaschists and a dolomite-rich Paleozoic sedimentary cover inherited from the Variscan evolution. The time span of this hydrothermal activity along a major Cretaceous normal fault responsible for part of the crustal thinning of the massif is estimated at about 15 Ma between 112 Ma and 97 Ma (Schärer et al. 1992; Fallourd et al. 2014; Boutin et al. 2015; Quesnel et al. 2019).

At the same time, during the Middle Cretaceous, the Variscan basement of the NPZ and the edge of the future Axial Zone underwent widespread penetrative albitization. The example of the Salvezines massif shows that the fluids responsible for desilicification, K loss and Na and Al enrichment of the Paleozoic rocks actively circulated during 20 Ma between 117 Ma and 98 Ma (Boulvais et al. 2006, 2007; Poujol et al. 2010). The regional scale of these metasomatic reactions and their synchronous character with the Trimouns-type talc-chlorite mineralization should be emphasized (Chapter 7).

### 5.2.5. *Lherzolites*

The marbles and flyschs of the NPZ are associated with about 40 outcrops of lherzolites. These rocks originate from the Earth's mantle and normally underlie the continental crust at depths of 30 km (Figures 5.2 and 5.8). The lherzolite massifs are small bodies (less than 3 km long) that outcrop in association with Mesozoic sedimentary series flanking the North Pyrenean fault from the east (Salvezines massif) (Monchoux 1970) to the west in the Basque Region (very small Leiza body) (DeFelipe et al. 2017) (Figure 5.2). These spinel lherzolites are different from one massif to another and have the great interest of displaying the whole diversity the mantle facies encountered separately within the ultrabasic enclaves sampled by the basalts across entire Europe.

As early as 1880, the mineralogist Alfred Lacroix devoted several publications to the description of these ultrabasic rocks. He tried to understand the interactions between lherzolites and the surrounding sediments or metasediments and discussed their origin (Lacroix 1894, 1895a, 1895b, 1897). In 1917, he published the results of a first complete study of the Pyrenean peridotites in which he defined the major petrographic types: *lherzolites, cortlandites, ariegites, hornblendites, lherzites* and *avézacites*. Then the lherzolites of the NPZ were the subject of several exhaustive mineralogical studies in the framework of doctoral theses. Avé-Lallemant

(1968) devoted his works to the rocks of Etang de Lherz and to the study of their petrofabrics, then Monchoux (1970) was interested in the totality of the NPZ and provided a detailed sheet dedicated to each peridotite body, finally Conquéré (1977) concentrated on Ariège bodies, in particular on the massif of Etang de Lherz (Figures 5.9 and 5.10).

**Figure 5.8.** *Some emblematic lherzolite massifs of the NPZ (lhz: lherzolite outcrops) (images by Lagabrielle)*

Due to poor knowledge relative to the nature of the subcrustal envelopes and in the ignorance of the petrogenetic processes related to mantle lherzolites, harzburgites and pyroxenites, the first mineralogists of the 19th and 20th centuries encountered difficulties to explain the presence of the lherzolites within the NPZ. For Ravier (1959), the breccias accompanying the Lherz peridotites show affinities with the kimberlite pipes of South Africa. He deduced that the lherzolites of the NPZ are ultrabasic magmas emplaced in the sediments of the NPZ and responsible for the thermometamorphism. He related these eruptions to the Cretaceous volcanic and hypovolcanic events of the NPZ.

**Figure 5.9.** *The lherzolites of Lherz in their natural setting*

LEGEND FOR FIGURE 5.9.– *(a) In this picture of a countryside art installation in 2007, we can see in the frame the dark cliff of lherzolite (lhz) covered by scree at its feet, to the right, outside of the frame, the Etang de Lherz; (b) view on the scarp of the lherzolites. Looking south from the granulites of the Port de Lherz (Lerz path). On the right (west), the mount Béas, where the calcareous-lherzolitic breccias outcrop (images by Lagabrielle).*

The context changed with the advent of plate tectonics theory. As the debate around the significance of ophiolites gained momentum on a global scale, Avé-Lallemant (1968), Conquéré (1971a, 1971b, 1972, 1973, 1977) and Gaudichet (1974) accepted that the Pyrenean lherzolites were derived from the upper mantle and were emplaced in a solid state within the NPZ. For Avé-Lallemant (1968) and Conquéré (1973), these represented a case of *alpine-type peridotite* and thus of solid-state intrusion, in opposition to the ultramafic rocks of the Tethyan ophiolites regarded as part of large laccolites described in Greece by the geologists of the French school.

At the end of the 1990s, mantle rocks petrologists showed that the Pyrenean lherzolites correspond to an old, depleted subcontinental mantle that was reimpregnated by tholeitic magmas during a first uplift event under a thinning crust with lithospheric conditions and equilibration at 800–900°C (Bodinier et al. 1988, 1990, 2004; Downes et al. 1991; Fabriès et al. 1991; Reisberg and Lorand 1995; Le Roux et al. 2007). For some authors, this reequilibration is contemporary with the genesis of the granulites paragenesis in the North Pyrenean massifs during the Permian (Pin and Vielzeuf 1983). The peridotites are often highly serpentinized (except at Etang de Lherz) and are crossed by veins of various types: garnet pyroxenites, gabbros, amphibolites, albitites (Conquéré 1971a, 1971b; Conquéré and Fabriès 1984a, 1984b; Bodinier et al. 1987; Vétil et al. 1988). Some of these veins were emplaced during the Mid-Cretaceous, as revealed by several chronological studies. Indeed, Ar/Ar and Sm/Nd ages obtained on amphiboles from the Lherz and Caussou bodies cluster around 110–105 Ma (Henry et al. 1998). Very similar ages have been obtained for corundum albitites from the Urdach–Les Pernes massif (Pin et al. 2001, 2006; Monchoux et al. 2006).

**Figure 5.10.** *The lherzolites of Etang de Lherz seen from the top of Mount Béas (image from Clerc)*

| The data acquired in the 1970s and 1980s were gathered by Vielzeuf and Kornprobst in a seminal article published in 1984 in which they proposed a geodynamical link between the lherzolites and granulites emplacement. This new concept is based on the geology of the Cretaceous Aulus Basin, which contains the Lherz body and was formed during a strong crustal thinning between the future Axial Zone and the continental block of the Trois-Seigneurs massif. The Cenozoic inversion of this basin brought together the deep-seated units exhumed during extension (peridotites and granulites) and allowed their solid-state tectonic intrusion into the metamorphic sediments of the basin floor. This paper synthesizes the concepts circulating in the 1980s–1990s on the meaning of both the NPZ and IMZ. |
|---|

**Box 5.3.** *Integration of geological constraints on crustal thinning and lherzolite uplift (Vielzeuf and Kornprobst 1984)*

Fabriès et al. (1991, 1998) provide detailed petrological and geothermobarometric data of the Pyrenean lherzolites. They propose a catalogue of the constraints imposed in order to build models of mantle exhumation at the Iberian-European boundary. In particular, they show that spinel lherzolites in the west and east of the range differ by their texture and by the relative abundance of harzburgite and garnet pyroxenite layers. In the west, the peridotites reached lithospheric conditions by passive cooling followed by a period of equilibration at the base of the lithosphere (1,050°C and 15–18 kbar), and then by a nonadiabatic ascent in two stages: from 60 km to 25 km deep, then from 25 km to 15 km and less. The first stage is related to Permian postorogenic extension and the second stage, which is accompanied by cooling to 600°C, is attributed to the opening of the Albian-Cenomanian Basins and associated tectonic denudation. Vissers et al. (1997) described low-temperature (600°C) shear bands in the poorly serpentinized lherzolites of the Turon de la Técouère (NPZ of Béarn, see Figure 5.8), which, according to the authors, reflect plastic deformation related to the upwelling of the lherzolites below the axis of the Albian Basins.

### 5.2.6. *The issue of polygenic breccias*

Breccias, corresponding to clastic formations with angular elements, are very often associated with the lherzolite outcrops of the NPZ. They have been the subject of numerous descriptions by pioneer geologists such as Lacroix and Ravier in particular. Curiously enough, however, apart from a few articles in the 1980s and 2000s, they have been little described and the exhaustive petrological and mineralogical analysis of these formations has not been completely carried out.

At Etang de Lherz (or Lers), the main peridotite body is located within a clastic sequence rich in breccia levels reworking peridotites, scapolite-foliated marbles, and rare gabbro and continental basement clasts. These are the internal and external breccias of Conquéré (1977) and the calcareous-lherzolitic breccias of Ravier (1959) and of Ternet et al. (1997) (Figures 5.11 and 5.12). This breccia complex is cartographically associated with an E–W band of ductilely deformed Jurassic and lower Cretaceous marbles (calcareous mylonites), in tectonic contact with the continental basement of the Trois-Seigneurs massif.

Lacroix proposed two interpretations of the Lherz breccias. In 1892, he pointed out the presence of Jurassic carbonate and lherzolite elements and on the existence of lherzolite gravels of detrital origin, demonstrating the anteriority of the lherzolites with respect to the breccias and the sedimentary nature of the latter. In 1895, he prefers to consider the intrusive origin of the lherzolites and the action of fumaroles to explain the appearance of specific mineralogical assemblages. He proposes then that the breccias result from "friction".

**Figure 5.11.** *A common outcrop of calcareo-lherzolitic breccias visible in the valley of Paumères near Etang de Lherz*

LEGEND FOR FIGURE 5.11.– *The progressive interbedding between two clastic materials of exclusively ultrabasic composition on one side and exclusively carbonate on the other is clearly visible. This association is possible if we consider the mixing of two debris-flows from sources of contrasted lithology (image from Lagabrielle).*

For Ravier (1959) and for Avé-Lallemant (1968), the Lherz breccias result from explosions due to gases released by the mantle during the solid intrusion phase. This hypothesis is supported in the work of Ravier (1959) by a series of rigorous descriptions of the Lherz outcrops and is based on a comparison with the kimberlite breccias in South Africa. In his interpretation, Ravier considered the lherzolites as "eruptive" rocks, which implied crystallization contemporaneous with the brecciation of the host rock. He emphasized the relationships between the lherzolites and the silica-poor alkaline magmas emplaced during the Cretaceous in the NPZ. He did not recognize the mantle origin and solid state of the lherzolites during emplacement; he was therefore in favor of a magmatic emplacement and wrote, among other things, that "in the collapse zones in the core of the chain, the suite of peridotites is introduced into the sediments during burial" (Ravier 1959, p. 211, translated by the author). In this line, the genesis of the Lherz breccias in relation to an intrusive phenomenon is taken up by Minnigh et al. (1980). These authors propose for the Lherz body an intrusive emplacement in a solid and hot state causing the explosive decarbonation of its sedimentary carbonate environment.

**Figure 5.12.** *The Lherz breccias*

LEGEND FOR FIGURE 5.12.– *(a) Typical association of lherzolite fragments and marbles (metamorphic foliation sometimes with scapolites). The cement of the breccia is a sandstone consisting of mineral debris resulting from the desaggregation of the lherzolite and the various marbles (see microphotograph d); (b) breccias with dominance of lherzolite, the presence of rare elements of marble suggests; (c) graded-bedded marble clasts in a lherzolite sandstone matrix; d) thin-section of a calcareous-lherzolitic microbreccia showing isolated marble grains in an ultrabasic matrix with olivines, clino- and orthopyroxene, spinel and rare serpentinite grains (crossed nicols) (images (a), (b) and (d) from Lagabrielle and (c) from Clerc).*

South of the Lherz massif, analysis of the contact with the sedimentary host reveals that the calcareous-lherzolitic breccias were deposited on the peridotites, as reported three times by Choukroune (1973, 1976, 1980) (Figure 5.13). These observations which represented important constraints on any model of Pyrenean subcontinental mantle emplacement were not universally accepted at the time. However, they were consistent with the report of the presence of reworked lherzolites in the Cretaceous flysch of the western NPZ. Hence, in his thesis, Roux (1983) relates the 1964 discovery by Villanova and coworkers of reworked lherzolite blocks in the Cenomanian flysch north of the Urdach massif (Béarn). He also described ultrabasic sandstones interbedded in the Upper Albian flysch of the same area. In the same period, Duée et al. (1984) and Fortané et al. (1986) described metric and plurimetric blocks of reworked lherzolites in the flysch, west of the Urdach lherzolite. It was thus demonstrated by several independent geologists that, in the western part of the NPZ at least, the lherzolites have been exposed to the seafloor at the bottom or along the borders of the Albian Basins. This fact has never been questioned subsequently (see, for example, Debroas et al. 2010; Canérot 2017a).

### 5.2.7. Conclusion of section 5.2

The low amount of convergence between the Iberian and Eurasian plates is one of the major interests of the Pyrenees. It favors the study of terrains belonging to the edges of basins opened during the pre-orogenic period and weakly inverted, in which the witnesses of extreme crustal thinning and mantle exhumation remain preserved (see the summary in Teixell et al. 2018). Such a situation represents a rare opportunity worlwide.

We recall that at the beginning of the 2000s, geological information on the Pyrenean lherzolite and IMZ issues obtained from stratigraphy, structural geology, mantle mineralogy, alkaline magmatism and thermometamorphism chronology were perfectly consistent with each other. These collectively accounted for the importance of events associated with extensional and transcurrent lithospheric movements occurring at the plate margin. However, the processes linking all of these these events remained understood, including the behavior of the mantle and deep crust at the end of uplift as well as the relationship between these deep layers and the sedimentary formations deposited on the basin floors.

The revival of the geology of the Pyrenees from 2008 onwards occurs when the processes of mantle exhumation at the axis of slow-spreading ridges and at the foot of hyper-extended passive margins around the world are compared to the case of the NPZ. This recent evolution of ideas concerning the mobility of lherzolites is reported in section 5.3. The works dedicated to the Pyrenees represent important contributions to the understanding of crustal hyperextension and subcontinental mantle exhumation processes along the nonvolcanic passive margins of continents. The access to these margins in the current oceans is possible thanks to considerable technological and financial efforts and the presence of a geological analog in mountain belt onshore offers structural and petrological references at much lower costs.

## 5.3. Mechanisms of crustal thinning and mantle denudation in the NPZ – recent field data and modern models

### 5.3.1. *Years 2008–2010: outcrop of mantle rocks at the axis of the future NPZ – evidence and preliminary models*

An important evolution in the concepts of Cretaceous Pyrenean rifting occurred between 2008 and 2010, as a result of independent fieldwork to the east of the NPZ in the Aulus Basin and to the west, in the Mauléon Basin.

In 2008, Lagabrielle and Bodinier proposed that the marble clast and mantle breccias outcropping in the core of the Aulus Basin, in the immediate vicinity of the lherzolites at Etang de Lherz of the pond (so-called calcareous-lherzolitic breccias of Ternet et al. (1997), see above) result from the desaggregation of submarine reliefs formed partly from an ultrabasic basement exposed, along a major shallow dipping normal fault (Figure 5.13). It must be pointed out that Choukroune (1973, 1980) had already mentioned that the lherzolites are reworked within detrital sedimentary formations in the form of debris flows, gravels and graded-bedded sands (Figures 5.11–5.13).

For Lagabrielle and Bodinier (2008), the ultrabasic rocks represent the subcontinental mantle exposed at the bottom of the Cretaceous Aulus Basin during the final stages of rifting. The main evidence resides in the mapping of the formations surrounding the lherzolite coupled with microscopic observation of the numerous exposed breccia-types. The model referred to a mechanism that had been evidenced 20 years before at the axis of slow-spreading oceanic ridges (see the synthesis in Lagabrielle et al. (1998)) and previously proposed for the Alpine Ocean as early as in the late 1980s (Lemoine et al. 1987; Lagabrielle and Cannat 1990). For the first time, the term "detachment fault" was used to characterize the major structure along which the mantle rises up from beneath the continental crust. In the hypothesis of Lagabrielle and Bodinier (2008), the exposure of the lherzolites results from the extensional Cretaceous event. This was in contrast to the hypothesis of Choukroune (1976) where the exposed mantle might be part of ophiolites included in the Variscan basement.

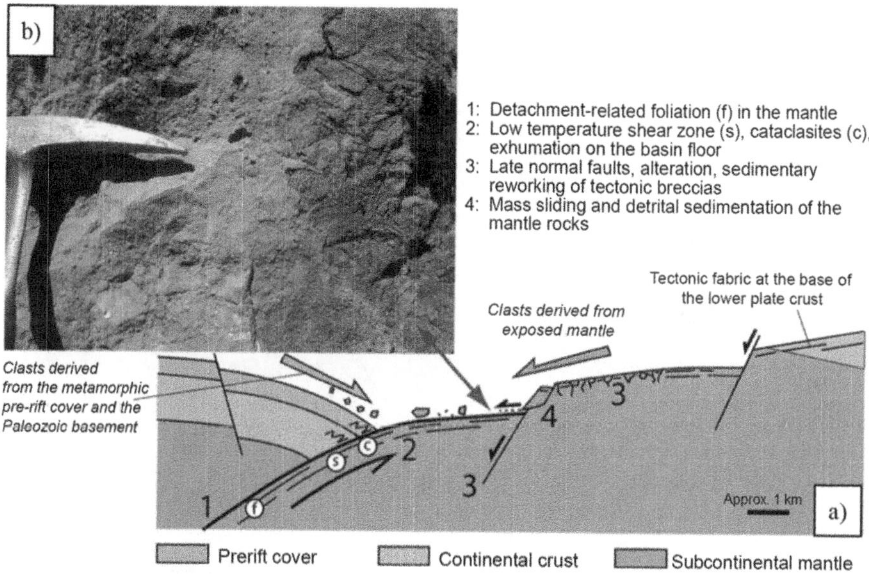


**Figure 5.13.** *Exposure of the mantle of the NPZ through a major shallow dipping normal fault (extensional detachment)*

LEGEND FOR FIGURE 5.13.– *(a) The first diagram showing the role played by a crust-mantle detachment to account for the geological setting around Etang de Lherz lherzolite body (modified from Lagabrielle and Bodinier 2008); (b) photograph of the stratigraphic contact between the lherzolites and the polymict microbreccias in the southern part of the Lherz massif (image by Lagabrielle).*

Jammes et al. (2009) proposes a viewpoint very similar to that of Lagabrielle and Bodinier (2008), and also make use of a detachment fault to explain the extreme thinning of the continental crust and the upwelling of the mantle under the Mauléon Basin. The study is based on analysis of seismic profiles and mapping of units from the Cretaceous Basin that opened from the Aptian to the Cenomanian and that were tectonically inverted during the compression initiated in the Santonian (Andrieu et al. 2021; Issautier et al. 2022). According to Jammes et al. (2009), tectonic evolution during crustal "hyperextension" (this term appears here for the first time in the NPZ literature) brings prerift sedimentary units into direct tectonic contact over the mantle. These units represent *extensional allochthons*, that is extensional klippes, which may evolve during the tectonic inversion into fold-thrust structures corresponding to the future Béarn Ranges.

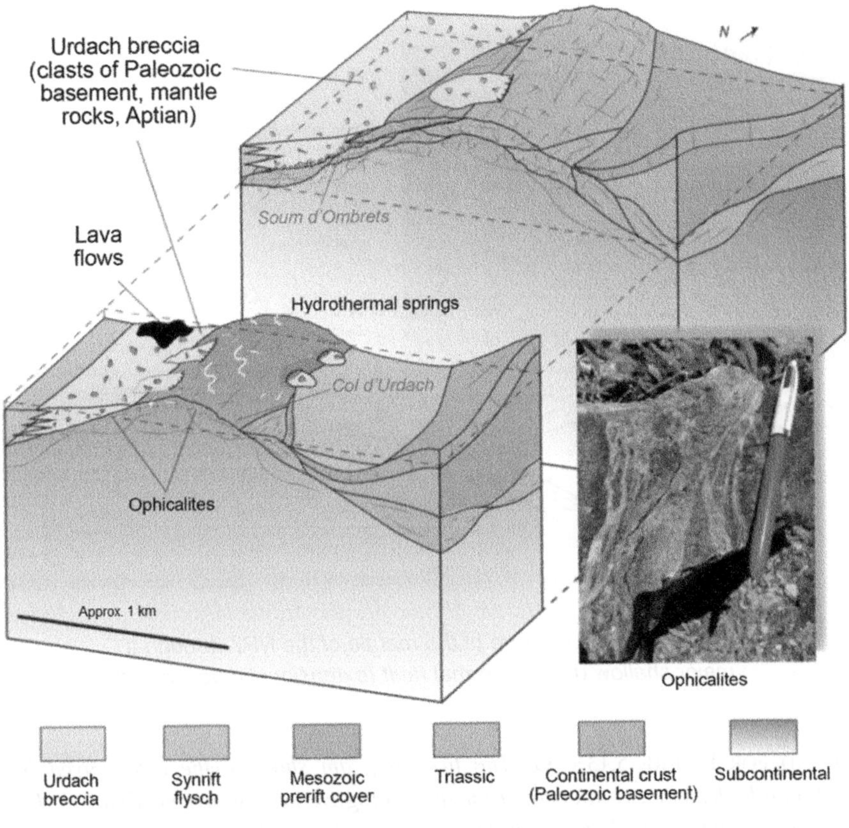


**Figure 5.14.** *Reconstruction of the ultramafic seafloor of the eastern edge of the Mauléon Basin during the Albian times according to the geology of the Urdach–Les Pernes area (modified from Corre 2017; Lagabrielle et al. 2019a)*

Jammes et al. (2009) describes an outcrop now considered as an emblematic witness of the exhumation and denudation of the NPZ mantle. This outcrop corresponds to the ophicalcites of the Bilâtre quarry, north of the Urdach lherzolites (Figure 5.14). These ultrabasic breccias are made up of serpentinites cemented by carbonates and are symptomatic of surficial interaction between the exhumed mantle rocks and seawater during the serpentinization of an ultrabasic seafloor. These are known in the ocean along slow ridges, at low pH and moderate temperature hydrothermal sites, where the mantle is exposed (e.g. Lost City site on the Atlantic ridge, see Früh-Green et al. 2003). At the Urdach site, mantle exposure on the floor of Cretaceous Basins is also confirmed by the presence of a cover of bedded sedimentary breccias reworking Paleozoic basement and mantle rocks, the Urdach Breccia (Debroas et al. 2010; Lagabrielle et al. 2019a) (Figure 5.15). The basal breccias are cemented by silica-rich deposits deriving from fluids expelled from the Triassic evaporites (Ntémé et al. 2020) (see also Figure 5.27). For Debroas et al. (2010), mantle outcrop does not occur along a detachment as proposed by Jammes et al. (2009) and Lagabrielle et al. (2010, 2019a), but instead along a N–S vertical fault that separated this area from the Mauléon Basin, the so-called Barlanès fault.

Lagabrielle et al. (2010) completed the model developed at Etang de Lherz by adding constraints obtained around various lherzolite massifs in the central and western Pyrenees. They thus defined S-type lherzolites, whose contacts with associated sediments are of sedimentary origin, and T-type lherzolites, whose contacts with the Mesozoic series and Paleozoic basement are tectonic in origin. They highlight for the first time the mobility of the prerift cover and emphasize the major role of the Triassic as a *décollement* level of the Mesozoic cover during rifting.

The Lherz breccias, to date, have continued to attract attention due to the presence of clasts of Mesozoic metasediments, mainly marbles, that have undergone thermometamorphism and ductile deformation. The exposure of the mantle rocks in the NPZ during the Cretaceous times was accompanied by the desaggregation of the entire prerift metamorphic cover, from the Triassic to the Cretaceous layers, including the Jurassic and Liassic metamorphic marbles. However, the works in the years 2008–2010 stumbled on this question of thermometamorphism and did not integrate it into the models and interpretations of the evolution of the Pyrenean rift. Therefore, in their preliminary drawing, Lagabrielle and Bodinier (2008) (Figure 5.13) simply point out that ductile deformation affects the prerift cover during crust-mantle detachment activity, but they do not propose any explanation to the presence of marble mineral clasts in the Lherz breccias. The model of Lagabrielle et al. (2010) also did not address this issue, as it envisioned unit slip in an "empty" basin without flysch cover. Jammes et al. (2009) did not consider the issue of the Pyrenean metamorphism.

In order to better constrain mantle exhumation processes, Clerc et al. (2012) focused on the detailed lithostratigraphy of outcrops south of Port de Lherz. Around the Lherz body, the composition of the breccia clasts varies from a 100% carbonate pole and a 100% ultrabasic pole (Figures 5.11 and 5.12). The breccia matrix is invariably composed of grains deriving from the desaggregation and reworking of the mantle rocks, serpentinized or not (olivine, pyroxenes, spinels, serpentines). This ultramafic assemblage is associated with fragments of marble, often foliated, bearing imprint of the Pyrenean HT-LP metamorphism (Figure 5.12). Ultramafic microclasts sometimes form millimeter-thick beds, showing slumps and internal laminations. By place, at the edge of the Lherz massif, the carbonates invade the serpentinized peridotite. The latter then displays the typical aspect of ophicalcites as shown by Clerc et al. (2014). The observation lherzolite-rich clastic levels at significant distances from any ultrabasic rock body implies a transport far from the source, by sedimentary processes, as already suggested by Choukroune (1976). Note that clasts of peridotite breccias and ultramafic litharenites are also reworked in these formations (sedimentary cannibalism), indicating sedimentation, followed by rapid diagenesis in an active tectonic environment.

**Figure 5.15.** *The Urdach breccia*

LEGEND FOR FIGURE 5.15.– *(a) Clastic level interbedded in the Upper Albian flysch. It is an assemblage of poorly sorted angular clast of continental basement mylonites and serpentinized and talcified peridotites; (b) breccias at the contact of the Paleozoic mylonite units. These consist exclusively of clasts derived from the disaggregation of the middle and upper continental crust, in this case Ordovician quartzites. The siliceous cement of these breccias is rich in fluid inclusions allowing the study of fluid-rock relationships during mantle exhumation (see Figure 5.28) (images by Lagabrielle (a) and Bergamini (b)).*

Observations very similar to those made around the Lherz ultramafic body are reported from the environment of small lherzolite bodies scattered in the NPZ south of the Saint-Barthélémy massif at Lordat, Bestiac and Caussou (de Saint Blanquat et al. 2016). The authors highlight the presence of peridotites in the form of clasts of all sizes, from millimeters to hectometers, embedded in debris flows also reworking fragments of the prerift cover. The degree of serpentinization is highly variable from grain to grain, implying that sources evolved differently during mantle uplift as well as on the floor of the Cretaceous Basins. Large amounts of small phyllites of either detrital or metamorphic origin are found in the Lherz and Bestiac breccias. This suggests that the breccia underwent a late stage of Cretaceous metamorphism. This crucial issue was addressed by Ravier (1959), who concluded that the Bestiac breccia cement is indeed metamorphic. De Saint Blanquat et al. (2016) indicate that this feature can be demonstrated in the brecciated series associated with the Bestiac lherzolites. However, this question remains difficult to resolve only by field or thin section observation. For de Saint Blanquat et al. (2016), the polygenic breccias are not related to the Pyrenean compression episodes, but were emplaced during mantle exhumation, or during the last stages of extension, just before the onset of inversion.

At several localities in the immediate vicinity of the Lherz and Freychinède bodies and in the veins crossing the Berqué lherzolites, clastic saphirine-bearing rocks described and analyzed by Monchoux (1972a, 1972b) are exposed. These rocks are made up of well-sorted mineral beds and microbreccias whose composition, surprisingly, associates an ultramafic assemblage with a singular assemblage of probable meta-Keuper debris (phlogopite, anthophyllite, enstatite, hornblende, etc.) and meta-ophites (scapolite and green hornblende). Uzel et al. (2020) demonstrated the presence of anhydrite and apatite inclusions rich in F, Cl and Sr within the saphirines and enstatites. They propose that the saphirine-bearing rocks were most likely detrital sediments deposited at the base of the breccia assemblage, and fed by the desaggregation of meta-evaporites that reached the maximum temperatures of thermometamorphism during the exhumation of the mantle (Figure 5.16).

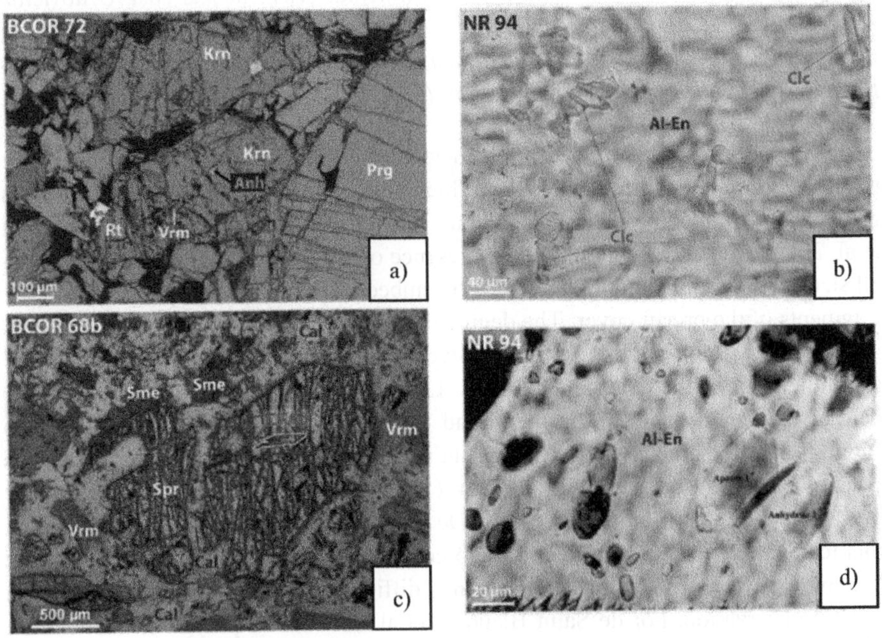


**Figure 5.16.** *Microphotographs of some saphirine rocks from Etang de Lherz (Uzel et al. 2020)*

LEGEND FOR FIGURE 5.16.– *The examples show solid inclusions suggesting that these rocks are the product of high-temperature metamorphism of Keuper evaporite sediments. (a) Anhydrite (Anh) inclusion within a kornerupine (Krn) crystal (Prg: paragonite; Vrm: vermiculite; Rt: rutile; backscattered electron image); (b) inclusions of clinochlore (Clc) in an aluminous enstatite (Al-En) (crossed nicols); (c) inclusion of chlorapatite (Cl-Ap, highlighted in blue) in a saphirine crystal (Spr) (Sme: smectite; Cal: calcite; reflected light); (d) coalescing inclusions of anhydrite and chlorapatite within an aluminous enstatite crystal (crossed nicols).*

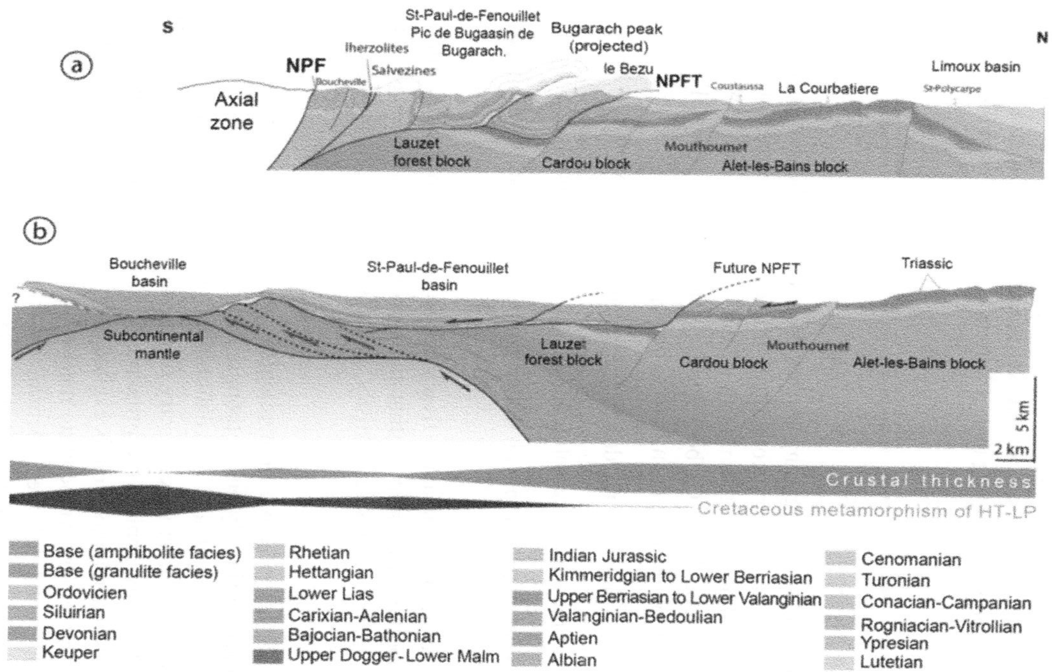


**Figure 5.17.** *A mantle exhumation model at the axis of the East Pyrenean rift according to Clerc (2012) and Clerc et al. (2016). This model (in b) is built on geological data from the NPZ section at the level of the synclines of Boucheville and Saint-Paul-de-Fenouillet shown in (a)*

Based on the data recalled earlier, successive geological models of Cretaceous Basin evolution have been proposed by Clerc et al. (2012, 2016) (see Figure 5.17), Clerc and Lagabrielle (2014), de Saint Blanquat et al. (2016), Lagabrielle et al. (2016), Saspiturry et al. (2020b) and Uzel et al. (2020). In these models, mantle was exhumed on the floor or along the edges of many small basins during the rapid Albian to Cenomanian Iberian drift. In section 5.3.5, these geological models are compared to digital models, including that of Duretz et al. (2019) (see Figure 5.33).

### 5.3.2. *Mantle exhumation and thermal evolution in the core of the Pyrenean rift*

As reported previously, petrography studies show that the thermometamorphism of the IMZ results from strong geothermal gradients concentrated right above ongoing crustal and lithospheric thinning (Vielzeuf and Kornprobst 1984; Golberg 1987; Dauteuil and Ricou 1989; Golberg and Leyreloup 1990). The following key points regarding thermometamorphism have to be kept in mind:

1) it runs along the transtensional domain represented by the NPZ;

2) it is coeval in time and space with alkaline magmatism (about 110–85 Ma);

3) it displays close relationship with the lherzolite massifs and the granulite crust scales;

4) it affected the Albo-Cenomanian flysch deposits during, or shortly after, their formation. An important constraint is that the Cenomanian and younger material exhibits metamorphism of variable intensity, but always less than that of the older sediments it overlies.

In the absence of effective mineralogical markers, accurate estimates of temperatures reached during Pyrenean thermometamorphism are difficult to obtain using the classical method of calibration by geochemical equilibria. Paleo-thermometry by Raman spectrometry of carbonaceous material (RSCM) fills this gap. This technology applies the principle of Raman Stokes diffusion, which shows light changes when a molecule is excited by a monochromatic laser. It allows us to estimate the maximum temperature reached at the peak of the metamorphism by considering the width of the spectrum of the carbonaceous material. This technique is particularly adapted to carbonate metasediments, a dominant lithology in the NPZ. Beyssac et al. (2002) proposed the first calibration of this geothermometer with a maximum temperature range of 330–650°C. Later on, Lahfid et al. (2010) extended its applicability to the maximum temperatures reached by rocks during diagenesis and low-grade metamorphism.

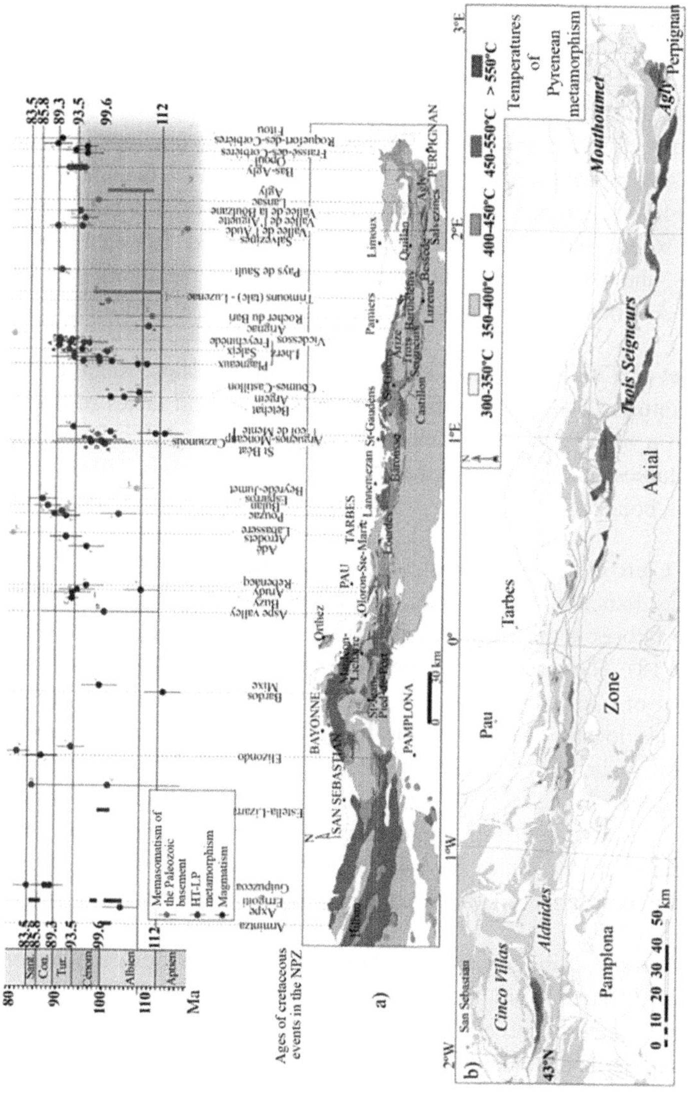


**Figure 5.18.** *(a) Age distribution of Pyrenean metamorphism and magmatism from Clerc et al. (2015); (b) Pyrenean metamorphism map from Ducoux (2017) based on the map from Clerc et al. (2015)*

In the NPZ, RSCM geothermometry was implemented by Lahfid and Clerc (Clerc 2012; Clerc et al. 2015). All maximum temperatures were qualitatively estimated by comparing the obtained spectra to those of the Glarus Region (Lahfid et al. 2010). Clerc et al. (2015) compiled a first updated map of metamorphism at the scale of the entire NPZ based on more than 100 RSCM measurements and a compilation of classical mineralogical equilibrium data (Figure 5.18). Obtaining new radiometric ages allowed authors to confirm that thermometamorphism developed during the 110–90 Ma interval (Albian to Turonian). Clerc et al. (2015) note the clear link between high-temperature zones and mantle outcrops and confirm the long-established zonation showing, (1) in the west, a low metamorphic area (Béarn) with widely developed flysch basins, (2) in the center, a domain of strong metamorphism where basins are narrower and thightened between Paleozoic basement blocks, and (3) in the east, a domain of very strong metamorphism largely affecting synrift sediments of the black flysch.

Lagabrielle et al. (2016) provided a detailed map of the Lherz Region showing that high temperatures (500–600°C) affected all lithologies and facies of the prerift sequence, but in a heterogeneous manner. This is inconsistent with homogeneous extensional tectonics and suggests that the prerift sediments were cut into independent thin, boudin-like units.

The map by Clerc et al. (2015) was completed by Ducoux (2017) with 200 new temperature measurements in the carbonates of the IMZ. This map also integrates the Nappe des Marbres south of the Cinco Villas massif and west of the Aldudes massif (Figure 5.18). The maximum temperature value (TRSMC) at the scale of the IMZ remains that of 626 ± 4°C measured by Clerc et al. (2015) in the heart of the Aulus trough. This location represents the highest grade metamorphic basin with numerous TRSMCs above 600°C. The highest value obtained by Ducoux (2018), 625 ± 17°C, comes from the western part of the Mesozoic Sarrancolin sedimentary Basin, south of the Barousse massif where many temperatures are above 575°C. The highest temperatures of the entire IMZ are finally located between the axial zone and the northern Pyrenean massifs in the eastern half of the Pyrenees.

The metamorphic distribution map of Ducoux (2018) provides isometamorphic zones somewhat different from those of Clerc et al. (2015), as they are more homogeneous in the central and eastern part of the IMZ. These zones correspond to a set of basins that have undergone identical metamorphism under a range of temperature between 400 and 630°C. We also see the 400–450°C isograde ends abruptly at the western tip of the Sarrancolin Basin, and only reappears in the Nappe des Marbres Basin. Ducoux (2018) further shows that some areas described as non-metamorphic actually have temperatures between 300 and 350°C, notably in the

Camarade Basin and in the sedimentary covers of the Milhas, Bessède-de-Sault and Cinco Villas massifs.

The Cretaceous thermicity of the Mauléon Basin has been reconstructed by Saspiturry (2019). The compilation of more than 150 TRSMC measurements allows him to evaluate the 3D distribution of the isotherms and their evolution during the inversion phase. The thermal anomaly of the Mauléon Basin is acquired during the Cretaceous hyperextension phase and lasts until the Campanian as evidenced by the Upper Santonian series that suffer relatively low metamorphic temperatures, but which align with this synrift gradient.

Saspiturry et al. (2019b, 2020a) show that the maximum temperature reached by the prerift and synrift series are controled by two independent factors:

1) the heat flow at the base of the hyper-stretched domain;

2) burial under the thick sedimentary pile of the flysch (*burial metamorphism*: as already envisioned by Ravier in 1959).

Temperature paleogradients for the Mauléon Basin were calculated by Saspiturry et al. (2019b, 2020a) from borehole temperature data (Figure 5.19). For the Iberian and European margins, they obtain values between 39°C/km and 46°C/km, slightly higher than those for non-stretched continental domains. The paleogradient of the hyper-extended domain was calibrated from three wells giving maximum values of 57°C/km and 60°C/km. These values are in agreement with those generally accepted in similar rift systems. They are lower than the 80°C/km accepted by Vacherat et al. (2014), but close to those proposed by Chelalou et al. (2016) (70°C/km) for the Boucheville Basin east of the IMZ.

The evolutionary model of Saspiturry et al. (2019b) (Figure 5.19) refers to motion along a large detachment fault with different kinematic and chronological characteristics, in a way similar to the models of Masini et al. (2014) and Hart et al. (2017). Increasing geothermal gradient toward the center of the basin affects the rheological behavior of the crust with predominantly brittle deformation at the edges of the system shifting to ductile deformation at its center, as claimed by the Clerc and Lagabrielle (2014) model. The study by Saspiturry (2019) also demonstrates that the anomalously high geothermal paleogradient coincides with the positive gravity anomaly present beneath the Mauléon Basin. This anomaly reflects the presence of the subcontinental mantle at shallow depth (~10 km), a result of mantle exhumation processes during the Cretaceous hyperextension phase, according to the models by Wang et al. (2016) and Lehujeur et al. (2021).

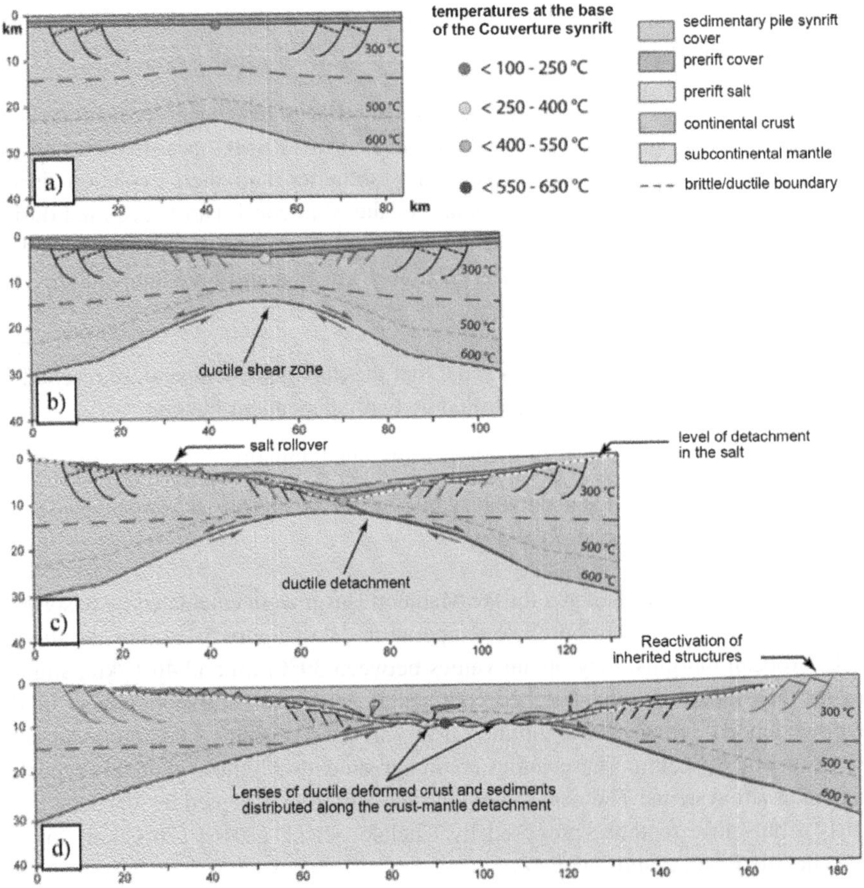


**Figure 5.19.** *Structural and thermal evolution of a Pyrenean type rift with "smooth slopes". This model shows the intervention of steep faults on the external zones of the rift and large detachments (in red) in the center, but it also implies a homogeneous ductile thinning of the crust. It is based on the example of the Mauléon Basin (modified from Saspiturry et al. 2021b)*

### 5.3.3. *Behavior of the cover, crust and mantle during exhumation*

#### 5.3.3.1. *Extensive S0/S1 foliation, detachment of the cover on the Keuper and lateral extraction of the continental crust*

Some attempts to apply classical models of hyper-stretched margin formation based on the study of the Iberian-Newfoundland conjugate margins to the Pyrenean rift have been proposed by Masini et al. (2014), Tugend et al. (2014, 2015) and Lescoutre (2019). In these models, the mechanisms of crustal thinning involve the dismembering of the crust into independent rigid blocks separated by listric faults and detachments (Péron-Pinvidic and Manatschal 2009; Mohn et al. 2012). In the necking zone, where crustal thinning is maximal, tilted blocks slide against each other and occupy the entire slope. In the distal margin zone, the mantle rocks are exposed between rigid blocks of continental crust representing *extensional allochthons*. The prerift cover remains welded to these blocks and undergoes tilting with them. We show below that this "Iberian-Newfoundland" rifting model cannot apply completely to the NPZ case.

Clerc and Lagabrielle (2014) emphasize an important originality of the IMZ. They show that the synmetamorphic S0/S1 foliation well developed in the IMZ marbles has an extensional character. In the ductilely deformed marbles, foliation is marked by the flattening-stretching of objects (fossils, clasts, minerals) and the development of shear zones showing high rates of extension (Figure 5.7). Parallelism between this foliation and the stratigraphy (S0) is better explained by the flattening of the sedimentary pile accompanying its horizontal extension. This is a rare case of coaxial extension parallel to S0 in a mountain range. In the study of the Saraillé massif (Bearnais Ranges), Corre et al. (2016) further show that the marbles and dolomites of the prerift cover behave in a ductile to brittle way and that the S0/S1 synextensional foliation is also parallel to the contact between the mantle basement and the prerift cover. This suggests that the prerift sedimentary pile remained in the center of the basin during extension and that the continental crust was removed horizontally beneath this cover that became mobile relatively and thus appears in an allochthon position (notion of lateral extraction). This concept is crucial to understand the interpretations developed later (Figures 5.20 and 5.21).

Following Clerc (2012), Vauchez et al. (2013), and Clerc and Lagabrielle (2014), the most recent models based on geological studies around the Paleozoic massifs, lherzolites, and metamorphic units of the NPZ and Basque Region account for the absence of brittle structures in the prerift sediments and basement in the vicinity of the exhumed mantle (Mouthereau et al. 2014; Vacherat et al. 2014; Corre et al. 2016; Teixell et al. 2016; Ducoux et al. 2019; Espurt et al. 2019; Saspiturry et al. 2019b; Lagabrielle et al. 2020) along this diffuse plate boundary (Asti et al. 2022).

In this sense, the Pyrenean models contrast with the "brittle" Iberian-Newfoundland-type models discussed above. Considering the differences existing in the thermic evolutions between the east and west of the NPZ in the Cretaceous, Clerc and Lagabrielle (2014) propose reconstructions of the Pyrenean Basins at three longitudes corresponding to three increasing thermal states. They show the existing correlation between zones of high metamorphic grade and extreme stretching of the prerift cover. Exposure of the mantle rocks on the floor of the rift basins together with the evidence of their direct contact (tectonic underplating) at the base of the prerift cover are seen as the result of horizontal lateral pulling of a ductilely thinning crust (lateral extraction). For the first time, the authors emphasize the fundamental role of the Triassic layers in allowing complete decoupling of the cover during rift formation. Subsequently, Clerc et al. (2016) completely improved this model by taking into account the coeval character of Black Flysch Group sedimentation and crustal extension (see Figures 5.32 and 5.33). They explain that ductile synextensional deformation occurred under a cover of sediments deposited in basins with hyper-thinned or absent crust. Therefore, during the Cenomanian-Turonian times, the Upper Albian synrift sediments actively deform as widening of the basins and related sedimentary infill continue.

**Figure 5.20.** *Boudinage of the Lower Jurassic dolomites of the prerift cover of the Agly massif. These structures mark the flattening of the sedimentary pile accompanying its horizontal extension during rifting under temperature conditions of the order of 350°C (Vauchez et al. 2013; Clerc and Lagabrielle 2014). The maximum height of boudins is in the order of 3 m*

Figure 5.22 shows a compilation of geological sections reconstructed across the different basins of the NPZ and the Basque-Cantabrian domains by different authors. All the profiles show clear similarities. Their central part is characterized by the absence of normal fault scarp and smooth slopes. This characteristic profile has been recognized in all basins opened during the Mid-Cretaceous times along the Iberian-Eurasian border (Lagabrielle et al. 2020; Saspiturry et al. 2020b). The smooth slope profile of these basins is consistent with the notion of a sag basin (i.e. lacking steep faults cutting the basin basement).

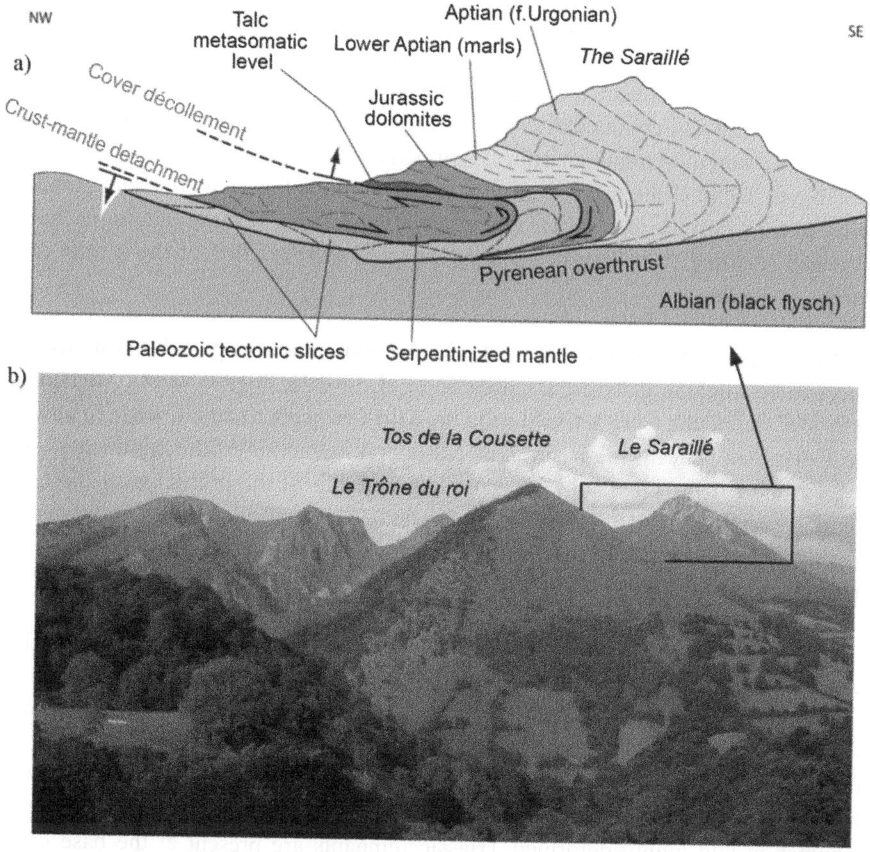


**Figure 5.21.** *The NPZ in the western Pyrenees: The "Chaînons Béarnais" (Bearn Ranges)*

LEGEND FOR FIGURE 5.21.– *Photograph (b) shows the Aptian pyramid of the Saraillé lherzolite (and Paleozoic) massif in the background on the right and the dolomite and limestone scarps of the prerift series (le Trône du roi) on the left. The Saraillé geological section (a) shows the pre-orogenic ductile boudinage of the Paleozoic crust and prerift cover (modified from Corre et al. 2016).*

#### 5.3.3.2. *Crustal lenses, ductile deformation and detachments*

Detailed mapping studies in the western NPZ massifs exposing both lherzolites and small Variscan basement slices welded to the lherzolites are prime targets for testing the Pyrenean rifting model of Clerc et al. (2016). Studies of the Urdach-Les Pernes and Saraillé massifs in the Béarn Ranges by Corre et al. (2016), Asti et al. (2019) and Lagabrielle et al. (2019a, 2019b) show that primary major contacts between mantle and continental crust are accessible to observation. In these massifs where Pyrenean thermometamorphism temperatures did not exceed 350–400°C, it is possible to decipher the present geometric relationships between crust and mantle in order to constrain the mechanical (rheological) behaviors of these lithologies during Cretaceous rifting. It is also possible to describe the evolution of the prerift cover during extension.

Based on descriptions of geologic section, microscope observations and microprobe mineral analyses, Lagabrielle et al. (2019a, 2019b) identify two major extensive faults that were active during the Mid-Cretaceous and allowed (1) uplift of the lherzolites beneath the basin axis, (2) lateral extraction of the continental crust beneath the rift margins and (3) decoupling and slip of the prerift cover along the Keuper evaporites and clays. These two fault zones, located at the crust-mantle boundary and at the base of the allochthonous prerift series, have been named crust-mantle detachment and cover *décollement* (Figure 5.23).

During the Upper Albian, the Urdach peridotites were exposed on the seafloor where they underwent partial carbonation accompanied by calcite crystallization in a network of orthogonal veins (ophicalcites). The serpentinites were partially overlain by the Urdach breccia formed by flows of ultramafic and Paleozoic debris. The mantle rocks and thin lenses of crustal mylonites form the core of a westward dipping fold. The prerift cover has undergone severe basal truncation during mantle exhumation and highly deformed Triassic remnants are present at the base of the detached prerift series. An alkaline spherulitic lava flow is present in the flysch close to the exhumed mantle (Figure 5.14).

In the Saraillé massif, the prerift cover and thin lenses of Paleozoic mylonites are involved in a fold having lherzolites in its core (Figure 5.21). Unlike the Urdach massif, these peridotites were never exposed on the floor of the Cretaceous Basins and always remained under a thin layer of crustal mylonites. Field data allow for the reconstruction of the state of deformation of the lherzolites in the crust-mantle detachment (Figures 5.21 and 5.23). A 20–50 m thick level of tectonic lenses of serpentinites separated by anastomosing shear zones is overlain by an endommagement zone of pyrite-invaded talcchists. The overburden is a multi-decameter thick zone resulting from tectonic brecciation of the Upper Triassic and the site of intense metasomatic transformations.

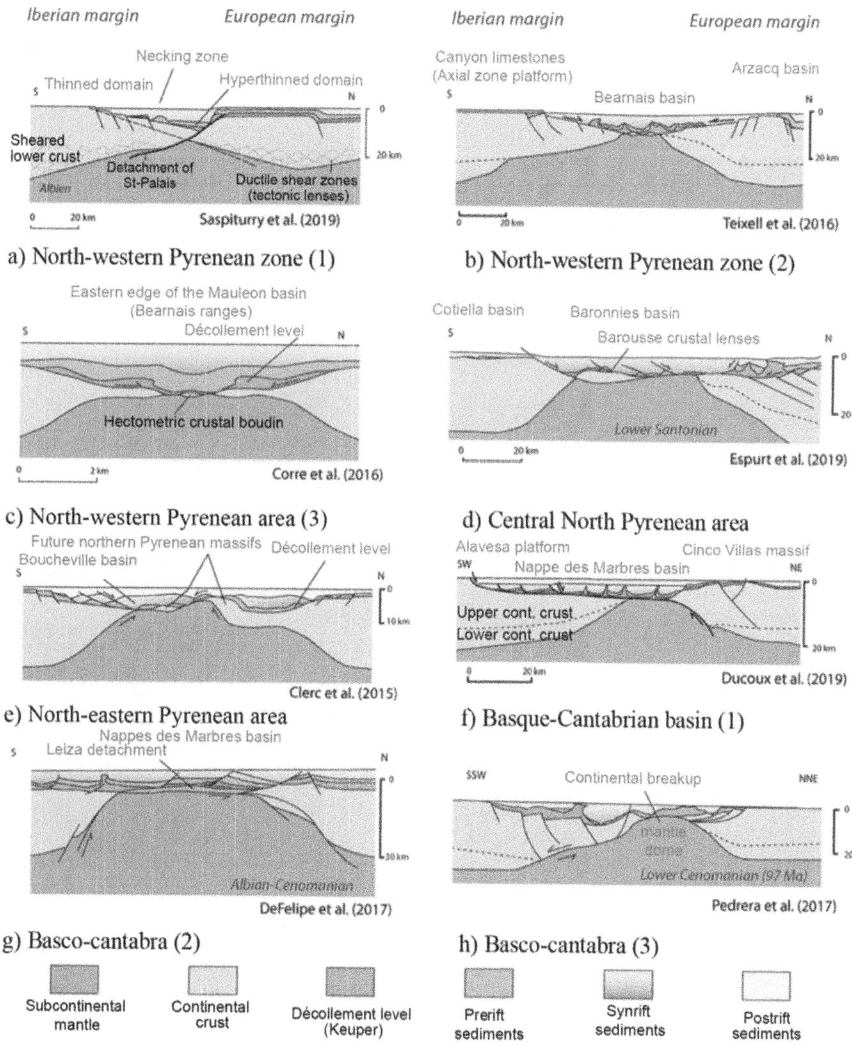


**Figure 5.22.** *Compilation of various profiles reconstructed across the Cretaceous Basins distributed along the NPZ and in the Basque-Cantabrian chain*

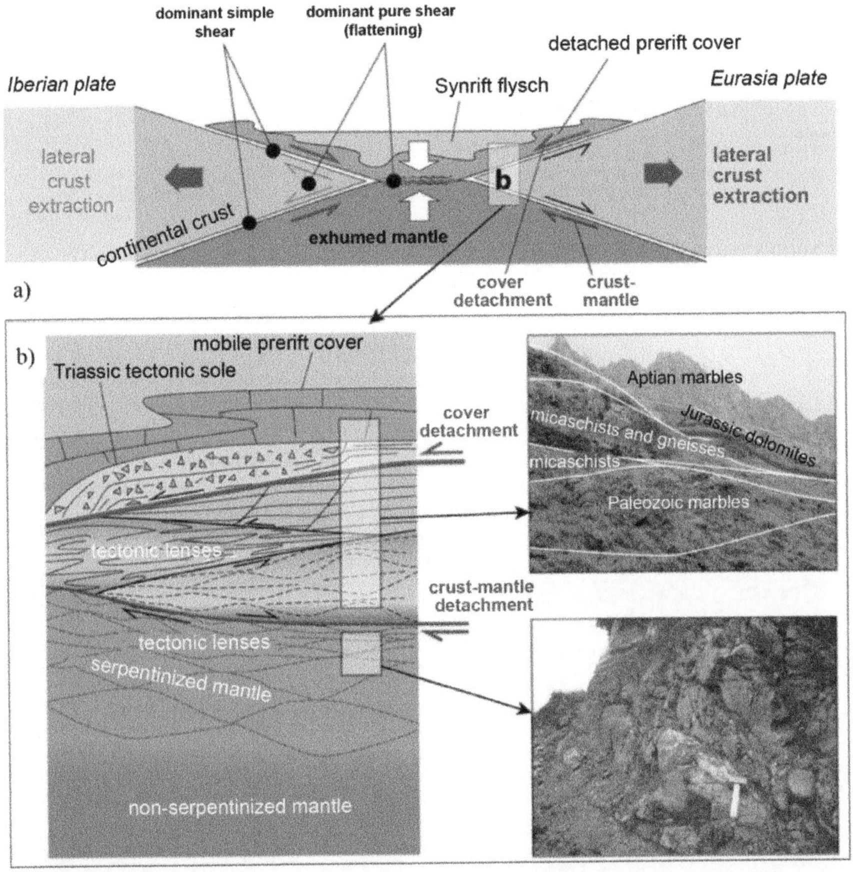


**Figure 5.23.** *Lateral extraction and crustal thinning. (a) Theoretical diagram showing the geometry of major faults allowing lateral extraction of hyper-thinned crust from the center of the Cretaceous Basins of the future NPZ; (b) detail of the area shown in (a) and corresponding photographs taken in the Saraillé massif (modified from Lagabrielle et al. 2019a, 2019b)*

**Figure 5.24.** *Deformation of the continental basement of the NPZ during the Cretaceous*

LEGEND FOR FIGURE 5.24.– *(a) Dated muscovite (red circle) from a mylonitic leucogranite sample in the Urdach massif showing that the Variscan crust underwent ductile deformation during the Albian-Cenomanian rifting as suggested by structural field data; (b) natural light microscopic section of the dated sample showing mylonitic deformation around a feldspar eye (modified from Asti et al. 2019).*

According to Corre et al. (2016), the thin lens-shaped crustal units still welded to the Saraillé and Urdach mantle are remnants of Paleozoic material strongly stretched and dismembered during the Cretaceous extension. In a comprehensive study of this material, Asti et al. (2019) first show that the abundant crustal mylonites in these lenses all exhibit foliation parallel to the contact with the underlying mantle, suggesting ductile deformation contemporary with exhumation. They then show that mylonitization occurred in a temperature range of 350–450°C (greenschist facies) consistent with a structural level in the middle crust. Finally, they demonstrate through Ar/Ar dating of a muscovite from the Urdach mylonites (Monié, Géosciences Montpellier) that this deformation occurred at 105.2 ± 0.5 Ma, that is, during the Albian. These results are consistent at the scale of the NPZ with the dating of the Saint-Barthélémy mylonite zone (Odlum and Stockli 2020) and the shear bands of the Agly massif (Odlum and Stockli 2019; Aumar et al. 2022).

Finally, crustal thinning of the distal margin may have occurred totally during the Cretaceous, but it is very likely that the Paleozoic crust underwent pre-thinning episodes during the Late Variscan Orogenic period. These issues are discussed by Asti et al. (2019) and summarized in Figure 5.25.

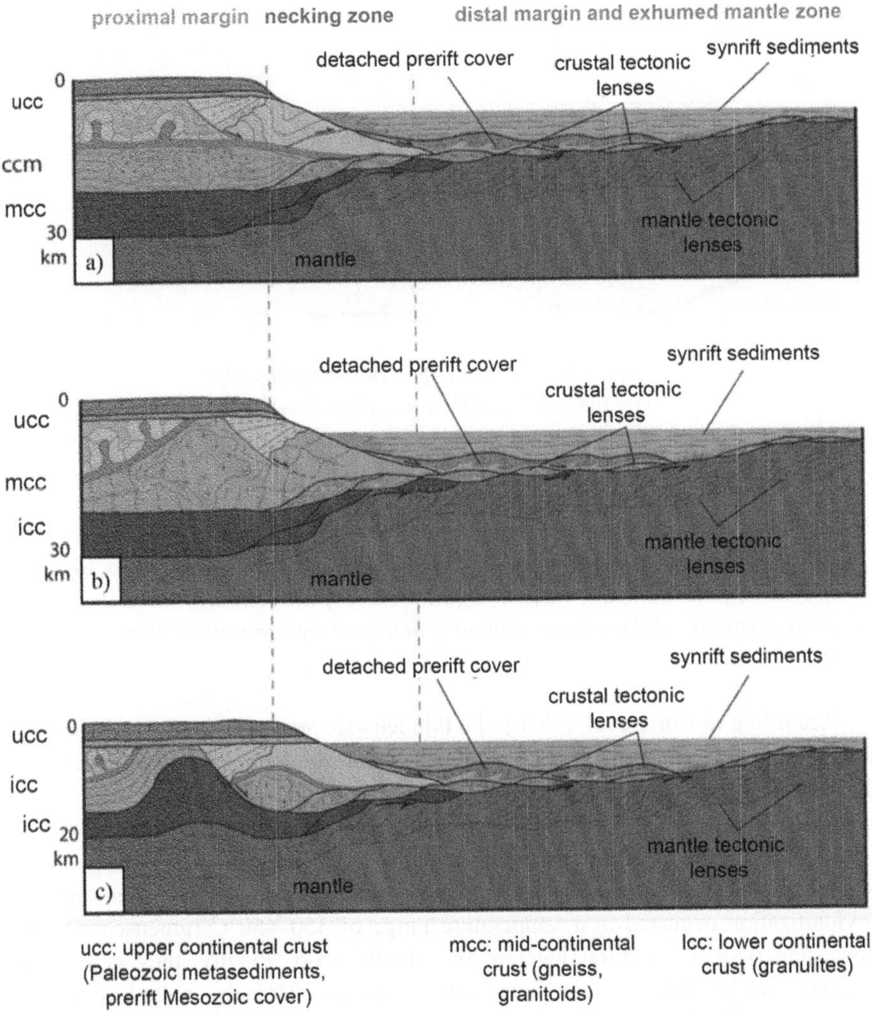


**Figure 5.25.** *Three hypotheses on the behavior of the Variscan basement and Permian tegument of the Iberian margin during Cretaceous rifting of the NPZ (adapted from Asti et al. 2019)*

LEGEND FOR FIGURE 5.25. – *In (a) and (b), the crust has a standard thickness of 30 km at the end of the Permian. In (b), the middle crust was exhumed and denudated at the end of the Variscan orogenesis. In (c), the crust underwent a 10 km thinning at the end of the Paleozoic.*

#### 5.3.3.3. *The outer edges of the Cretaceous Basins of the NPZ*

In the previous sections, we focused on processes active in the deepest, axial part of the Cretaceous Basins, emphasizing ductile synextensional tectonics. This process does not apply to the outer parts of the basins, which remain relatively cold. The top of the margins is preserved in some areas, notably in the west of the NPZ where the Iberian margin outcrops south of the Mauléon Basin, and in the center of the NPZ where breccias exposures are the mark of steep paleo-fault scarps (e.g. Castel Nérou breccias of the Baronnies Basin) (Figure 5.2). In the western part of the range, the European margin is not accessible in the field, but is particularly well known from boreholes and numerous seismic profiles. In contrast, in the eastern Pyrenees, the European margin is exposed south of the Mouthoumet massif where cover décollements and unit truncations accompany the deepening of Cretaceous Basins (Clerc and Lagabrielle 2014).

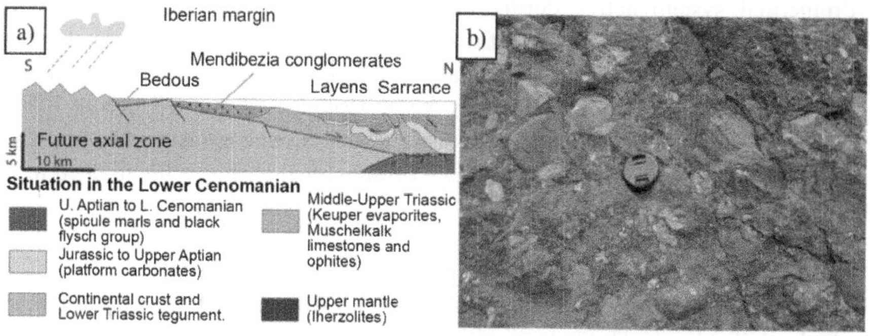


**Figure 5.26.** *(a) Position on the Iberian margin (diagram adapted from a figure by Labaume and Teixell (2020)); (b) photograph of the Mendibelza conglomerates (image by Lagabrielle)*

The works by Saspiturry (2019) and Saspiturry et al. (2021a) show that the upper Iberian margin of the Mauléon Basin is affected by differential vertical movements from the Albian to the Lower Cenomanian times. First, there is tectonic subsidence of the central part of the basin and uplift of its southern margin with emplacement of the Mendibelza conglomerates (Figure 5.26). The sedimentation of the Mendibelza conglomerates is controlled by a 15°–30° northwestward tilt of the base of this margin, causing the cover to slide toward the basin along a *décollement* on the Upper Triassic salt and clays. The European margin, on the other hand, is little affected by Mesozoic cover slip and records less vertical movement. The Middle

Cenomanian–Upper Santonian period is marked by a widening of the deep basin. A shallow carbonate platform develops on both the European and Iberian margins of the Mauléon Basin (canyon limestones). For Saspiturry et al. (2019b), the process of successive tilts results from flip-flop detachment tectonics and confers a final pseudosymmetric geometry to the Mauléon rift (Figure 5.19).

### 5.3.4. *Fluid flow, metamorphism and metasomatism during the formation of Cretaceous Basins in the NPZ*

#### 5.3.4.1. *Crustal thinning, metasomatism and primary brines*

As seen earlier, Boulvais et al. (2007) and Poujol et al. (2010) emphasized the importance of fluid circulations affecting the edge of the Axial Zone and the entire NPZ; they characterize the corresponding metasomatic events as a "huge" hydrothermal system active during the relative drift of Iberia against Eurasia. Subsequent studies have largely confirmed this view and have clarified the origin of the fluids while expanding the time window of their action. The role of the Keuper evaporites was again highlighted and the inheritance of pre-Cretaceous rifting episodes was revealed. This very large-scale flow system contrasts with local fluid circulations, which are possible at the scale of individual basins as highlighted in the Boucheville Basin (Boulvais 2016).

Boutin et al. (2015) provided important chronological constraints on fluid circulation in the continental crust during the Mesozoic using in situ laser ablation methods (ICP-MS, U-Th-Pb laser ablation) on a large assemblage of rare earth minerals from talc-chlorite deposits in the eastern Pyrenees, including the Trimouns site (Chapter 7). They confirm that talc formation did occur during the Aptian-Cenomanian period, but indicated that chlorite formation occurred over a larger span time, from the Permian to the Cretaceous with an episode in the Mid-Jurassic, coeval with the Tethyan rifting and the opening of the Ligurian Tethys. A link between the major Cretaceous talcification event in the Trimouns fault and the calco-sodic metasomatosis episodes already described in the NPZ massifs is again outlined (Boulvais et al. 2007).

In the study of the metasomatism of the northern Pyrenean massifs of Agly and Saint-Barthélémy, Fallourd et al. (2014) complete the studies of calco-sodic metasomatism. They confirm that it is a complementary process to the hydrothermal activity leading to albitization and talc-chlorite mineralization, and affecting metasedimentary rocks as well as Paleozoic granites. In situ laser ablation dating (*ICP-MS, U-Th-Pb* laser ablation) of titanite (sphene) grains formed in albitites show that this episode took place between 110 and 92 Ma. The temperature of the fluids circulating in the crust is estimated at about 550°C, which is consistent with the data obtained in the IMZ in the vicinity of the studied massifs. The study of Fallourd et al. (2014) allow us, once again, to link calco-sodic metasomatism with Pyrenean thermometamorphism and thus with the process of crustal extension and thinning. Fallourd et al. (2014) thus confirm all previous dating, notably that of Schärer et al. (1992) on the Trimouns site.

Quesnel et al. (2019) in turn complete the study of the Cretaceous metasomatism of the NPZ by considering the mineralogy of the Trimouns deposits and their fluid inclusion evolution. Talc-chlorite mineralization is characterized by the presence of fluids trapped in microscopic inclusions within dolomites, quartz and calcites. Stable isotope ratios of the elements C, O, D and Cl and the Cl/Br ratio in the fluid inclusions give results compatible with an origin by primary brines confirming the major role played by the Keuper during the Cretaceous hyperextension episode.

The abundant data concerning Triassic fluids collected between the years 2000 and 2020 in the Pyrenees are gathered in a synthesis by Cathelineau et al. (2021). These authors show the originality of the Pyrenean system where the Triassic constitutes a huge reservoir of brines, mobilized during the rifting and collision cycles (Figure 5.27).

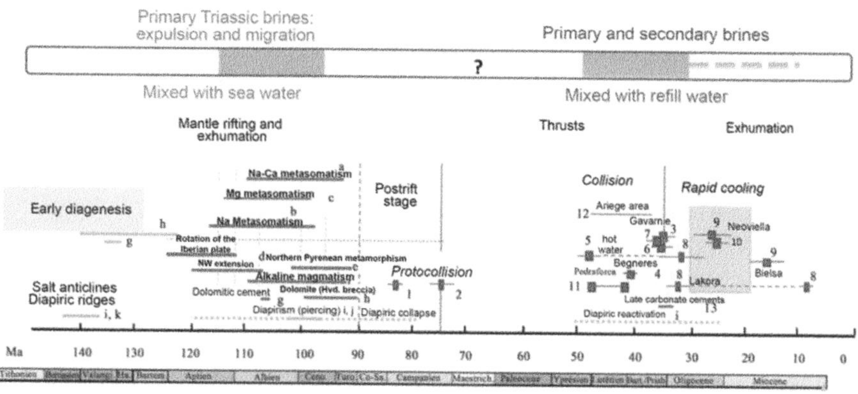


**Figure 5.27.** *Chronological table showing the various types of Triassic-related fluids involved in fluid-rock interactions during the evolution of the Pyrenees (modified from Cathelineau et al. 2021)*

### 5.3.4.2. *Exhumation of the mantle and circulation of fluids*

In the previous section, we focused on fluids that flowed through the continental crust during thinning. Here, we consider fluids that flowed through the two fault zones located at the crust-mantle boundary and at the base of the detached prerift series, respectively: the crust-mantle detachment and the cover *décollement* (Figure 5.23).

Salardon et al. (2016) conducted one of the first comprehensive isotopic and paleothermometric studies of fluids that circulated in the NPZ cover during and after Pyrenean metamorphism. They highlight the influence of Triassic evaporites in the carbonate veins of the Bearnais Ranges, but favor a signature of Cretaceous magmatism. Incerpi et al. (2020), on the other hand, emphasize the role of fluids responsible for the dolomitization and albitization of metasediments in the same region. They emphasize the progressive passage of fluids precipitating carbonates, then silica during the hyper-extensional evolution of this distal paleomargin. In the vicinity of the Saint-Barthélémy massif, Odlum and Stöckly (2020) show the possible links between the crustal fluids at the origin of the talc-chlorite mineralization of Trimouns (along a major normal fault of the European margin) and the mantle fluids resulting from the serpentinization of the exhumed mantle on the floor of the rift of the future NPZ (Figure 5.28).

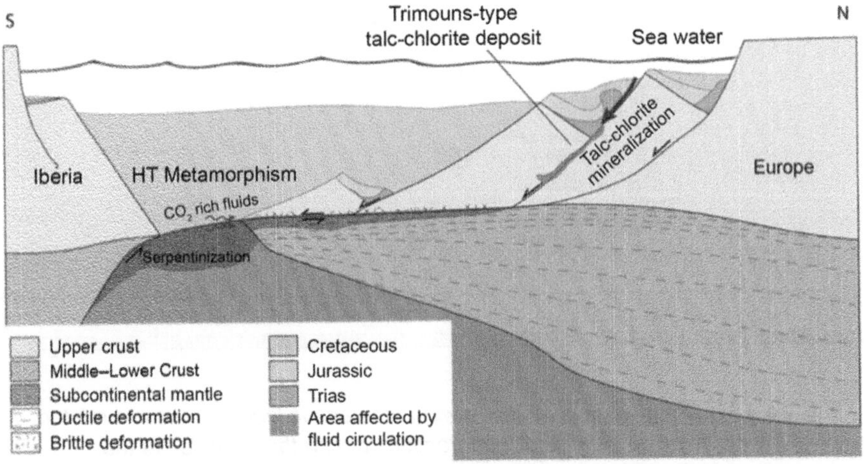


**Figure 5.28.** *The fluid system during the Albian extension in the region of the Saint-Barthélémy massif and the Trimouns talc-chlorite deposit (modified from Odlum and Stöckly 2020)*

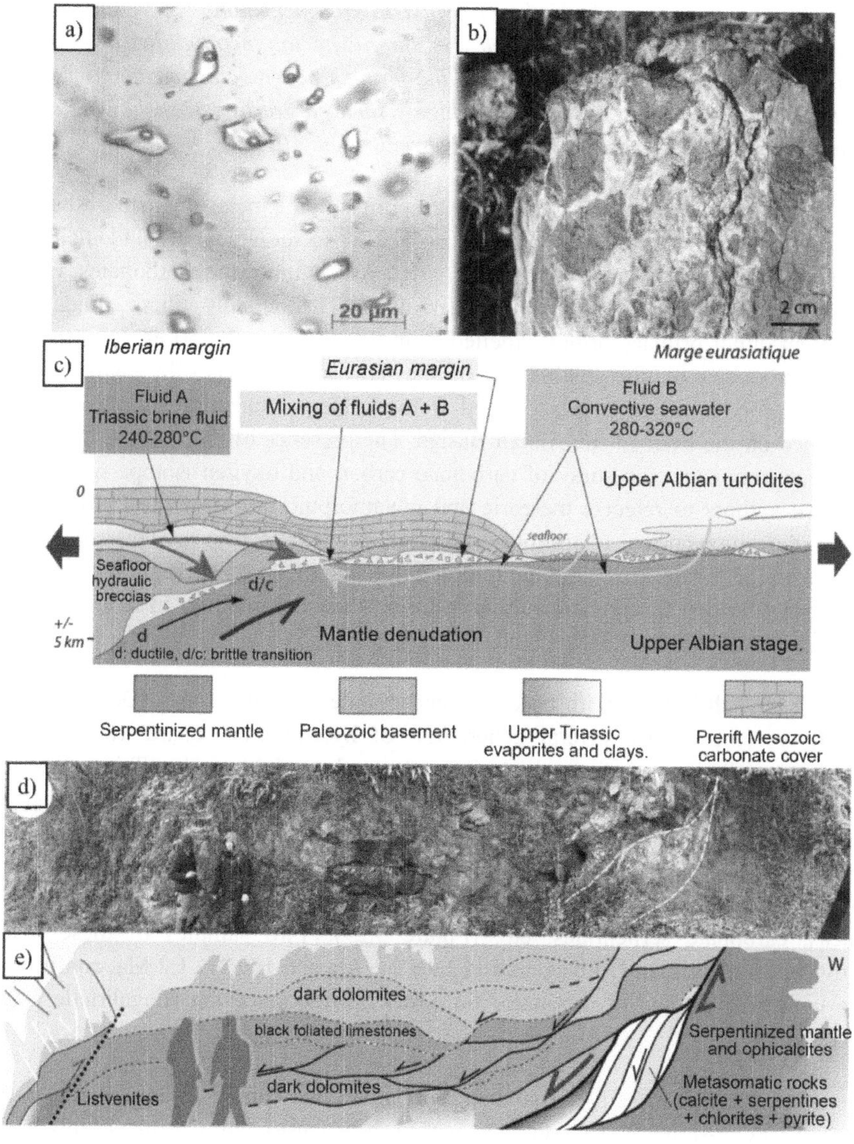


**Figure 5.29.** *Fluid involved in the Albian-Cenomanian crust-mantle detachment in the Urdach massif*

LEGEND FOR FIGURE 5.29. – *(a) Two-phase fluid inclusions in the siliceous matrix of continental basement breccias; (b) monogenic tectonic breccia of Ordovician quartzites cemented by hydrothermal quartz; (c) schematic representation of fluid-rock interactions in the Urdach lherzolite exhumation (modified from Ntemé et al. 2020); (d and e) cross-section of the Urdach cover detachment showing the contact between the lherzolites and the highly sheared prerift series showing strong tectonic lenticulation (modified from Lagabrielle et al. 2019a).*

The major role played by the Keuper evaporites in the metasomatism of the NPZ was initially demonstrated by Corre et al. (2018) in the study of fluid inclusions contained in calcites and quartz in veins cross-cutting the carbonates of the metamorphic cover of the Saraillé lherzolites. Theoretically, these fluids may represent either primary brines expelled from the Triassic evaporites, or secondary brines produced by the dissolution of the Triassic evaporites by diagenetic or metamorphic fluids, or brines derived from the serpentinization of mantle rocks exhumed on the floor of the Albian Basins. The presence of three-phase inclusions with salt cubes and the study of carbonate carbon and oxygen isotope ratios have allowed authors to reject a meteoric and metamorphic origin of the fluids and to favor the intervention of primary or even secondary brines. This study has demonstrated for the first time the remote influence of Triassic evaporites and their migration through the prerift sediments during their deformation in relation to the Cretaceous extension.

The Urdach breccia with basement and mantle mylonite clasts (Debroas et al. 2010) is cemented by several generations of hydrothermal quartz rich in fluid inclusions. Studies of these inclusions show that highly salty Triassic fluids contributed to the precipitation of the cement during several episodes (Ntémé et al. 2020). Fluids of mantle and continental origin circulated actively along the crust-mantle detachment and allowed the crystallization of abundant hydrothermal quartz and lenticular deposits containing quartz, calcite, pyrite, chlorite and Cr-rich micas, carbonate serpentines, etc. (Figure 5.29). Two samples of these rocks provided in situ zircon U/Pb ages of 112.9 ± 1.6 Ma and 109.4 ± 1.2 Ma, confirming the upper Albian age of the major fluid-rock interaction event (Lagabrielle et al. 2019a).

In the Urdach massif, the cover *décollement* also contains metasomatic rocks such as calcite-serpentine associations within shear lenses and listvenites (ultramafic rocks that underwent serpentinization and carbonation with crystallization of chromium-bearing micas). These rocks show large-scale fluid-rock exchange from mantle and crustal material (Figure 5.29).

In the crust-mantle detachment of the Saraillé massif, Lagabrielle et al. (2019b) highlight strong metasomatic alteration by multicomponent fluids that led to the precipitation of quartz, dolomite, talc, Cr-rich chlorite, amphiboles, magnesite and pyrite. These data show the involvement of the mantle in the fluids circulating the base of the prerift cover. Here, the lower Triassic levels are completely transformed into dolomite breccias (cargneules) and chlorite schists, rich in mica and basement clasts. The breccias also contain small bipyramid quartz from the evaporites, sometimes mixed with serpentinized mantle clasts (Lagabrielle et al. 2019b).

### 5.3.5. *Exhumation of the mantle: synthesis of geological data and modeling*

Data collected since 2010 on the lherzolite massifs of the NPZ allow us to propose a reconstruction of the distal domain architecture of the inverted North Iberian margin in the Late Cretaceous. This reconstruction, which applies to all basins up to the European margin, shows a very symmetrical rift domain, dominated by ductile processes in their center as represented by many authors in Figure 5.22 and on the synthesis section in Figure 5.30.

All mineralogical and geochronological data acquired in recent years in the IMZ of the NPZ confirm the strong volume reduction of the Triassic by mechanical crushing and dissolution of evaporites whose chemical elements participate in the massive metasomatism of the continental cover and crust, up to the Trimouns-type outer margin major faults, along which temperatures reached 550°C (Figure 5.28). In the west, the temperatures of the fluids responsible for the metamorphic crystallizations within the extensive detachments do not exceed those of the greenschist facies (about 350°C) and the Paleozoic basement deforms ductilely at an average maximum temperature of 450°C under the prerift cover. This hyper-stretched cover is locally discontinuous and reveals the underlying mantle, as observed at the Urdach site. The mantle undergoes strong hydrothermal alteration with serpentinization and carbonation and receives debris derived from the exhumed rocks around (mantle, basement and rare sediments). The breccias are cemented by quartz having interacted with primary brines.

In the central NPZ, thermal conditions are more severe, as neocrystallization at the expense of Triassic material occurs at temperatures that allow crystallization of assemblages including anthophyllites, spinels, clinopyroxenes and saphirine instead of the chlorite and talc assemblage. In these regions, the mantle rocks were not exposed everywhere on the seafloor.

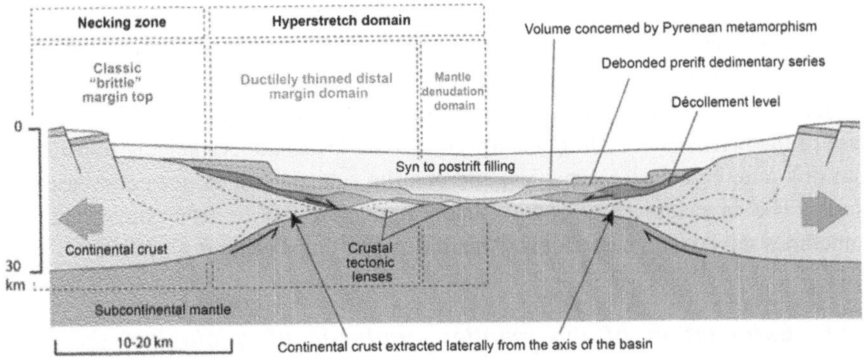


**Figure 5.30.** *A model of the architecture of the Cretaceous Basins of the NPZ. The rift is symmetrical, dominated by ductile processes in its center and the development of boudins (or tectonic lenses) of continental basement lying over the exhumed mantle. This architecture led to the concept of smooth-slope basins (after Lagabrielle et al. 2020)*

At Lherz, the presence of ophicalcites demonstrates that the mantle formed the bottom of the Aulus Basin, which is confirmed by the composition of the Lherz breccias rich in lherzolite debris. Note, however, that the model of exhumation with complete mantle denudation is not absolutely demonstrated at Lherz, unlike at Urdach, where geological and micropaleontological data are unavoidable. Indeed, there is still uncertainty about the age of the Lherz breccias, which may date from the very early stages of compression as well. The fact that they apparently do not rework post-Albian flysch is consistent with early emplacement, prior to compression and most probably during Cretaceous extension as proposed by Clerc et al. (2013) and Lagabrielle et al. (2016). However, this point is not fully resolved in 2020.

The basis of the new concepts underlying the models presented in this chapter is, on the one hand, the lateral extraction of the continental crust allowing denudation of the mantle and, on the other hand, the extraction of the metasediments during further sedimentation of the flysch. The particularity of the models lies in the presence of a thick layer with very low resistance constituting a detachment level allowing the pre- and synrift crust to remain in the center of the basin and to maintain the thermal anomaly. There is thus a strong rheological contrast between the brittle and ductile parts within the model. The continental crust is ductilely deformed in the central zone, allowing the individualization of crustal blobs that will form the future North Pyrenean massifs. The mantle is partially denuded at the basin axis, only covered by very discontinuous basement lenses. The prerift cover is plastically deformed, individualizing very stretched lenses affected by thermometamorphism.

These concepts are taken up by Aumar et al. (2022) and applied to the case of Cretaceous deformation of the Agly massif, which they confirm by dating monazites and micas at 127–94 Ma. Their study emphasizes the crustal deformation in a series of tectonic lenses at all scales, making these structures a universal model for the thinned crust of the NPZ (Figure 5.31).

Finally, by contrast to the last federating model of Vielzeuf and Kornprobst (1984), in these recent models, the emplacement of the peridotites is no longer regarded as the result of a tectonic intrusion (cold or hot) within the Mesozoic sedimentary sequences during a transpressive or compressive phase. Recent works concerning the Pyrenees on a larger scale are based, partly or totally, on this modern concept. For example, Tugend et al. (2014) consider the issue at the scale of the Bay of Biscay Region and propose that the mantle is exposed over domains representing considerable areas at the foot of the Armorican and Cantabrian margins (Chapter 6).

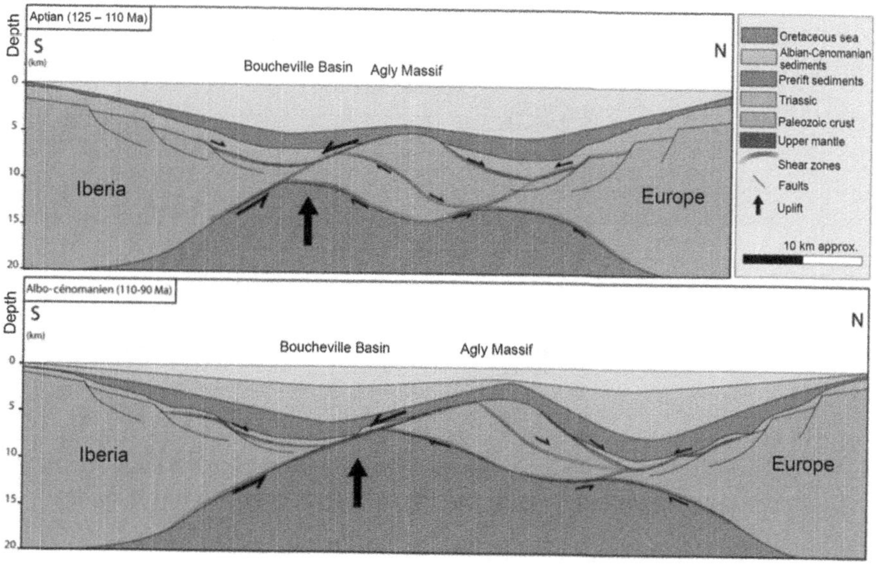


**Figure 5.31.** *Deformation model of the continental crust of the NPZ in the Agly massif during Cretaceous extension. Note the importance of tectonic lenticulation (modified from Aumar et al. 2022)*

Thermomechanical models applied directly to the Pyrenean rifting case are not numerous. In contrast, models related to the evolution of passive continental margins, such as those of Huismans and Beaumont (2011) and Brune et al. (2014) (Figure 5.32), have been widely discussed and applied to the NPZ case, with varying degrees of success.

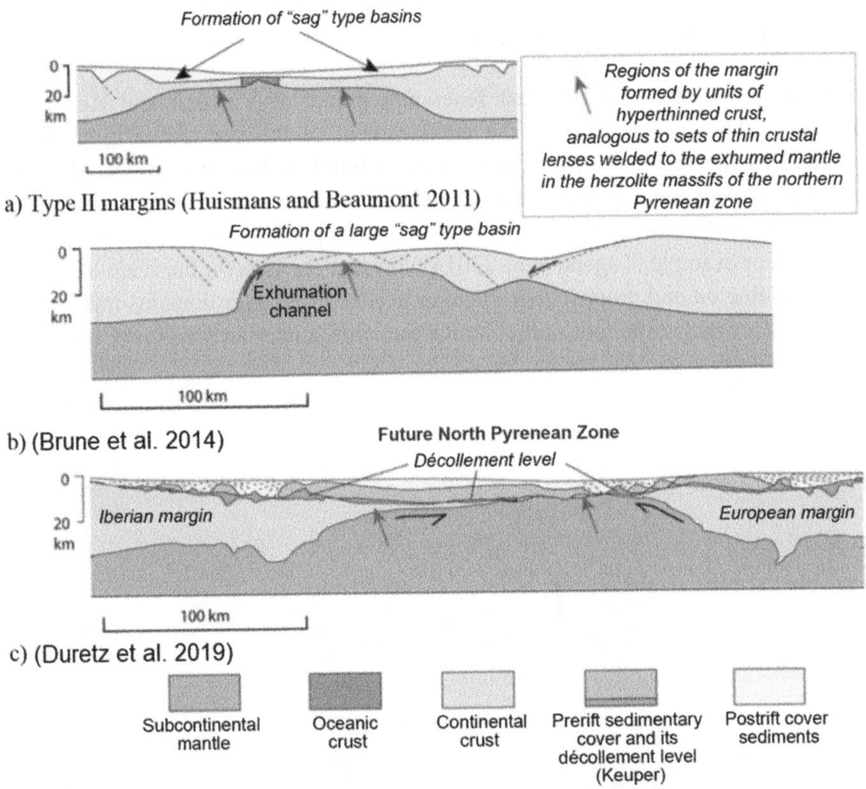


**Figure 5.32.** *Three thermomechanical models describing the evolution of continental margins (a and b) and the NPZ (c). Comments are in the text*

The first model to faithfully use the geological conditions of the Cretaceous Pyrenean rifting is that of Duretz et al. (2019) (Figure 5.34). The starting conditions are those of a 30-km-thick continental crust including at its top a *décollement* level made of 1 km thick evaporites and clays (the Upper Triassic) uniformly overlain by carbonates representing the 2 km thick prerift series. The thicknesses of these layers are consistent with stratigraphic data from the western Pyrenees (see James and Canérot 1988; Canérot 2008, 2017a, 2017b).

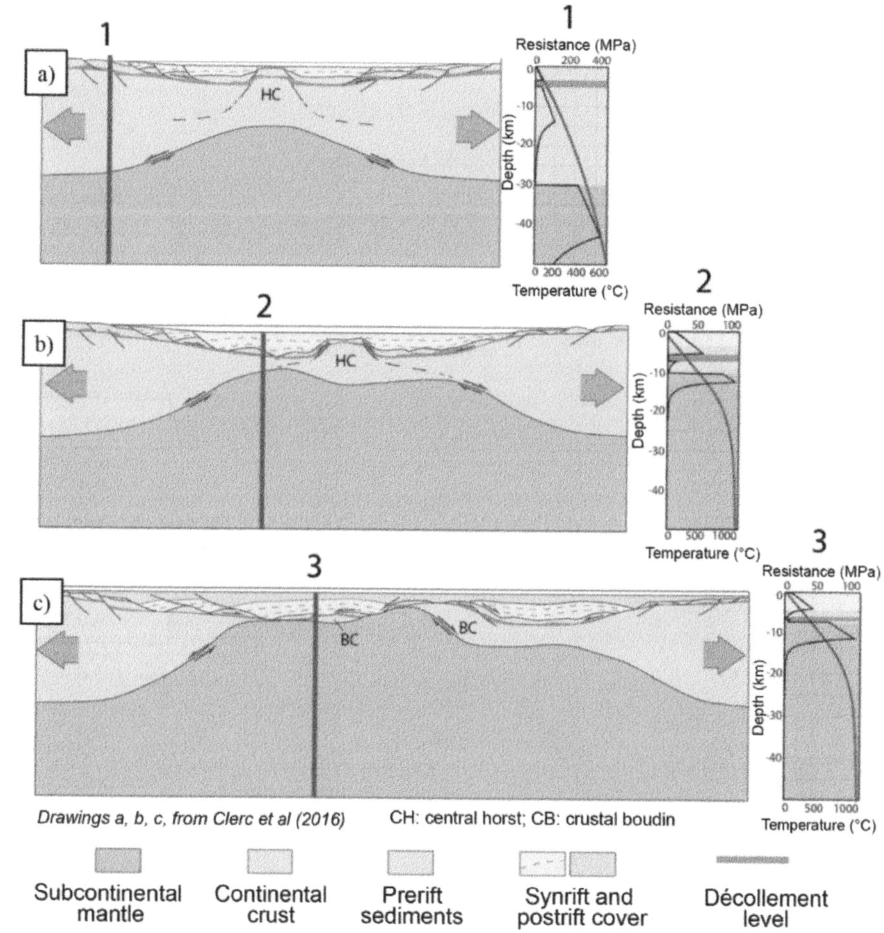


**Figure 5.33.** *Mechanical behavior of various layers involved in the evolution of the Cretaceous Pyrenean rifting (modified from Lagabrielle et al. 2020). Rheological profiles 1–3 show the resistance of the envelopes to extensional forces (comments in the text in sections 5.3.2 and 5.3.5)*

During its development, the model predicts the sedimentation of flysch by filling the basins that are created. It also accounts for the thermal evolution of the whole system, including that of the sedimentary cover. From the earliest stages of extension, the role of the Triassic *décollement* is highlighted by the stagnation of the cover in the basin created by crustal thinning. Decoupling along the salt and salt clays thus allows the cover to slide relatively inward the model, creating two zones of denudation on the rift shoulders.

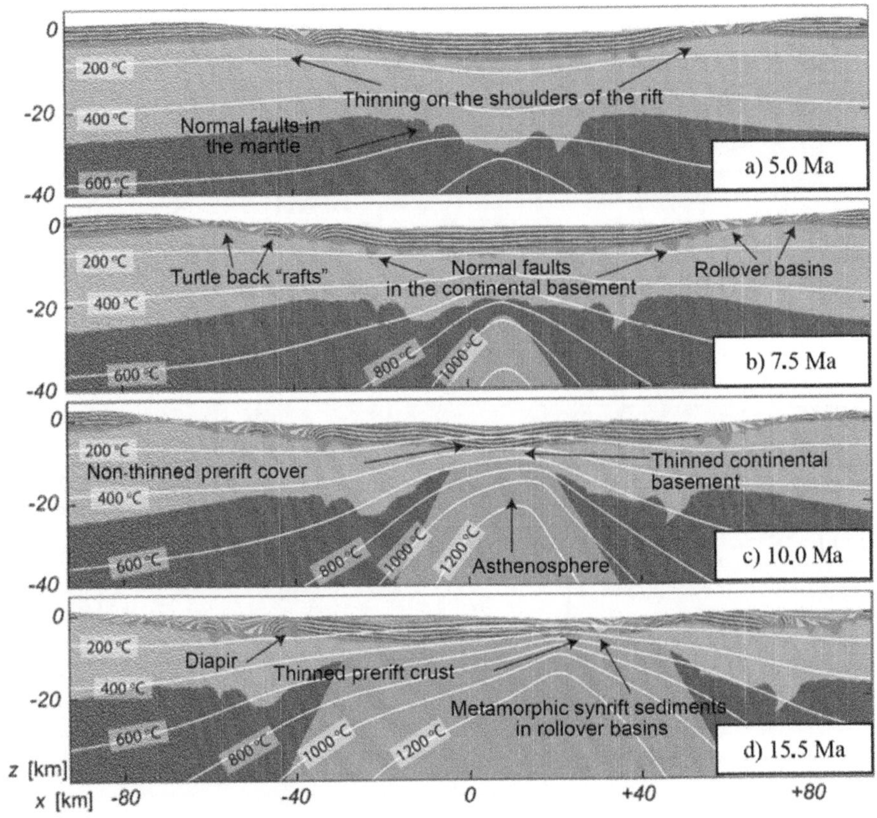


**Figure 5.34.** *The thermomechanical model of Duretz et al. (2019). Explanations are in the text*

The temperature increases under the prerift cover in the center of the basin. The thermal cover effect is enhanced by the deposition of the synrift flysch layers. This results in ductile deformation of the base of the cover and the continental crust, which do not form tilted blocks in the center of the system, but tectonic lenses (boudins) elongated above the mantle (Figures 5.32 and 5.33). This model therefore reproduces the geological evolution of the NPZ quite accurately and once again highlights the role of the Triassic layers in the rifting processes.

## 5.4. Conclusion

The numerous works carried out since 2010 in the framework of large governmental and industrial research programs on the Pyrenees have promoted a better understanding of the mechanisms leading to the NPZ Cretaceous rifting. Geodynamic processes at a lithospheric scale are now proposed, which account for crustal hyperthinning and exhumation of the subcontinental mantle. These processes integrate the consistent set of sedimentological, structural, geochronological and metamorphic constraints acquired over 150 years in the NPZ.

The new models of NPZ Cretaceous rifting highlight the following facts: (1) the uplift of the lherzolites that allows direct tectonic contact of the mantle rocks with the base of the prerift sediments, (2) the withdrawal of the continental crust under the rift margins by lateral extraction and (3) the decoupling and relative sliding of the prerift cover along the Keuper evaporites and clays. The notion of relative allochthony of the mobile prerift series on the Upper Triassic evaporites (soapstone layer) is fundamental and represents a major contribution to understanding the evolution of the NPZ. Due to constant occurrence at the bottom of the widening and deepening basin, the detached prerift cover allows the heat due to mantle ascent to be preserved and maintains the thermal anomaly for at least 20 Ma. Thermicity is also increased by the burial effect under the depositing flysch. For this reason, the thinning continental crust is not cooled below 350–400°C. It deforms ductilely and homogeneously as it is gradually extracted laterally from the center of the basins. However, at these temperatures, the crust never reaches the conditions of partial melting. Therefore, there is no granitization related to the "hot" Pyrenean rifting.

The mechanisms of Pyrenean rifting proposed in these recent models involve the individualization of thin hectometric crustal units that remain welded to the exhumed mantle. These units are systematically observed in most of the lherzolitic massifs of the NPZ. Crustal deformation is homogeneous and occurs as a combination of symmetrical anastomosing mylonitic shear zones and ductile flattening within the tectonic lenses. Geochronological dating in the Urdach massif has confirmed the Albian age of this deformation.

## 5.5. Acknowledgments

The concepts gathered in this chapter are the result of discussions and exchanges with many geologists since 2006. I would like to thank all the colleagues, post-doctoral fellows, PhD students and Master 2 students with whom I worked during the rich years following my first missions in the Pyrenees. I also thank the leaders of the various governmental and industrial programs (ANR Pyrope, RGF of

BRGM and OROGEN of TotalEnergies-BRGM-CNRS-INSU) who trusted us and financially supported our research. I would also like to mention the inhabitants of the Bearn Region who warmly welcomed us and gave us shelter during our field missions in these regions. Finally, I would like to thank the young editors, N. Saspiturry, J. Uzel and A. Ortiz, of the two Pyrenean volumes of this SCIENCES collection for ISTE and Wiley for their dynamic involvement in the project.

## 5.6. References

Albarède, F. and Michard-Vitrac, A. (1978a). Datation du métamorphisme des terrains secondaires des Pyrénées par des méthodes Ar-Ar et Rb-Sr. Ses relations avec les péridotites associées. *Bulletin de la société géologique de France*, 20(7), 681–688.

Albarède, F. and Michard-Vitrac, A. (1978b). Age and significance of the North Pyrenean metamorphism. *Earth and Planetary Science Letters*, 40(3), 327–332.

Andrieu, S., Saspiturry, N., Lartigau, M., Issautier, B., Lasseur, E. (2021). Sedimentary evolution and vertical movements during the Cenomanian to Santonian interval on the Iberian margin of the Pyrenees: Comparison with the southern Aquitaine Basin. *BSGF – Earth Sciences Bulletin*, 192, 19 [Online]. Available at: https://doi.org/10.1051/bsgf/2021011.

Asti, R., Lagabrielle, Y., Fourcade, S., Corre, B., Monié, P. (2019). How do continents deform during mantle exhumation? Insights from the northern Iberia inverted paleo-passive margin, Western Pyrenees (France). *Tectonics*, 38, 1666–1693.

Asti, R., Saspiturry, N., Angrand, P. (2022). The Mesozoic Iberia-Eurasia plate boundary: A wide domain of distributed transtensional deformation rather than a single wrench fault. *Earth-Science Reviews*, 230, 104040 [Online]. Available at: https://doi.org/10.1016/j.earscirev.2022.104040.

Aumar, C., Bosse, V., Merle, O., Monié, P. (2022). Syn-rift Cretaceous deformation in the Variscan Agly massif (Eastern Pyrenees, France). *BSGF – Earth Sciences Bulletin*, 193(6), 2–27.

Avé-Lallemant, H.G. (1968). Structural and petrofabric analysis of an “alpine type” peridotite: The lherzolite of the French Pyrenees. *Leidse Geol. Med.*, 42, 1–57.

Azambre, B. (1967). Sur les roches intrusives sous-saturées du Crétacé des Pyrénées. PhD Thesis, Université Paris 6.

Azambre, B. and Monchoux, P. (1998). Métagabbros amphiboliques et mise en place crustale des lherzolites des Pyrénées (France). *Comptes rendus de l'académie des sciences*, 327(1), 9–15 [Online]. Available at: https://doi.org/10.1016/S1251- 8050(98)80012-8.

Azambre, B. and Rossy, M. (1976). Le magmatisme alcalin d'âge crétacé dans les Pyrénées occidentales ; ses relations avec le métamorphisme et la tectonique. *Bulletin de la société géologique de France*, 7(18), 1725–1728.

Azambre, B. and Rossy, M. (1981). Caractère alcalin du magmatisme triasique des Corbières orientales. *Bulletin de la société géologique de France*, 23(3), 253–262.

Azambre, B., Ravier, J., Thiébault, J. (1971). À propos du phénomène de dipyrisation des roches éruptives d'âge secondaire des Pyrénées. *Comptes rendus de l'académie des sciences Paris*, 272, 2137–2139.

Azambre, B., Rossy, M., Elloy, R. (1981). Les dolérites triasiques (ophites) des Pyrénées : données nouvelles fournies par les sondages pétroliers en Aquitaine. *Bulletin de la société géologique de France*, 23(3), 263–269.

Azambre, B., Rossy, M., Lago, M. (1987). Caractéristiques pétrologiques des dolérites tholéitiques d'âge triasiques (ophites) du domaine pyrénéen. *Bulletin de minéralogie*, 110, 379–96.

Azambre, B., Rossy, M., Albarède, F. (1992). Petrology of the alkaline magmatism from the Cretaceous North-Pyrenean rift zone (France and Spain). *Eur. J. Mineral.*, 4(4), 813–834.

Bernus-Maury, C. (1984). Étude des paragenèses caractéristiques du métamorphisme mésozoïque dans la partie orientale des Pyrénées. PhD Thesis, Université Paris 6.

Beyssac, O., Goffé, B., Chopin, C., Rouzaud, J.N. (2002). Raman spectra of carbonaceous material in metasediments: A new geothermometer. *Journal of Metamorphic Geology*, 20, 859–871.

Béziat, D., Joron, J.L., Monchoux, P., Treuil, M., Walgenwitz, F. (1991). Geodynamic implications of geochemical data for the Pyrenean ophites (Spain-France). *Chem. Geol.*, 89, 243–262 [Online]. Available at: https://doi.org/10.1016/0009-2541(91)90019-N.

Biteau, J.-J., Le Marrec, A., Le Vot, M., Masset, J.-M. (2006). The Aquitaine Basin. *Pet. Geosci.*, 12, 247–273.

Bodinier, J.-L, Fabriès, J., Lorand, J.P., Dupuy, C. (1987). Geochemistry of amphibole pyroxenite veins from the Lherz and Freychinède ultramafic bodies (Ariège, French Pyrenees). *Bulletin de minéralogie*, 110, 345–358.

Bodinier, J.-L., Dupuy, C., Dostal, J. (1988). Geochemistry and petrogenesis of Eastern Pyrenean peridotites. *Geochimica et Cosmochimica Acta*, 52, 2893–2907.

Bodinier, J.-L., Vasseur, G., Vernières, J., Dupuy, C., Fabriès, J. (1990). Mechanisms of mantle metasomatism: Geochemical evidence from the Lherz Orogenic peridotite. *J. Petrol.*, 31, 597–628.

Bodinier, J.-L., Menzies, M.A., Shimizu, N., Frey, F.A., McPherson, E. (2004). Silicate, hydrous and carbonate metasomatism at Lherz, France: Contemporaneous derivatives of silicate melt-harzburgite reaction. *J. Petrol.*, 45(2), 299–320 [Online]. Available at: https://doi.org/10.1093/petrology/egg107.

Boillot, G., Capdevilla, R., Hennequin-Marchand, I., Lamboy, M., Lepretre, J.P. (1973). La zone nord-pyrénéenne, ses prolongements sur la marge continentale nord-espagnole et sa signification structurale. *Comptes rendus de l'académie des sciences Paris*, 227, 2629–2632.

Boillot, G., Recq, M., Winterer, E.L., Meyer, A.W., Applegate, J., Baltuck, M., Williamson, M. (1987). Tectonic denudation of the upper mantle along passive margins: A model based on drilling results (ODP leg 103, western Galicia margin, Spain). *Tectonophysics*, 132, 335–342 [Online]. Available at: https://doi.org/10.1016/0040-1951(87)90352-0.

Boirie, J.-M. (1981). Étude sédimentologique des Poudingues de Mendibelza (P.A.). PhD Thesis, Université Paul Sabatier de Toulouse.

Boirie, J.-M. and Souquet P. (1982). Les poudingues de Mendibelza : dépôts de cônes sousmarins du rift Albien des Pyrénées. *Bulletin centres de recherches exploration-production Elf- Aquitaine*, 6(2), 405–435.

Boulvais, P. (2016). Fluid generation in the Boucheville Basin as a consequence of the North Pyrenean metamorphism. *Comptes rendus geosci: Rifting to Mountain BuilDing: The Pyrenean Belt*, 348, 301–311 [Online]. Available at: https://doi.org/10.1016/j.crte.2015.06.013.

Boulvais, P., de Parseval, P., D'Hulst, A., Paris, P. (2006). Carbonate alteration associated with talc-chlorite mineralization in the eastern Pyrenees, with emphasis on the St. Barthelemy Massif. *Mineralogy and Petrology*, 88(3), 499–526 [Online]. Available at: https://doi.org/10.1007/s00710-006-0124-x.

Boulvais, P., Ruffet, G., Cornichet, J., Mermet, M. (2007). Cretaceous albitization and dequartzification of Hercynian peraluminous granite in the Salvezines Massif (French Pyrénées). *Lithos*, 93(1/2), 89–106 [Online]. Available at: https://doi.org/10.1016/j.lithos.2006.05.001.

Boutin, A., de Saint Blanquat, M., Poujol, M., Boulvais, P., de Parseval, P., Rouleau, C., Robert, J.F. (2015). Succession of Permian and Mesozoic metasomatic events in the eastern Pyrenees with emphasis on the Trimouns talc-chlorite deposit. *International Journal of Earth Sciences*, 105, 747–770 [Online]. Available at: https://doi.org/10.1007/s00531-015-1223-x.

Brune, S., Heine, C., Pérez-Gussinyé, M., Sobolev, S.V. (2014). Rift migration explains continental margin asymmetry and crustal hyper-extension. *Nat. Commun.*, 5, 1–9 [Online]. Available at: https://doi.org/10.1038/ncomms5014.

Brunet, M.F. (1991). Subsidence et géodynamique du Bassin d'Aquitaine. Relations avec l'ouverture de l'Atlantique. PhD Thesis, Université Paris VI.

Canérot, J. (1988). Manifestations de l'halocinèse dans les Chaînons Béarnais (Zone Nord-Pyrénéenne) au Crétacé inférieur. *Comptes rendus de l'académie des sciences Paris*, 306, 1099–1102.

Canérot, J. (1989). Rifting eocrétacé et halocinèse sur la marge ibérique des Pyrénées Occidentales (France). Conséquences structurales. *Bulletin centres de recherches exploration-production Elf- Aquitaine*, 13, 87–99.

Canérot, J. (2008). *Les Pyrénées : histoire géologique*. Atlantica, Paris.

Canérot, J. (2017a). The pull apart-type Tardets-Mauléon Basin, a key to understand the formation of the Pyrenees. *Bulletin de la société géologique de France*, 188, 35 [Online]. Available at: https://doi.org/10.1051/bsgf/2017198.

Canérot, J. (2017b). Origine de la chaîne des Pyrénées : collision entre les plaques ibérique et européenne ou inversion d'un ancien rift continental avorté ? *Bulletin de la société d'histoire naturelle de Toulouse*, 153, 95–110.

Canérot, J. and James, V. (1999). Diapirism and post-Triassic structural development of the western Pyrenees and southern Aquitaine. *Eclogae Geol. Helv.*, 92(1), 63–72.

Carracedo, M., Larrea, F.J., Alonso, A. (1999). Estructura y organización de las coladas submarinas: características de las lavas almohadilladas de edad cretácica que afloran en la Cordillera Vasco-Cantábrica. *Estudios Geológicos*, 55(5/6), 45–53.

Castanares, L.M., Robles, S., Gimeno, D., Vicente Bravo, J.C. (2001). The submarine volcanic system of the Errigoiti Formation (Albian-Santonian of the Basque-Canta-brian Basin, Northern Spain): Stratigraphic framework, facies, and sequences. *Journal of Sedimentary Research*, 71(2), 318–333 [Online]. Available at: https://doi.org/10.1306/080700710318.

Castéras, M. (1933). Recherches sur la structure du versant nord des Pyrénées centrales et orientales [Bull. Serv. Carte Géol. France, 27 (189)]. Paris.

Castéras, M. (1953). Plissement d'âge crétacé dans la région du massif d'Igounce (B.P.). *Comptes rendus de l'académie des sciences Paris*, 257, 1729.

Castéras, M. and Souquet, P. (1964). Sur la constitution et sur la structure de la couverture crétacée de la Zone primaire axiale pyrénéenne à l'ouest du pic d'Anie. *Comptes rendus de l'académie des sciences Paris*, 259(17), 2881–2886.

Castéras, M., Canérot, J., Paris, J., Tisin, D., Azambre, B., Alimen, H. (1970). Carte géologique de la France au 1/50000, feuille d'Oloron Sainte Marie. BRGM, Orléans.

Castéras, M., Gottis, M., Clin, M., Guignard, J., Paris, J., Galharague, J. (1971). Carte géologique de la France au 1/50 000, feuille de Tardets-Sorholus. BRGM, Orléans.

Cathelineau, M., Boiron, M.C., Jakomulski, H. (2021). Triassic evaporites: A vast reservoir of brines mobilized successively during rifting and thrusting in the Pyrenees. *Journal of the Geological Society of London* [Online]. Available at: https://doi.org/10.1144/jgs2020-259.

Chelalou, R., Nalpas, T., Bousquet, R., Prevost, M., Lahfid, A., Poujol, M., Ringenbach, J.-C., Ballard, J.-F. (2016). New sedimentological, structural and paleo-thermicity data in the Boucheville Basin (eastern North Pyrenean Zone, France). *Comptes rendus géoscience*, 348, 312–321.

Choukroune, P. (1972). Relation entre tectonique et métamorphisme dans la zone nord-pyrénéenne centrale et orientale. *Bulletin de la société géologique de France*, XIV(7), 3–11.

Choukroune, P. (1973). La brèche de Lherz dite "d'explosion liée à la mise en place des lherzolites" est une brèche sédimentaire d'âge cénozoïque. *Comptes rendus de l'académie des sciences Paris*, 277, 2621–2624.

Choukroune, P. (1976). *Structure et évolution tectonique de la zone Nord-Pyrénéenne. Analyse de la déformation dans une portion de chaîne à schistosité subverticale*. Société géologique de France, Paris.

Choukroune, P. (1980). Comment and reply on "Quenching: An additional model for emplacement of the lherzolite at Lers (French Pyrenees)". *Geology*, 8(11), 514–515 [Online]. Available at: https://doi.org/10.1130/00917613.

Choukroune, P. and Mattauer, M. (1978). Tectonique des plaques et Pyrénées : sur le fonctionnement de la faille transformante nord-Pyrénéenne ; comparaisons avec les modèles actuels. *Bulletin de la société géologique de France*, 20, 689–700.

Choukroune, P. and Séguret, M. (1973). *Carte structurale des Pyrénées*. ELF-ERAP, Paris.

Claude, D. (1990). Étude stratigraphique, sédimentologique et structurale des dépôts mésozoïques au nord du massif du Labourd : rôle de la faille de Pamplona (Pays Basque). PhD Thesis, Université de Bordeaux 3.

Clerc, C. (2012). Évolution du domaine nord-pyrénéen au Crétacé. Amincissement crustal extrême et thermicité élevée : un analogue pour les marges passives. PhD Thesis, Université Pierre et Marie Curie, Paris.

Clerc, C. and Lagabrielle, Y. (2014). Thermal control on the modes of crustal thinning leading to mantle exhumation. Insights from the Cretaceous Pyrenean hot paleomargins. *Tectonics*, 33, 1340–1359 [Online]. Available at: https://doi.org/doi:10.1002/2013TC003471.

Clerc, C., Lagabrielle, Y., Neumaier, M., Reynaud, J.-Y., de Saint Blanquat, M. (2012). Exhumation of subcontinental mantle rocks: Evidence from ultramafic-bearing clastic deposits nearby the Lherz peridotite body, French Pyrenees. *Bulletin de la société géologique de France*, 183(5), 443–459 [Online]. Available at: https://doi.org/doi:10.2113/gssgfbull.183.5.443.

Clerc, C., Boulvais, P., Lagabrielle, Y., de Saint Blanquat, M. (2014). Ophicalcites from the Northern Pyrenean Belt: A field, petrographic and stable isotope study. *International Journal of Earth Sciences*, 103(21), 141–163.

Clerc, C., Lahfid, A., Monié, P., Lagabrielle, Y., Chopin, C., Poujol, M., Boulvais, P., Ringenbach, J.-C., Masini, E., de Saint Blanquat, M. (2015). High-temperature metamorphism during extreme thinning of the continental crust: A reappraisal of the North Pyrenean passive paleomargin. *Solid Earth*, 6, 643–668 [Online]. Available at: https://doi.org/doi:10.5194/se-6-643-2015.

Clerc, C., Lagabrielle, Y., Labaume, P., Ringenbach, J.-C., Vauchez, A., Nalpas, T., Bousquet, R., Ballard, J.-F., Lahfid, A., Fourcade, S. (2016). Basement – Cover decoupling and progressive exhumation of metamorphic sediments at hot rifted margin. Insights from the Northeastern Pyrenean analog. *Tectonophysics*, 686, 82–97.

Cochelin, B., Chardon, D., Denèle, Y., Gumiaux, C., Le Bayon, B. (2017). Vertical strain partitioning in hot Variscan crust: Syn-convergence escape of the Pyrenees in the Iberian-Armorican syntax. *Bulletin de la société géologique de France*, 188(6), 39 [Online]. Available at: https://doi.org/doi:10.1051/bsgf/2017206.

Cochelin, B., Lemirre, B., Denèle, Y., de Saint Blanquat, M., Lahfid, A., Duchêne, S. (2018). Structural inheritance in the central Pyrenees: The variscan to Alpine tectonometamorphic evolution of the axial zone. *J. Geol. Soc.*, 175, 336–351 [Online]. Available at: https://doi.org/doi:10.1144/jgs2017-066.

Colchen, M., Ternet, Y., Debroas, E.J. (1997). Carte géologique de la France au 1/50 000, feuille 1086. BRGM, Orléans.

Combes, P.J. and Peybernès, B. (1989). Tectonique albienne dans les gisements de bauxites des Pyrénées ariégeoises (France) en relation avec l'évolution géodynamique de la marge passive européenne. *Comptes rendus de l'académie des sciences Paris*, 308(2), 1005–1010.

Conquéré, F. (1971). Les pyroxénolites à amphibole et les amphibolites associées aux lherzolites du gisement de Lherz (Ariège, France) : un exemple du rôle de l'eau au cours de la cristallisation fractionnée des liquides issus de la fusion partielle de lherzolites. *Contributions to Mineralogy and Petrology*, 33(1), 32–61 [Online]. Available at: https://doi.org/doi:10.1007/BF00373793.

Conquéré, F. (1972). Caractères distinctifs et genèse des péridotites associées aux pyroxénolites dans les lherzolites des Pyrénées (Ariège, France). *Comptes rendus de l'académie des sciences Paris*, 274, 1244–1247.

Conquéré, F. (1973). L'origine des lherzolites et roches associées de l'étang de Lherz (Ariège, France). *Comptes rendus de l'académie des sciences Paris*, 277, 1589–1592.

Conquéré, F. (1977). Pétrologie des pyroxénites litées dans les complexes ultramafiques de l'Ariège (France) et autres gisements de lherzolite à spinelle. Compositions minéralogiques et chimiques, évolution des conditions d'équilibre des pyroxénites. *Bulletin de minéralogie*, 100, 42–80.

Conquéré, F. and Fabries, J. (1984a). Chemical disequilibrium and its thermal significance in spinel-peridotites from the Lherz and Freychinède ultramafic bodies (Ariège; French Pyrenees). In *Kimberlites II: The Mantle and Crust-Mantle Relationships*, Kornprobst J. (ed.). Elsevier, Amsterdam [*Developments in Petrology*, 11(2), 319–331].

Conquéré, F. and Fabries, J. (1984b). Caractères pétrographiques et chimiques des péridotites à spinelle des gisements ultramafiques de Lherz et de Freychinède (Ariège, Pyrénées françaises). *Annales scientifiques de l'Université de Clermont-Ferrand II*, 2(74), 55–83.

Corre, B. (2017). La bordure nord de la plaque ibérique à l'Albo-Cénomanien. Architecture d'une marge passive de type ductile (Chaînons Béarnais, Pyrénées occidentales). PhD Thesis, Université de Rennes 1.

Corre, B., Lagabrielle, Y., Labaume, P., Fourcade, S., Clerc, C., Ballèvre, M. (2016). Deformation associated with mantle exhumation in a distal, hot passive margin environment: New constraints from the Saraillé Massif (Chaînons Béarnais, North-Pyrenean Zone). *Comptes rendus géoscience*, 348, 279–289.

Corre, B., Boulvais, P., Boiron, M.C., Lagabrielle, Y., Marasi, L., Clerc, C. (2018). Fluid circulations in response to mantle exhumation at the passive margin setting in the North Pyrenean Zone, France. *Mineralogy and Petrology*, 112(5), 647–670 [Online]. Available at: https://doi.org/doi:10.1007/s00710-018-0559-x.

Costa, S. and Maluski., H. (1988). Use of the $^{40}$Ar-$^{39}$Ar stepwise heating method for dating mylonite zones: An example from the St. Barthélémy massif (Northern Pyrenees, France). *Chemical Geology: Isotope Geoscience Section*, 72(2), 127–144 [Online]. Available at: https://doi.org/doi:10.1016/0168-9622(88)90061-9.

Dauteuil, O. and Ricou, L.E. (1989). Une circulation de fluides de haute température à l'origine du métamorphisme crétacé nord-pyrénéen. *Geodinamica Acta*, 3(3), 237–250.

Dauteuil, O., Raymond, D., Ricou., L.E. (1987). Brèches de fracturation hydraulique dans la zone métamorphique des Pyrénées, exemple à l'Est du Saint-Barthélémy. *Comptes rendus de l'académie des sciences Paris, série II*, 304, 1025–1028.

Debeaux, M. and Thiébaut, J. (1958). Les affleurements du socle paléozoïque entre les massifs de la Barousse et de Milhas. *Bulletin de la societe d'histoire naturelle de Toulouse*, 93, 522–528.

Debroas, E.J. (1976a). Les flyschs crétacés pyrénéens de la Ballongue, du Castillonnais et de la région d'Arbas (Ariège, Haute-Garonne). PhD Thesis, Université de Toulouse.

Debroas, E.J. (1976b). Nouvelles observations sur les relations entre métamorphisme et tectonique dans les terrains secondaires des Pyrénées. *Comptes rendus de l'académie des sciences Paris*, 283(D), 1707–1710.

Debroas, E.J. (1977). Identification d'un paléocône sous-marin dans le Flysch ardoisier des Pyrénées ariégeoises. *Bulletin de la société géologique de France*, 7(19), 267–274.

Debroas, E.J. (1978). Évolution de la fosse du flysch ardoisier de l'Albien au Sénonien inférieur (zone interne métamorphique des Pyrénées navarro-languedociennes). *Bulletin de la société géologique de France*, 7(20), 639–648.

Debroas, E.J. (1987). Modèle de bassin triangulaire à l'intersection de décrochements divergents pour le fossé albo-cénomanien de la Ballongue (Zone nord-pyrénéenne, France). *Bulletin de la société géologique de France*, 8(5), 887–898.

Debroas, E.J. (1990). Le flysch noir albo-cénomanien témoin de la structuration albienne à sénonienne de la Zone nord-pyrénéenne en Bigorre (hautes Pyrénées, Fr.). *Bulletin de la société géologique de France*, 8, 273–285.

Debroas, E.J. and Souquet, P. (1972). Contribution à la connaissance du flysch ardoisier nord-pyrénéen : les schistes métamorphiques du synclinal de la Ballongue. *Comptes rendus de l'académie des sciences Paris*, 275, 2813–2816.

Debroas, E.J. and Souquet, P. (1976). Sédimentogenèse et position structurale des flyschs crétacés du versant nord des Pyrénées centrales. *Bull. BRGM*, 2, 305–320.

Debroas, E.J., Mediavila, F., Souquet, P. (1978). La zone des failles transcurrentes de Bigorre dans la région de Lourdes et Bagnères. *Bulletin de la société géologique de France*, 7(20), 639–648.

Debroas, E.J., Lagier, Y., Souquet, P. (1983). Turbidites calcaires exceptionnelles dans le flysch turono-coniacien du versant nord des Pyrénées occidentales. *Bulletin de la société géologique de France*, 7(25), 911–919.

Debroas, E.J., Canérot, J., Bilotte, M. (2010). Les Brèches d'Urdach, témoins de l'exhumation du manteau pyrénéen dans un escarpement de faille Vraconnien-Cénomanien inférieur (zone nord-pyrénéenne, Pyrénées-Atlantiques, France). *Géologie de France*, 2, 53–63.

DeFelipe, I., Pedreira, D., Pulgar, J.A., Iriarte, E., Mendia, M. (2017). Mantle exhumation and metamorphism in the Basque-Cantabrian Basin (N Spain): Stable and clumped isotope analysis in carbonates and comparison with ophicalcites in the North-Pyrenean Zone (Urdach and Lherz). *Geochem. Geophys. Geosyst.*, 18(2), 631–652.

Deramond, J., Souquet, P., Fondecave-Wallez, M.J., Spetch, M. (1993). Relationships between thrust tectonics and sequence stratigraphy surfaces in foredeeps: Model and examples from the Pyrenees (Cretaceous–Eocene, France, Spain). *Geological Society, London, Special Publications*, 71(1), 193–219.

Downes, H., Bodinier, J.-L., Thirlwall, M.F., Lorand, J.-P., Fabriès, J. (1991). REE and Sr-Nd isotopic geochemistry of the Eastern Pyrenean peridotite massifs: Sub-continental lithospheric mantle modified by continental magmatism. Orogenic Lherzolites and Mantle Processes. *J. Petrol.*, 2, 97–115.

Ducasse, L. and Velasque, P.C. (1988). Géotraverse dans la partie occidentale des Pyrénées, de l'avant-pays aquitain au bassin de l'Ebre : effet d'une inversion structurale sur l'édification d'une chaîne intracontinentale. PhD Thesis, Université Paul Cézanne, Aix-Marseille.

Ducasse, L., Velasque, P.C., Muller, J. (1986). Glissement de couverture et panneaux basculés dans la région des Arbailles (Pyrénées occidentales) : un modèle évolutif crétacé de la marge nord-ibérique à l'Est de la transformante de Pamplon. *Comptes rendus de l'académie des sciences Paris*, 303(16), 1477–1482.

Ducoux, M. (2017). Structure, thermicité et évolution géodynamique de la Zone Interne Métamorphique des Pyrénées. PhD Thesis, Université d'Orléans.

Ducoux, M., Jolivet, L., Cagnard, F., Gumiaux, C., Baudin, T., Masini, E., Callot, J.P., Aubourg, C., Lahfid, A., Homonnay, E. (2019). The Nappe des Marbres unit of the Basque-Cantabrian basin: The tectono-thermal evolution of a fossil hyperextended rift basin. *Tectonics*, 38(11), 3881–3915 [Online]. Available at: https://doi.org/10.1029/2018TC005348.

Duée, G., Lagabrielle, Y., Coutelle, A. Fortané, A. (1984). Les lherzolites associées aux chaînons béarnais (Pyrénées Occidentales) : mise à l'affleurement anté-dogger et resédimentation albo-cénomanienne. *Comptes rendus de l'académie des sciences Paris*, 299(17), 1205–1209.

Durand-Delga, M. (1964). Remarques sur la stratigraphie et la structure du Mésozoïque situé entre Estagel et Perpignan (Pyrénées-Orientales). *Comptes rendus de l'académie des sciences Paris*, 259, 837–840.

Durand-Delga, M. (1965). Au sujet de la phase anté-cénomanienne à l'Est de Quillan. *Comptes rendus somm. soc. géol. France*, 2, 61–62.

Durand-Delga, M. (1980). La chaîne des Pyrénées et son avant-pays aquitain-languedocien. In *Itinéraires géologiques, excursions dans le "Grand Sud-Ouest", 26e C.G.I. Mém. Centre rech. explor. prod. Elf-Aquitaine*. Paris.

Duretz, T., Asti, R., Lagabrielle, Y., Brun, J.P., Jourdon, A., Clerc, C., Corre, B. (2019). Numerical modelling of syn-rift salt tectonics during Cretaceous Pyrenean rifting. *Basin Res.*, 32, 1–16 [Online]. Available at: https://doi.org/10.1111/bre.12389.

Espurt, N., Angrand, P., Teixell, A., Labaume, P., Ford, M., de Saint Blanquat, M., Chevrot, S. (2019). Crustal-scale balanced cross-section and restaurations of the Central Pyrenean belt (Nest-Cinca transect): Superimposed orogenesis and Pyrenean rift system evolution. *Tectonophysics*, 764, 25–45 [Online]. Available at: https://doi.org/10.1016/j.tecto.2019.04.026.

Fabriès, J., Lorand, J.-P., Bodinier, J.L., Dupuy, C. (1991). Evolution of the Upper Mantle beneath the Pyrenees: Evidence from Orogenic Spinel Lherzolite Massifs. *Journal of Petrology*, 2, 55–76 [Online]. Available at: https://doi.org/10.1093/petrology/Special_Volume.2.55.

Fabriès, J., Lorand, J.-P., Bodinier, J.L. (1998). Petrogenetic evolution of orogenic lherzolite massifs in the central and western Pyrenees. *Tectonophysics*, 292, 145–167 [Online]. Available at: https://doi.org/doi:10.1016/S0040-1951(98)00055-9.

Fallourd, S., Poujol, M., Boulvais, P., Paquette, J.L., de Saint Blanquat M., Rémy, P. (2014). In situ LA-ICP-MS U–Pb titanite dating of Na–Ca metasomatism in orogenic belts: The North Pyrenean example. *International Journal of Earth Sciences*, 103(3), 667–682 [Online]. Available at: https://doi.org/doi:10.1007/s00531-013-0978-1.

Fonteilles, M. (1970). Géologie des terrains métamorphiques et granitiques du massif hercynien de l'Agly (Pyrénées orientales). *Bull. bur. rech. géol. min. Paris*, 4, 21–72.

Fortané, A., Duée, G., Lagabrielle, Y., Coutelle, A. (1986). Lherzolites and the western "Chaînons Béarnais" (French Pyrénées): Structural and paleogeographical pattern. *Tectonophysics*, 129, 81–98.

Früh-Green, G.L., Kelley, D.S., Bernasconi, S.M., Karson, J.A., Ludwig, K.A., Butterfield, D.A., Boschi, C., Proskurowski, G. (2003). 30,000 years of hydrothermal activity at the Lost City vent field. *Science*, 301, 495–498.

Gaudichet, A. (1974). Étude pétrographique des lherzolites de la région d'Oloron-Ste Marie (Pyrénées-Atlantiques). PhD Thesis, Université de Paris 6.

Golberg, J.M. (1987). Le métamorphisme mésozoïque dans la partie orientale des Pyrénées : relation avec l'évolution de la chaîne au Crétacé. Thesis, Centre géologique et géophysique, Université des sciences et techniques du Languedoc, Montpellier.

Golberg, J.M. and Leyreloup, A.F. (1985). Mise en évidence de fluides mantelliques dans la zone nord pyrénéenne (région de Lherz) : la formation bréchique du col d'Agnes. *Comptes rendus de l'académie des sciences Paris*, 300, 71–74.

Golberg, J.M. and Leyreloup, A.F. (1990). High temperature-low pressure Cretaceous metamorphism related to crustal thinning (Eastern North Pyrenean Zone, France). *Contributions to Mineralogy and Petrology*, 104, 194–207.

Golberg, J.M. and Maluski, H. (1988). Données nouvelles et mise au point sur l'âge du métamorphisme pyrénéen. *Comptes rendus de l'académie des sciences Paris*, 306, 429–435.

Golberg, J.M., Maluski, H., Leyreloup, A.-F. (1986). Petrological and age relationship between emplacement of magmatic breccia, alkaline magmatism, and static metamorphism in the North Pyrenean Zone. *Tectonophysics*, 129, 275–290 [Online]. Available at: https://doi.org/10.1016/0040-1951(86)90256-8.

Gong, Z., Langereis, C.G., Mullender, T.A.T. (2008). The rotation of Iberia during the Aptian and the opening of the Bay of Biscay. *Earth and Planetary Science Letters*, 273, 80–93.

Hart, N.R., Stockli, D.F., Lavier, L.L., Hayman, N.W. (2017). Thermal evolution of a hyperextended rift basin, Mauléon Basin, western Pyrenees. *Tectonics*, 36, 1103–1128 [Online]. Available at: https://doi.org/doi:10.1002/2016TC004365.

Henry, P., Azambre, B., Montigny, R., Rossy, M., Stevenson, R.K. (1998). Late mantle evolution of the Pyrenean sub-continental lithospheric mantle in the light of new $^{40}$Ar- $^{39}$Ar and Sm-Nd ages on pyroxenites and peridotites (Pyrenees, France). *Tectonophysics*, 296, 103–123 [Online]. Available at: https://doi.org/doi:10.1016/S0040-1951(98)00139-5.

Huismans, R.S. and Beaumont, C. (2011). Depth-dependent extension, two-stage breakup and cratonic underplating at rifted margins. *Nature*, 473, 74–78 [Online]. Available at: https://doi.org/10.1038/nature09988.

Incerpi, N., Manatschal, G., Martire, L., Bernasconi, S.M., Gerdes, A., Bertok, C. (2020). Characteristics and timing of hydrothermal fluid circulation in the fossil Pyrenean hyperextended rift system: New constraints from thc Chaînons Béarnais (W Pyrenees). *International Journal of Earth Sciences*, 109, 1071–1093 [Online]. Available at: https://doi.org/doi.org/10.1007/s00531-020-01852-6.

Issautier, B., Lasseur, E., Saspiturry, N., Andrieu, S., Angrand, A., Serrano, O. (2022). Onset of the Iberian-European plates convergence: Late Cretaceous flexural response of a hot lithosphere. *Tectonophysics*, 843, 229504 [Online]. Available at: https://doi.org/10.1016/j.tecto.2022.229504.

James, V. and Canérot, J. (1988). Diapirisme et structuration post-triasique des Pyrénées occidentales et de l'Aquitaine méridionale (France). *Eclogae Geol. Helveticae*, 92, 63–72.

Jammes, S., Manatschal, G., Lavier, L., Masini. E. (2009). Tectonosedimentary evolution related to extreme crustal thinning ahead of a propagating ocean: Example of the western Pyrenees. *Tectonics*, 28, 4 [Online]. Available at: https://doi.org/10.1029/2008TC002406.

Jammes, S., Lavier, L., Manatschal, G. (2010a). Extreme crustal thinning in the Bay of Biscay and the Western Pyrenees: From observations to modeling. *Geochem. Geophys. Geosystems*, 11 [Online]. Available at: https://doi.org/10.1029/2010GC003218.

Jammes, S., Tiberi, C., Manatschal, G. (2010b). 3D architecture of a complex transcurrent rift system: The example of the Bay of Biscay-Western Pyrenees. *Tectonophysics*, 489, 210–226 [Online]. Available at: https://doi.org/10.1016/j.tecto.2010.04.023.

Jammes, S., Manatschal, G., Lavier, L. (2010c). Interaction between pre-rift salt and detachment faulting in hyperextended rift systems: The example of the Parentis and Mauléon basins (Bay of Biscay and western Pyrenees). *AAPG Bull.*, 94, 957–975 [Online]. Available at: https://doi.org/10.1306/12090909116.

Lacroix, A. (1892). Sur les modifications minéralogiques effectuées par la lherzolite sur les calcaires du Jurassique de l'Ariège. *Comptes rendus de l'académie des sciences Paris*, 115, 974–976.

Lacroix, A. (1894). Étude minéralogique de la lherzolite des Pyrénées et de ses phénomènes de contact. *Nouv. Archives Muséum*, 6, 209–308.

Lacroix, A. (1895a). Sur les phénomènes de contact de la lherzolite des Pyrénées. *Comptes rendus de l'académie des sciences Paris*, 120, 339–342.

Lacroix, A. (1895b). Considérations sur le métamorphisme de contact, auxquelles conduit l'étude des phénomènes de contact de la lherzolite des Pyrénées. *Comptes rendus de l'académie des sciences Paris*, 120, 388–391.

Lacroix, A. (1895c). Les phénomènes de contact de la lherzolite et de quelques ophites des Pyrénées. *Bulletin des services de la carte géologique de la France et des topographies souterraines*, 6, 307–446.

Lacroix, A. (1897). Phénomènes de contact de la lherzolite et des diabases ophitiques (ophites) (Pyrénées). *Bulletin des services de la carte géologique de la France et des topographies souterraines*, 8, 131–133.

Lacroix, A. (1900). Sur l'origine des brèches calcaires secondaires de l'Ariège : conséquences à en tirer au point de vue de l'âge de la lherzolite. *Comptes rendus de l'académie des sciences Paris*, 131, 96–98.

Lacroix, A. (1917). Les péridotites des Pyrénées et les autres roches intrusives non feldspathiques qui les accompagnent. *Comptes rendus de l'académie des sciences Paris*, 165, 381–387.

Lacroix, A. (1920). Les roches éruptives du Crétacé des Pyrénées et la nomenclature des roches éruptives modifiées. *Comptes rendus de l'académie des sciences Paris*, 170, 685–690.

Labaume, P. and Teixell, A. (2020). Evolution of salt structures of the Pyrenean rift (Chaînons Béarnais, France): From hyper-extension to tectonic inversion. *Tectonophysics*, 785, 228451 [Online]. Available at: https://doi.org/10.1016/j.tecto.2020.228451.

Lagabrielle, Y. and Bodinier, J.L. (2008). Submarine reworking of exhumed subcontinental mantle rocks: Field evidence from the Lherz peridotites, French Pyrenees. *Terra Nova*, 20, 11–21 [Online]. Available at: https://doi.org/doi:10.1111/j.1365-3121.2007.00781.x.

Lagabrielle, Y. and Cannat, M. (1990). Alpine jurassic ophiolites resemble the modern central Atlantic basement. *Geology*, 18, 319–322.

Lagabrielle, Y., Bideau, D., Cannat, M., Karson, K.A., Mével, C. (1998). Ultramafic-mafic plutonic rock suites exposed along the Mid-Atlantic Ridge (10°N-30°N). Symmetrical-asymmetrical distribution and implications for seafloor spreading processes. In *Faulting and Magmatism at Mid-Ocean Ridges*, Buck, W.R., Delaney, P.T., Karson, J.A., Lagabrielle, Y. (eds). American Geophysical Union, Washington, D.C.

Lagabrielle, Y., Labaume, P., de Saint Blanquat, M. (2010). Mantle exhumation, crustal denudation, and gravity tectonics during Cretaceous rifting in the Pyrenean realm (SW Europe): Insights from the geological setting of the lherzolite bodies. *Tectonics*, 29(4), 1–26.

Lagabrielle, Y., Clerc, C., Vauchez, A., Lahfid, A., Labaume, P., Azambre, B., Fourcade, S., Dautria, J.-M. (2016). Very high geothermal gradient during mantle exhumation recorded in mylonitic marbles and carbonate breccias from a Mesozoic Pyrenean palaeomargin (Lherz area, North Pyrenean Zone, France). *Comptes rendus géoscience*, 348, 257–267.

Lagabrielle, Y., Asti, R., Fourcade, S., Corre, B., Poujol, M., Uzel, J., Labaume, P., Clerc, C., Lafay, R., Picazo, S. et al. (2019a). Mantle exhumation at magma-poor passive continental margins. Part I. 3D architecture and metasomatic evolution of a fossil exhumed mantle domain (Urdach lherzolite, north-western Pyrenees, France). *BSGF – Earth Sciences Bulletin*, 190(8), 1–26 [Online]. Available at: https://doi.org/10.1051/bsgf/2019007.

Lagabrielle, Y., Asti, R., Fourcade, S., Corre, B., Labaume, P., Uzel, J., Clerc, C., Lafay, R., Picazo, S. (2019b). Mantle exhumation at magma-poor passive continental margins. Part II: Tectonic and metasomatic evolution of large-displacement detachment faults preserved in a fossil distal margin domain (Saraillé lherzolites, northwestern Pyrenees, France). *BSGF – Earth Sciences Bulletin*, 190(14), 26–54.

Lagabrielle, Y., Asti, R., Duretz, T., Clerc, C., Fourcade, S., Teixell, A., Labaume, P., Corre, B., Saspiturry, N. (2020). A review of cretaceous smooth-slopes extensional basins along the Iberia-Eurasia plate boundary: How pre-rift salt controls the modes of continental rifting and mantle exhumation. *Earth-Science Reviews*, 201, 103071.

Lahfid, A., Beyssac, O., Deville, E., Negro, F., Chopin, C., Goffé, B. (2010). Evolution of the Raman spectrum of carbonaceous material in low-grade metasediments of the Glarus Alps (Switzerland). *Terra Nova*, 22(5), 354–360.

Le Roux, V., Bodinier, J.L., Tommasi, A., Alard, O., Dautria, J.M., Vauchez, A., Riches, A.J.V. (2007). The Lherz spinel lherzolite: Refertilized rather than pristine mantle. *Earth and Planetary Science Letters*, 259(3/4), 599–612.

Lefebvre, T. (1928). La structure de la moitié occidentale des Pyrénées françaises d'après P. Viennot. *Annales de géographie*, 37(209), 465–469.

Lehujeur, M., Chevrot, S., Villaseñor, A., Masini, E., Saspiturry, N., Lescoutre, R., Sylvander, S. (2021). Three-dimensional shear velocity structure of the Mauleon and Arzacq basins (Pyrenees). *BSGF – Earth Sciences Bulletin*, 192, 47 [Online]. Available at: https://doi.org/10.1051/bsgf/2021039.

Lemoine, M., Tricart, P., Boillot, G. (1987). Ultramafic and gabbroic ocean floor of the Ligurian Tethys (Alps, Corsica, Apennines): In search of a genetic imodel. *Geology*, 15(7), 622–625.

Lescoutre, R. and Manatschal, G. (2020). Role of rift-inheritance and segmentation for orogenic evolution: Example from the Pyrenean-Cantabrian system. *BSGF – Earth Sciences Bulletin*, 191, 18.

Lescoutre, R., Manatschal, G., Muñoz, J-A. (2021). Nature, origin and evolution of the Pyrenean-Cantabrian junction. *Tectonics*, 40, 1–40.

Masini, E., Manatschal, G., Tugend, J., Mohn, G., Flament, J.M. (2014). The tectonosedimentary evolution of a hyper-extended rift basin: The example of the Arzacq-Mauléon rift system (Western Pyrenees, SW France). *International Journal of Earth Sciences*, 103, 1–28.

Mattauer, M. (1964). Sur les schistosités d'âge tertiaire de la zone axiale des Pyrénées. *Comptes rendus de l'académie des sciences Paris*, 259(2891), 2891–2892.

Mattauer, M. (1968). Les traits structuraux essentiels de la chaîne des Pyrénées. *Revue de géologie dynamique et de géographie physique*, 10, 3–12.

Mattauer, M. and Proust, M. (1965). Sur la présence et la nature de deux importantes phases tectoniques dans les terrains secondaires des Pyrénées orientales. *Comptes rendus somm. soc. géol. France*, 4, 132–133.

Minnigh, L.D., van Calsteren, P.W.C., den Tex, E. (1980). Quenching: An additional model for emplacement of the lherzolite at Lers (French Pyrenees). *Geology*, 8(1), 18–21.

Moine, B., Fortune, J.P., Moreau, P., Viguier, F. (1989). Comparative mineralogy, geochemistry, and conditions of formation of two metasomatic talc and chlorite deposits; Trimouns (Pyrenees, France) and Rabenwald (Eastern Alps, Austria). *Economic Geology*, 84(5), 1398–1416.

Monchoux, P. (1970). Les lherzolites pyrénéennes : contribution à l'étude de leur minéralogie, de leur genèse et de leurs transformations. Thesis, Université de Toulouse.

Monchoux, P. (1972a). Roches à sapphirine au contact des lherzolites pyrénéennes. *Contributions to Mineralogy and Petrology*, 37, 47–64.

Monchoux, P. (1972b). Description des gisements de sapphirine associés aux lherzolites pyrénéennes I: Gisements de l'étang de Lherz. *Bulletin de la société d'histoire naturelle de Toulouse*, 410–419.

Monchoux, P., Fontan, F., de Parseval, P., Martin, R.F., Wang, R.C. (2006). Igneous albitite dikes in orogenic lherzolites, western Pyrénées, France: A possible source for corundum and alkali feldspar xenocrysts in basaltic terranes I: Mineralogical associations. *The Canadian Mineralogist*, 44, 811–836.

Monié, P., Soliva, J., Brunel, M., Maluski, H. (1994). Les cisaillements mylonitiques du granite de Millas (Pyrénées, France) ; âge Crétacé $^{40}Ar/^{39}Ar$ et interprétation tectonique. *Bulletin de la société géologique de France*, 165, 559–571.

Montigny, R., Azambre, B., Rossy, M., Thuizat, R. (1982a). Étude K/Ar du magmatisme basique lié au Trias supérieur des Pyrénées – Conséquences méthodologiques et pétrographiques. *Bulletin de minéralogie*, 105, 673–680.

Montigny, R., Azambre, B., Rossy, M., Thuizat, R. (1982b). K-Ar study of Cretaceous magmatism and metamorphism in the Pyrenees: Age and length of rotation of the Iberian Peninsula. *Tectonophysics*, 129, 257–273.

Mouthereau, F., Filleaudeau, P.Y., Vacherat, A., Pik, R., Lacombe, O., Fellin, M.G., Castelltort, S., Christophoul, F., Masini, E. (2014). Placing limits to shortening evolution in the Pyrenees: Role of margin architecture and implications for the Iberia/Europe convergence. *Tectonics*, 33, 2283–2314.

Nteme, J., Boiron, M.C., Lagabrielle, Y., Cathelineau, M., Quesnel, B., Boulvais, P. (2020). Fluid-rock interactions during continental rifting and mantle exhumation: the case of the Urdach Lherzolites body (North Pyrenees). *Journal of the Geological Society, London*, 178, 1–20.

Odlum, M.L. and Stockli, D.F. (2019). Thermotectonic evolution of the North Pyrenean Agly Massif during Early Cretaceous hyperextension using multi-mineral U-Pb thermochronometry. *Tectonics*, 38, 1509–1531.

Odlum, M.L. and Stockli, D.F. (2020). Geochronologic constraints on deformation and metasomatism along an exhumed mylonitic shear zone using apatite U-Pb, geochemistry, and microtextural analysis. *Earth and Planetary Science Letters*, 538, 116177.

Olivier, P., Gleizes, G., Paquette, J.L. (2004). Gneiss domes and granite emplacement in an obliquely convergent regime: New interpretation of the Variscan Agly Massif (Eastern Pyrenees, France). In *Gneiss Domes in Orogeny*, Whitney, D.L., Teyssier, C., Siddoway, C.S. (eds). Geological Society of America, Boulder, CO.

Olivier, P., Gleizes, G., Paquette, J.-L., Muñoz Sáez, C. (2008). Structure and U–Pb dating of the Saint-Arnac pluton and the Ansignan charnockite (Agly Massif): A cross-section from the upper to the middle crust of the Variscan Eastern Pyrenees. *Journal of the Geological Society, London*, 165, 141–152.

Paquet, J. and Mansy, J.L. (1991). La structure de l'Est de Pyrénées (transversale du massif de l'Agly) : un exemple d'amincissement crustal. *Comptes rendus de l'académie des sciences Paris*, 312, 913–919.

Passchier, C.W. (1982). Mylonitic deformation in the St Barthélémy Massif, French Pyrenees, with emphasis on the generation relationship between ultramylonite and pseudotachylite. *GUA Pap. Geol.*, 1(16), 1–173.

Pedrera, A., García-Senz, J., Ayala, C., Ruiz-Constán, A., Rodríguez-Fernández, L.R., Robador, A., González-Menéndez, L. (2017). Reconstruction of the exhumed mantle across the North Iberian Margin by crustal-scale 3-D gravity inversion and geological cross section. *Tectonics*, 36(12), 3155–3177.

Péron-Pinvidic, G. and Manatschal, G. (2009). The final rifting evolution at deep magma-poor passive margins from Iberia-Newfoundland: A new point of view. *International Journal of Earth Sciences*, 98(7), 1581–1597.

Peybernès, B. (1976). Le Jurassique et le Crétacé inférieur des Pyrénées franco-espagnoles entre la Garonne et la Méditerranée. PhD Thesis, Université de Toulouse.

Peybernès, B. and Souquet, P. (1984). Basement blocks and tecto-sedimentary evolution in the Pyrenees during mesozoic times. *Geol. Mag.*, 121(4), 397–539.

Pin, C. and Vielzeuf, D. (1983). Granulites and related rocks in Variscan median Europe; a dualistic interpretation. *Tectonophysics*, 93, 47–74.

Pin, C., Paquette, J.-L., Monchoux, P., Hammouda, T. (2001). First field-scale occurrence of Si-Al-Na-rich low-degree partial melt from the upper mantle. *Geology*, 29, 451–454.

Pin, C., Monchoux, P., Paquette, J.-L., Azambre, B., Wang, R.C., Martin, R.F. (2006). Igneous albitite dikes in orogenic lherzolites, Western Pyrénées, France: A possible source for corundum and alkali feldspar xenocrysts in basaltic terranes. II. Geochemical and petrogenetic considerations. *The Canadian Mineralogist*, 44, 843–856.

Poujol, M., Boulvais, P., Kosler, J. (2010). Regional-scale Cretaceous albitization in the Pyrenees: Evidence from in situ U-Th-Pb dating of monazite, titanite and zircon. *J. Geol. Soc.*, 167(4), 751–767.

Quesnel, B., Boiron, M.C., Cathelineau, M., Truche, L., Rigaudier, T., Bardoux, G., Agrinier, P., de Saint Blanquat, M., Masini, E., Gaucher, E.C. (2019). Nature and origin of mineralizing fluids in hyper-extensional systems, the case of Cretaceous Mg-metasomatism in the Pyrenees. *Geofluids*, 7213050, 18.

Ravier, J. (1959). Le métamorphisme des terrains secondaires des Pyrénées. *Mémoires de la Société géologique de France*, 86, 1–250.

Ravier, J. and Thiébaut, J. (1982). Lagunar origin of mesozoic marbles and hornfelds at the Col d'Agnes (Ariège). *Comptes rendus de l'académie des sciences Paris*, 294(2), 127–130.

Reisberg, L. and Lorand, J.-P. (1995). Longevity of sub-continental mantle lithosphere from osmium isotope systematics in orogenic peridotite massifs. *Nature*, 376, 159–162.

Rossy, M. (1988). Contribution à l'étude du magmatisme mésozoïque du domaine Pyrénéen. I. Le Trias dans l'ensemble du domaine. II. Le Crétacé dans les provinces Basques d'Espagne. PhD Thesis, Université de Besançon.

Rossy, M., Azambre, B., Albarède, F. (1992). REE and Sr/1bNd isotope geochemistry of the alkaline magmatism from the Cretaceous North Pyrenean Rift Zone (France–Spain). *Chemical Geology*, 97(1/2), 33–46.

Roux, J.C. (1983). Recherches stratigraphiques et sédimentologiques sur les flyschs crétacés pyrénéens au sud d'Oloron (Pyrénées-Atlantiques). PhD Thesis, Université Paul Sabatier, Toulouse.

de Saint Blanquat, M. (1993). The ductile normal fault of the Saint Barthélémy Massif Variscan evolution of the north-pyrenean catazonal massifs considered from the viewpoint of their thermal history. *Geodin. Acta*, 59–77.

de Saint Blanquat, M., Brunel, M., Mattauer, M. (1986). Les zones de cisaillement du massif nord-pyrénéen du Saint-Barthélémy, témoins probables de l'extension crustale d'âge crétacé. *Comptes rendus de l'académie des sciences Paris*, 303(14), 1339–1344.

de Saint Blanquat, M., Lardeaux, J.M., Brunel, M. (1990). Petrological arguments for high-temperature extensional deformation in the Pyrenean Variscan crust (Saint Barthélémy Massif, Ariège, France). *Tectonophysics*, 177, 245–262.

de Saint Blanquat, M., Bajolet, F., Grand'Homme, A., Proietti, A., Zanti, M., Boutin, A., Clerc, C., Lagabrielle, Y., Labaume, P. (2016). Cretaceous mantle exhumation in the central Pyrenees: New constraints from the peridotites in eastern Ariège (North Pyrenean zone, France). *Comptes rendus géoscience*, 348, 268–278.

Salardon, R., Carpentier, C., Bellahsen, N., Pironon, J., France-Lanord, C. (2016). Interactions between tectonics and fluid circulations in an inverted hyper-extended basin: Example of mesozoic carbonate rocks of the western North Pyrenean Zone (Chaînons Béarnais, France). *Marine and Petroleum Geology*, 80, 563–586.

Saspiturry, N. (2019). Évolution sédimentaire, structurale et thermique d'un rift hyperaminci : de l'héritage post-hercynien à l'inversion alpine. Exemple du bassin de Mauléon (Pyrénées). PhD Thesis, Université de Bordeaux.

Saspiturry, N., Cochelin, B., Razin, P., Leleu, S., Lemirre, B., Bouscary, C., Allanic, C. (2019a). Tectono-sedimentary evolution of a rift system controlled by Permian postorogenic extension and metamorphic core complex formation (Bidarray Basin and Ursuya dome, Western Pyrenees). *Tectonophysics*, 768, 228180 [Online]. Available at: https://doi.org/10.1016/j.tecto.2019.228180.

Saspiturry, N., Razin, P., Baudin, T., Serrano, O., Issautier, B., Lasseur, E., Leleu, S. (2019b). Symmetry vs. asymmetry of a hyper-thinned rift: Example of the Mauléon Basin (Western Pyrenees, France). *Marine and Petroleum Geology*, 104, 86–105 [Online]. Available at: https://doi.org/10.1016/j.marpetgeo.2019.03.031.

Saspiturry, N., Lahfid, A., Baudin, T., Guillou-Frottier, L., Razin, P., Issautier, B., Le Bayon, B., Serrano, O., Lagabrielle, Y., Corre, B. (2020a). Paleogeothermal gradients across an inverted hyperextended rift system: Example of the Mauléon fossil rift (Western Pyrenees). *Tectonics*, 39, e2020TC006206 [Online]. Available at: https://doi.org/10.1029/2020TC006206.

Saspiturry, N., Razin, P., Allanic, C., Issautier, B., Baudin, T., Lasseur, E., Leleu, S. (2020b). Closure of a hyperextended system in an orogenic lithospheric pop-up, Western Pyrenees: The role of mantle buttressing and rift structural inheritance. *Terra Nova*, 32(4), 253–260 [Online]. Available at: https://doi.org.10.1111/ter.12457.

Saspiturry, N., Issautier, B., Razin, P., Andrieu, S., Lasseur, E., Allanic, C., Serrano, O., Baudin, T., Cochelin, B. (2021a). Review of the syn-rift to early post-rift depositional systems of the Cretaceous Mauléon rift: Sedimentary record of continental crust hyperextension and mantle denudation (Western Pyrenees). *Bulletin de la société géologique de France* [Online]. Available at: https://doi.org. 10.1051/bsgf/2021044.

Saspiturry, N., Issautier, B., Razin, P., Baudin, T., Asti, R., Lagabrielle, Y., Allanic, C., Serrano, O. (2021b). Review of Iberia-Eurasia plate-boundary basins: Role of sedimentary burial on depth-dependent continental crust ductile thinning during rifting and continental breakup. *Basin Research*, 1626–1661 [Online]. Available at: https://doi.org. 10.1111/bre.12529.

Schärer, U., de Parseval, P., Polvé, M., de Saint Blanquat, M. (1999). Formation of the Trimouns talc-chlorite deposit (Pyrenees) from persistent hydrothermal activity between 112 and 97 Ma. *Terra Nova*, 11(1), 30–37.

Souquet, P. and Peybernès, B. (1987). Allochtonie des massifs primaires nord-pyrénéens des Pyrénées centrales. *Comptes rendus de l'académie des sciences Paris*, 305, 733–739.

Souquet, P., Bilotte, M., Canérot, J., Debroas, E.J., Peybernes, B., Rey, J. (1975). Nouvelle interprétation de la structure des Pyrénées. *Comptes rendus de l'académie des sciences Paris*, 281, 609–612.

Souquet, P., Peybèrnes, B., Bilotte, M., Debroas, E.J. (1977). La chaîne alpine des Pyrénées. *Géologie Alpine*, 53(2), 193–216.

Souquet, P., Debroas, E.J., Boirie, J.M., Pons, P., Fixari, G., Dol, J., Thieuloy, J.P., Bonnemaison, M., Manivit, H., Peybernès, B. (1985). Le groupe du Flysch noir (albo-cénomanien) dans les Pyrénées. *Bulletin des centres de recherches exploration-production Elf-Aquitaine*, 9(1), 183–252.

Souquet, P., Peybernès, B., Combes, P.J. (1990). Corrélations séquentielles et modèles de séquences de dépôt dans l'Albien – Cénomanien basal des Pyrénées franco-espagnoles. *Comptes rendus de l'académie des sciences Paris*, 311(2), 1235–1241.

Teixell, A., Labaume, P., Lagabrielle, Y. (2016). The crustal evolution of the west-central Pyrenees revisited: Inferences from a new kinematic scenario. *Comptes rendus géoscience*, 348, 257–267.

Teixell, A., Labaume, P., Ayarza, P., Espurt, N., de Saint Blanquat, M., Lagabrielle, Y. (2018). Crustal structure and evolution of the Pyrenean-Cantabrian belt: A review and new interpretations from recent concepts and data. *Tectonophysics*, 724, 146–170.

Ternet, Y., Colchen, M., Debroas, E.J., Azambre, B., Debon, F., Bouchez, J.L., Gleizes, G. (1997). Notice explicative, Carte géologie France (1/50 000), feuille Aulus-les-Bains (1086). BRGM éditions, Orléans.

Thiébaut, J., Debeaux, M., Durand-Walkenheim, C., Souquet, P., Gourinard, Y., Bandet, Y., Fondecave-Wallez, M.J. (1988). Métamorphisme et halocinèse crétacés dans les évaporites de Betchat, le long du chevauchement frontal nord-pyrénéen (Haute-Garonne et Ariège, France). *Comptes rendus de l'académie des sciences Paris*, 307, 1135–1150.

Thiébaut, J., Durand-Walkenheim, C., Debeaux, M., Souquet, P. (1992). Métamorphisme des évaporites triasiques du versant nord des Pyrénées centrales et occidentales. *Bulletin de la société d'histoire naturelle de Toulouse*, 128, 77–84.

Tournaire-Guille, B., Olivier, P., Paquette, J.L., Bosse, V., Guillaume, D. (2019). Evolution of the middle crust of the Pyrenees during the Paleozoic: New data on the plutonic rocks from the North Pyrenean Agly Massif. *International Journal of Earth Sciences*, 108(1), 245–265.

Tugend, J., Manatschal, G., Kusznir, N.J., Masini, E., Mohn, G., Thinon, I. (2014). Formation and deformation of hyperextended rift systems: Insights from rift domain mapping in the Bay of Biscay-Pyrenees. *Tectonics*, 33, 1239–1276 [Online]. Available at: https://doi.org/10.1002/2014TC003529.

Tugend, J., Manatschal, G., Kusznir, N.J., Masini, E. (2015). Characterizing and identifying structural domains at rifted continental margins: Application to the Bay of Biscay margins and its Western Pyrenean fossil remnants. *Geol. Soc. Lond. Spec. Publ.*, 413, 171–203.

Uzel, J., Lagabrielle, Y., Fourcade, S., Chopin, C., Monchoux, P., Clerc, C., Poujol, M. (2020). The sapphirine-bearing rocks in contact with the Lherz peridotite body: New mineralogical data, age and attempt at interpretation. *BSGF – Earth Sciences Bulletin*, 191(5).

Vacherat, A., Mouthereau, F., Pik, R., Bernet, M., Gautheron, C., Masini, E., Lahfid, A. (2014). Thermal imprint of rift-related processes in orogens as recorded in the Pyrenees. *Earth Planet. Sci. Lett.*, 408, 296–306.

Vauchez, A., Clerc, C., Bestani, L., Lagabrielle, Y., Chauvet, A., Lahfid, A., Mainprice, D. (2013). Preorogenic exhumation of the North Pyrenean Agly Massif (Eastern Pyrenees-France). *Tectonics*, 32(2), 95–106.

Vétil, J.Y., Lorand, J.P., Fabriès, J. (1988). Conditions de mise en place des filons des pyroxénites à amphibole du massif ultramafique de Lherz (Ariège, France). *Comptes rendus de l'académie des sciences Paris*, 307, 587–593.

Vielzeuf, D. (1984). Relations de phase dans le faciès granulite et implications géodynamiques. L'exemple des granulites des Pyrénées. PhD Thesis, Université de Clermont II, Clermont-Ferrand.

Vielzeuf, D. and Kornprobst, J. (1984). Crustal splitting and the emplacement of Pyrenean lherzolites and granulites. *Earth Planet. Sci. Lett.*, 67, 87–96.

Viers, G. (1953). La structure du Pays Basque d'après les travaux récents. *Revue géographique des Pyrénées et du Sud-Ouest. Sud-Ouest Européen*, 24(2), 154–159.

Vissers, R.L.M., Drury, M.R., Newman, J., Fliervoet, T.F. (1997). Mylonitic deformation in upper mantle peridotites of the North Pyrenean Zone (France): Implications for strength and strain localization in the lithosphere. *Tectonophysics*, 279, 303–325.

Wang, Y., Chevrot, S., Monteiller, V., Komatitsch, D., Mouthereau, F., Manatschal, G., Martin, R. (2016). The deep roots of the western Pyrenees revealed by full waveform inversion of teleseismic P waves. *Geology*, 44, 475–478 [Online]. Available at: https://doi.org/10.1130/G37812.1.

# 6

# The Bay of Biscay Opening in the Frame of the Iberian-European Plate Boundary Formation

**Julie TUGEND[1,2] and Isabelle THINON[3]**

*[1] ISTeP, CNRS, INSU, Sorbonne Université, Paris, France*

*[2] GEC, CY Cergy Paris Université, France*

*[3] BRGM, Orléans, France*

This chapter aims to review the present-day structure of the Bay of Biscay oceanic basin and its continental margins, as well as the geodynamic framework of its opening. Located offshore France and Spain (western Europe), the Bay of Biscay corresponds to a Cretaceous V-shaped oceanic basin terminating to the east within the Parentis Basin. While the structure of the northern Bay of Biscay passive margin is well-preserved, sharing characteristic features with other magma-poor rifted margins, its southern counterpart has been inverted and integrated as part of the accretionary prism of a fossil proto-subduction trench during the Upper Cretaceous to Cenozoic Pyrenean orogeny. Continental breakup initiated during the latest Early Cretaceous time and propagated eastwards, resulting in the transient localization of a plate boundary between Europe and the fragmented Iberian plate. At the large scale, the formation and evolution of the Bay of Biscay relates to the northward propagation of the southern North Atlantic Ocean and plate kinematic evolution

---

For a color version of all the figures in this chapter, see www.iste.co.uk/saspiturry/evolution1.zip.

*Evolution of the Pyrenees during the Variscan and Alpine Cycles 1*,
coordinated by Nicolas SASPITURRY, Jessica UZEL, Alexandre ORTIZ. 

between Europe, Africa and fragmented Iberia, but also between North America and Europe.

## 6.1. Introduction

During the early stages of marine geosciences, numerous studies focused on the Bay of Biscay abyssal plain and its margins (Figure 6.1(a)), providing pioneer results on the structure of oceanic basins and their transition with continents. These first data sets highlighted the presumably oceanic nature of the V-shaped Bay of Biscay abyssal plain (Matthews and Williams 1968; Bacon et al. 1969; Damotte et al. 1971; Le Borgne et al. 1971; Montadert et al. 1971, 1979; Grau et al. 1973). They also showed it was bordered to the north by a passive continental margin and to the south by a fossil subduction trench (Montadert and Winnock 1971; Sibuet and Le Pichon 1971; Boillot et al. 1979; Montadert et al. 1979; Avedik et al. 1982; Derégnaucourt and Boillot 1982; Grimaud et al. 1982; Ginzburg et al. 1985; Whitmarsh et al. 1986; Sibuet et al. 1993). Though sparse at the time, seismic and gravity data also revealed the unexpected ambiguous structure, neither typically oceanic nor continental, of the wide Armorican basin at the toe of the northern continental passive margin of the Bay of Biscay (e.g. Bacon et al. 1969; Montadert et al. 1971, 1974, 1979; Avedik and Howard 1979; Le Pichon and Sibuet 1981; Derégnaucourt 1981; Derégnaucourt and Boillot 1982; Vaillant 1988; Figure 6.1(a)).

Following these early investigations, numerous data sets have progressively been collected and processed, including reflection and refraction seismic profiles (see compilations in Montadert et al. (1971, 1979); Thinon (1999); Fernández-Viejo et al. (2021)), rocks sampled during dredges and submersible dives (Pastouret et al. 1974; Pastouret and Auffret 1976; Auffret et al. 1979; Malod and Boillot 1980; Groupe CYMOR et al. 1981; Pastouret and Maury 1982; Boillot and Malod 1988), borehole data (Figure 6.2; DSDP leg 12, sites 118 and 119: Laughton and Berggren (1972) and DSDP leg 48, sites 400, 401 and 402: Montadert and Roberts (1979) and petroleum wells), as well as magnetic and gravity data (e.g. Matthews and Williams 1968; Bacon et al. 1969; Damotte et al. 1971; Galdeano et al. 1989). A shift in our knowledge of the Bay of Biscay margins and its ocean-continent transitions (OCTs) occurred in the 1980s and 1990s, with the acquisition of increasingly higher resolution petroleum (SNEAp, 1980–1981) and academic multichannel reflection and refraction seismic profiles (NORGASIS, IAM, ESCIN, MARCONI surveys; e.g. Avedik and Olivet 1994; Gallart et al. 1995; Alvarez-Marron et al. 1996, 1997; Pulgar et al. 1996; Fernandez-Viejo et al. 1998, 2011; Thinon 1999; Roca et al. 2011).

The northern Bay of Biscay became a laboratory to investigate the stratigraphic, structural and subsidence evolution of passive margins during rifting and continental breakup, but also during convergence (Avedik et al. 1982; Barbier et al. 1986; Whitmarsh et al. 1986; Thinon 1999; Thinon et al. 2001, 2002, 2003; Tugend et al. 2014, 2015b), while its southern counterpart became a key location to study the structure and formation processes of accretionary prisms upon subduction initiation (Gallart et al. 1995; Alvarez-Marron et al. 1996, 1997; Pulgar et al. 1996; Fernandez-Viejo et al. 1998, 2011; Gallastegui et al. 2002; Ferrer et al. 2008; Pedreira et al. 2015). Furthermore, being at the junction between the European, Iberian and North American plates, the characterization of the Bay of Biscay crustal structure and the understanding of the different steps of its formation are crucial to constrain plate kinematic reconstructions. Many different models have been proposed and remain debated, inferring a strike slip, transtensional and/or orthogonal opening of the Bay of Biscay (Carey 1958; Montadert and Winnock 1971; Le Pichon et al. 1971; Ferragne et al. 1978; Ries 1978; Olivet et al. 1984; Malod and Mauffret 1990; Roest and Srivastava 1991; Sibuet and Collette 1991; Olivet 1996; Vissers et al. 2012; Nirrengarten et al. 2018; King et al. 2021).

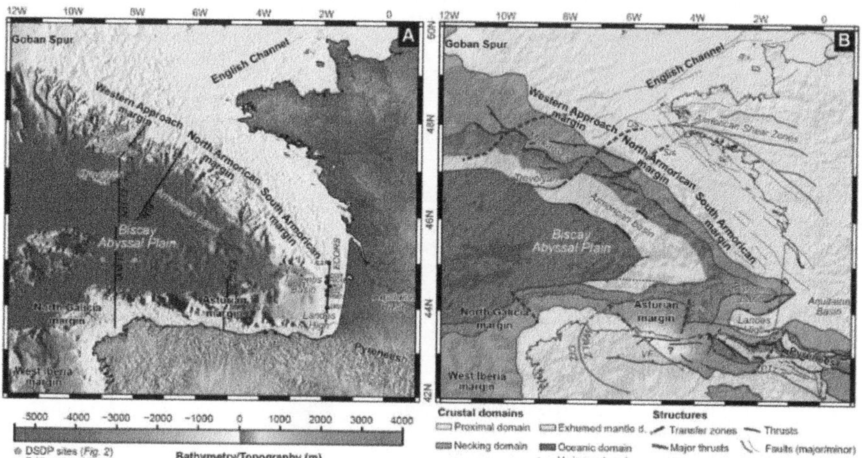


**Figure 6.1.** *Maps of the Bay of Biscay and the Pyrenees showing bathymetry and the main structural domains inherited from Cretaceous rifting*

LEGEND FOR FIGURE 6.1.– *(a) Bathymetric map of the Bay of Biscay with the main segments and morphological features indicated (bathymetry and topography from Becker et al. (2009)). The location of DSDP legs 12 and 48 and interpreted seismic reflection profiles presented in this chapter are also shown. (b) Structural map of the rift system formed during the opening of the Bay of Biscay (from Thinon (1999);*

*Thinon et al. (2003); Tugend et al. (2014)) with updates from Tugend et al. (2015b) (eastern oceanic domain); Druet et al. (2018) (North Galicia margin); Yang et al. (2020) (Goban Spur) and Le Maire et al. (2021) (Western Approaches, Armorican and Iberian continental shelves). Os: Ouessant system, Sis: Sizun system from Le Maire et al. (2021).*

Hereafter, we will present the results of pioneer and most recent investigations revealing the 3D structure of the Bay of Biscay and its continental margins, reviewing some of the most significant along strike variations of the stratigraphic and crustal structure. These observations and their interpretations are then integrated at a larger scale in the geodynamic framework of the formation of the Iberian-European plate boundary, highlighting its polyphase spatial and temporal evolution, and complex partitioning between transtensional and extensional deformation.

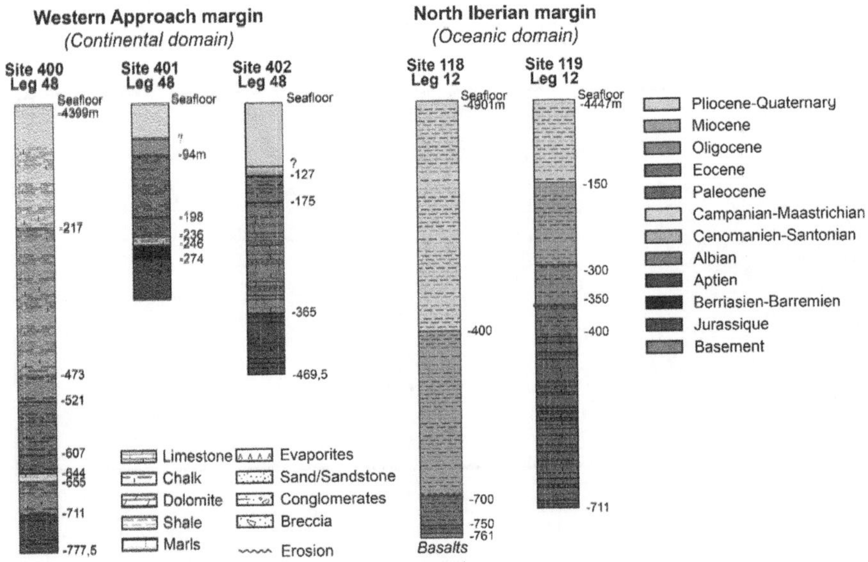


**Figure 6.2.** *Drill sites from the Deep Sea Drilling Program (DSDP) located in the Bay of Biscay (as compiled in Tugend (2013))*

LEGEND FOR FIGURE 6.2.– *Sites 400, 401 and 402 of leg 48 drilled the sedimentary cover of the Western Approach continental margin (Montadert and Roberts 1979; The Shipboard Scientific Party 1979a, 1979b, 1979c). Sites 118 and 119 of leg*

*12 drilled the sedimentary cover and uppermost part of the basement of the Bay of Biscay oceanic abyssal plain (Laughton et al. 1972a, 1972b; Laughton and Berggren 1972).*

## 6.2. Structure and age of the Bay of Biscay

The Bay of Biscay is a V-shape oceanic basin formed between the European and Iberian plates during the northward propagation of the Atlantic Ocean (Figure 6.1). It is bordered to the north by the North Biscay passive margin, where rift structures are well preserved (Figure 6.3; De Charpal et al. 1978; Montadert et al. 1979; Barbier et al. 1986; Thinon 1999; Thinon et al. 2003; Tugend 2013; Tugend et al. 2014). The North Iberian margin forms its conjugate; however, the former passive margin is not preserved anymore, and is now largely integrated in the E-W trending accretionary wedge framing a former incipient subduction zone related to the formation of the Pyrenean orogen (Figure 6.4; Boillot et al. 1979; Alvarez-Marron et al. 1997; Fernandez-Viejo et al. 1998, 2013; Gallastegui et al. 2002; Tugend et al. 2014, 2015; Pedreira et al. 2015; Cadenas et al. 2018, 2020). To the east, the two conjugate margins meet in the Aquitaine margin, where the V-shape Bay of Biscay terminates in the Parentis Basin (Figure 6.5).

The increasing acquisition of reflection and refraction seismic data combined with magnetic data, gravity modeling and inversions, progressively revealed the structure and distribution of continental and oceanic domains and in-between OCT in the Bay of Biscay and its continental margins (Figure 6.1(b); Thinon et al. 2003; Roca et al. 2011; Tugend et al. 2014, 2015b; Ruiz et al. 2017; Cadenas et al. 2018; Druet et al. 2018; Le Maire et al. 2021). A further distinction is made in the continental domain to highlight variations in crustal and lithospheric thinning during rifting, which can be sub-divided into the proximal, necking and hyper-thinned domains (see geological and geophysical mapping criteria in Tugend et al. (2015a)). The mapping of these structural and crustal domains hence shows the different stages of the Bay of Biscay passive margin formation in space (Figure 6.1(b)).

Along-strike variability of the Bay of Biscay margins crustal structure are observed and interpreted as the result of heterogeneities in the pre-rift basement, spatial variations in stress orientation during rifting, variable preservation to compressional deformation during the Pyrenean orogeny and variable pre-rift salt distribution (e.g. Thinon 1999; Thinon et al. 2003; Ferrer et al. 2008; Jammes et al. 2009, 2010b, 2010c; Roca et al. 2011; Tugend et al. 2014, 2015b; Cadenas et al. 2018, 2020; Le Maire et al. 2021). A set of accommodation zones or transfer zones has tentatively been identified, delimiting different margin segments: the Goban,

Western Approaches, North and South Armorican segments of the North Biscay passive margin and the Galician and Asturian segments of the North Iberian margin (Figure 6.1).

### 6.2.1. *The North Biscay passive margin*

The North Biscay rifted margin was initially considered as the archetype of a non-volcanic passive margin characterized by a series of titled blocs delimited by listric faults and the absence of any syn-rift volcanism (e.g. Montadert et al. 1974, 1979b; de Charpal et al. 1978; Avedik et al. 1979, 1982; Ginzburg et al. 1985; Barbier et al. 1986; Whitmarsh et al. 1986; Le Pichon and Barbier 1987). Evidence of local volcanic edifices emplaced during late rifting in the interpreted exhumed mantle of the OCT (e.g. Thinon et al. 2003; Tugend et al. 2014) indicates that it is best to refer to it as a magma-poor rifted margin.

The increasing acquisition of seismic data showed significant along strike changes of the North Biscay passive margin crustal structure with the Goban, Western Approaches, northern and southern Armorican passive margin segments being distinguished from west to east (Figure 6.1; Thinon et al. 2003; Bullock 2004; Bullock and Minshull 2005; Tugend et al. 2014), but only the structure of the Western Approaches and Armorican passive margin segments is presented hereafter (Figure 6.3). A set of transfer zones located in the continuity of major Variscan and post-Variscan structures has been interpreted to delimit these segments (Thinon 1999; Thinon et al. 2003; Tugend et al. 2014; Le Maire et al. 2021). However, the transition from one segment to the other is often progressive, suggesting that these transfer zones are complex accommodation zones, whose trend is at least partly controlled by crustal or even lithospheric scale pre-rift structures (Thinon 1999; Thinon et al. 2003; Tugend et al. 2014; Le Maire et al. 2021).

#### 6.2.1.1. *The Western Approach passive margin*

The Western Approaches passive margin segment is located in the south-westward continuation of the English Channel (Figure 6.1). The English Channel is structured by a series of NE-SW trending basins (Western Approaches Basins (WAB); e.g. Pinet et al. 1991) formed during earlier rift episodes in the Permo-Triassic and Late Triassic to Early Jurassic (Ruffell 1995). The Mesozoic structure of these WAB is therefore expected to interfere with that of the N105°-trending Western Approach margin (Montadert et al. 1979; Vaillant 1988; Thinon 1999). However, it remains poorly constrained.

The crustal structure of the Western Approach margin is imaged along the Norgasis 11 and 12 seismic reflection profiles (Figure 6.3) and constrained by refraction seismic data (Avedik 1975; Avedik and Howard 1979; Avedik et al. 1982; Ginzburg et al. 1985; Whitmarsh et al. 1986), as well as gravity inversion results (Tugend et al. 2014). Its sedimentary cover was sampled by several DSDP wells (sites 400, 401, 402, Figure 6.2; Montadert and Roberts et al. 1979) and dredged along escarpments (Pastouret et al. 1974; Pastouret and Auffret 1976; Auffret et al. 1979; Groupe CYAMOR et al. 1981; Pastouret and Maury 1982).

Landward of the Meriadzek Terrace, the proximal domain shows crustal thicknesses ranging from 30 to 25 km thick (e.g. Avedik and Howard 1979; Tugend et al. 2014). Seaward of a horst delimiting the proximal domain, crustal thinning from ~25 to ~15 km is progressive and seems mostly accommodated under the Meriadzek Terrace (Figures 6.1 and 6.3, necking domain), seaward of which the crust is thinned to less than 10 km (Figures 6.1 and 6.3, hyperthinned domain). The end of the continental crust is separated from oceanic domain by a nearly 50 km wide transitional domain identified in seismic refraction data (Avedik et al. 1982; Thinon 1999). Even if the nature of this domain is not constrained by boreholes, gravity inversion results are compatible with an OCT formed by exhumed serpentinized mantle (Figure 6.3; Whitmarsh et al. 1986; Tugend et al. 2014).

Early studies showed that the architecture of the Western Approach margin was characterized by a series of tilted blocs delimited by oceanward dipping listric faults soling out onto a remarkable reflector referred to as the S-reflector and commonly interpreted as the brittle-ductile transition (De Charpal et al. 1978, Barbier et al. 1986; Thinon 1999). The structure of the margin is not solely characterized by oceanward-rotated fault blocs (as already highlighted by Vaillant (1988)); a series of supra-detachment basins controlled by larger offset normal faults have also been identified in the necking domain or in the OCT (Tugend et al. 2014, 2015a). Even if the North Biscay margin escaped most of the Pyrenean deformation, evidence of compressional deformation is recorded at the edge of the continental crust forming the Trevelyan structure (Debyser et al. 1971; Montadert et al. 1971, 1974, 1979b; Deregnaucourt and Boillot 1982; Thinon 1999; Thinon et al. 2001), most likely using the serpentinized mantle of the OCT as a decollement layer (Figure 6.3; Tugend et al. 2014). Compressional deformation is also observed at the boundaries between crustal domains and is more diffuse in the oceanic domain (Thinon et al. 2001; Tugend et al. 2014), possibly resulting from the inversion of normal faults that initially structured the upper part of oceanic crust.

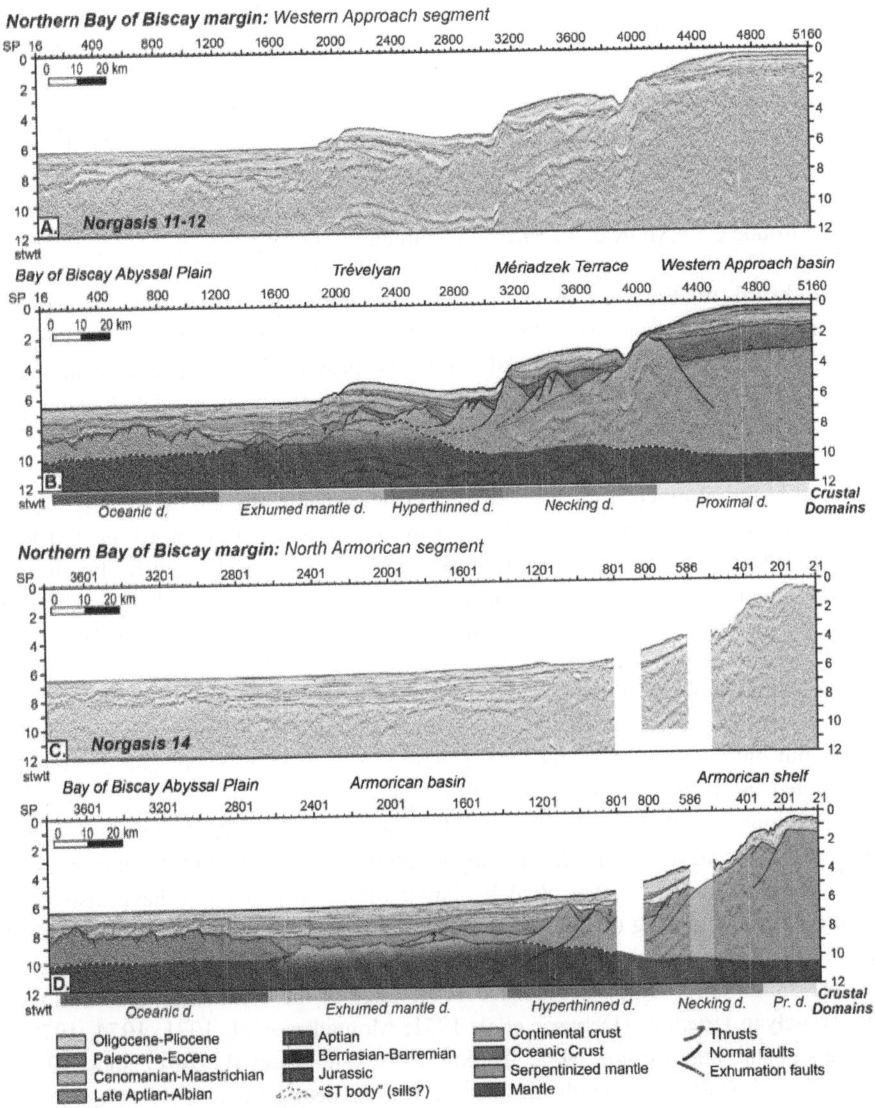


**Figure 6.3.** *Architecture of the North Biscay passive margin*

LEGEND FOR FIGURE 6.3.– *(a) Norgasis 11-12 seismic reflection profiles across the Western Approach passive margin segment (Avedik and Olivet 1994; Thinon 1999); (b) geological interpretation of the Norgasis 11-12 seismic reflection profiles*

*(Tugend et al. 2014); (c) Norgasis 14 seismic profile across the north Armorican passive margin segment (Avedik and Olivet 1994; Thinon 1999; Thinon et al. 2003); (d) geological interpretation of the Norgasis 14 seismic reflection profile (Thinon et al. 2003; Tugend 2013).*

Typical syn-rift geometries like growth strata are rarely observed (Vaillant 1988; Thinon et al. 2002), but seismic observations calibrated from the DSDP drilling results combined to subsidence analyses indicate that the margin formed subsequently to a latest Jurassic to Early Cretaceous rifting (Montadert et al. 1971, 1979; Derégnaucourt 1981; Derégnaucourt and Boillot 1982; Thinon et al. 2002). Even though we cannot exclude that previous rift events like the Late Triassic rifting contributed to shaping the present-day structure of the Western Approach (Vaillant 1988), such contributions remain to be observed and quantitatively evaluated. Despite little direct constraints on syn-rift sequences older than the Aptian, different stages have been distinguished. Rifting likely initiated just before or at the beginning of the Berriasian, but the main subsidence occurred from Barremian to Aptian time according to stratigraphic correlations (Thinon et al. 2002). Rift basins over the necking domain are filled by Berriasian to Barremian sediments, while the same sequence is tilted in the hyperthinned domain (Thinon et al. 2002), suggesting that extensional faulting was delayed to Aptian time in the distal margin (Figure 6.3; Thinon et al. 2002; Tugend et al. 2015a). The seismic facies of the first sequence deposited over the OCT is atypical and has been interpreted as allochthonous sediments emplaced as mega-slumps before oceanic accretion onset (3B sequence; Thinon et al. 2002) in latest Aptian to Albian time.

### 6.2.1.2. *Northern and southern Armorican margins*

The Armorican passive margin segment bridges the Western Approach segment to the north-west with the northern part of the Parentis Basin to the south-east (Figure 6.1). A northern and a southern segment of the Armorican passive margin are distinguished based on a change in the trend of the passive margin passing from N100°–N110° to N130°–N140° from north to south (Figure 6.1). The Armorican passive margin segment is characterized by a wide shelf. Though the outer structure of the Armorican shelf remains poorly constrained, continentward structures are dominated by Variscan and post-Variscan trends (e.g. Lefort 1975, 1980; Chantraine et al. 2003; Ballèvre et al. 2009, 2012; Le Maire et al. 2021).

The crustal structure of the Armorican margin is here imaged along the Norgasis 14 seismic reflection profiles (Figure 6.3). A coincident refraction line covers the distal part of the passive margin, from the edge of the continental crust to the first oceanic crust (Thinon et al. 2003). Complementary constraints on crustal thickness variations were determined from gravity inversion (Tugend et al. 2014) or using forward gravity modeling (Thinon et al. 2003). On the shelf, the sedimentary cover was sampled and dated because of petroleum wells located on the Armorican shelf (PENMA) and in the nearby English Channel (Thinon 1999; Thinon et al. 2002). In the abyssal plain, the sedimentary cover of the Armorican Basin was calibrated by seismic correlation with the DSDP-Leg 48 wells located in the adjacent Western Approaches segment (Thinon et al. 2002).

The architecture of the Armorican passive margin differs in many ways from that of the Western Approach. The thinning of the continental crust is more abrupt, resulting in a narrower necking domain, while only few grabens and tilted blocs occur in the hyperthinned domain (Figure 6.3; Thinon et al. 2003; Tugend et al. 2014). The OCT of the Armorican segment hosts a ~80 km-wide deep basin, the Armorican basin, whose basement shows atypical high velocity structure (7.4–7.5 km/s; Thinon et al. 2003). Based on analogies with the West Iberia OCT whose basement was drilled, the Armorican basin is interpreted to be floored by serpentinized exhumed mantle (Thinon et al. 2003), a hypothesis also compatible with gravity inversion results (Tugend 2013; Tugend et al. 2014). The Armorican passive margin segment escaped most of the Pyrenean deformation. Local inversions and reactivations are observed at the boundary between crustal domains, notably between the oceanic domain and the OCT (Figure 6.3; Thinon et al. 2001; Tugend et al. 2014). During Pyrenean convergence, the landward edge of the oceanic domain was uplifted creating a plateau (Thinon et al. 2001, 2003), while the OCT subsided relative to adjacent crustal domains (Thinon 1999).

The structure of the OCT varies between the northern and southern Armorican margin segment. The first sediments deposited in the proximal part of the northern Armorican OCT are latest Aptian in age (3B sequence; Thinon et al. 2002). Seaward, near the first oceanic crust, the first sequence on the basement shows chaotic facies and high velocities (6 km/s) and is interpreted as volcano-sedimentary sequences or as extrusive rocks (ST body in Figure 6.3; Thinon et al. 2003). The OCT of the southern Armorican segment shows an enigmatic seismic unit with chaotic facies (seismic unit 3C) that is mapped below the first sedimentary unit deposited on oceanic crust (Unit 3 in Thinon et al. (2003)). Though the nature of the 3C seismic unit is debated, recent interpretations suggest it is part of the OCT basement and could correspond to faulted serpentinized mantle with minor volcanism (Gillard et al. 2019). The distal parts of the southern Armorican OCT show several isolated seamounts that could correspond to volcanic edifices (Thinon et al. 2003). This observation is consistent with the southern Armorican OCT showing a series of strong intensity magnetic anomalies (Le Maire et al. 2021). The contrasting structure of the Armorican OCT could suggest that mantle exhumation was associated with more magmatic material in the southern part of the Armorican margin than in its northern part (Le Maire et al. 2021).

## 6.2.2. *North Iberia margin*

The southern border of the Bay of Biscay was initially interpreted as a former self-sustained subduction zone based on the recognition of morphological similarities with active subduction trenches (Sibuet and Le Pichon 1971). Attempts at documenting the deep structure of the trough revealed that only an incipient subduction stage was recorded (Boillot et al. 1979; Alvarez-Marron et al. 1997; Fernandez-Viejo et al. 1998) as result of a partial to complete overprint of the former rifted margin (Malod et al. 1982; Boillot et al. 1985; Boillot and Mallot 1988; Gallastegui et al. 2002; Roca et al. 2011; Tugend et al. 2014, 2015; Pedreira et al. 2015; Cadenas et al. 2018, 2020).

Along strike of the North Iberian margin, morphological changes are observed allowing the distinction between (1) the North Galicia margin segment that connects to the south with the West Iberia margin and (2) the Cantabrian margin segment that connects to the east with the Aquitaine margin. The change occurs near the Aviles Canyon located in the offshore prolongation of the Ventaniella fault (Figure 6.1). The details of the accommodation zone between these two segments are not known, but this change occurs at the transition between two Variscan zones (Central Iberian zone and West Asturian-Leonese zone), suggesting that the accommodation zone is at least partly controlled by pre-rift crustal to lithospheric heterogeneities.

### 6.2.2.1. *North Galicia margin*

The North Galicia margin connects the southern part of the Bay of Biscay with the West Iberian rifted margin (Figure 6.1). To the east, this WNW-ESE trending segment connects with the Cantabrian or Asturian margin separated by the Aviles Canyon, while west of the La Coruña canyon, the margin is shifted to the south and possibly segmented by NW-SE transfer faults (Boillot and Malod 1988).

The crustal structure of the margin is here illustrated along the IAM 12 seismic reflection profile (Figure 6.4) including knowledge of the velocity structure from a coincident refraction line (Fernandez-Viejo et al. 1998). Complementary information on crustal thickness variations comes from recent forward gravity modeling (Druet et al. 2017) and gravity inversion results (Tugend et al. 2014). A series of dives and samples collected during dredges provide constraints on the nature of the basement and sediments over the continental shelf and slope during the Mesozoic (Capdevila et al. 1974; Boillot et al. 1985; Boillot and Malod 1988). Two

DSDP sites (118 and 119, leg 13, Figure 6.2) were drilled in the abyssal plain north of the Galicia margin, recovering quaternary to Palaeocene sediments on top of altered basaltic rocks (Laughton et al. 1972a). The Iberian continental shelf is weakly thinned, suggesting that it corresponds to the former proximal margin domain (Figure 6.4). A set of inverted normal faults have been identified during dives conducted along the slope (Boillot et al. 1985; Boillot and Malod 1988), but they could not be observed on the IAM 12 seismic profile. At the base of the slope, a narrow accretionary wedge (<40 km) separates the continental margin from the oceanic domain of the Biscay abyssal plain (Fernandez-Viejo et al. 2012; Figure 6.4). The relatively high velocities modeled from the IAM 12 refraction experiment (Fernandez-Viejo et al. 1998; Fernandez-Viejo et al. 2012) suggest that the accretionary wedge includes sediments but also continental crust, hence possibly corresponding to the inverted former distal hyper-thinned domain (Tugend et al. 2014, 2015b). In contrast, to the north, the oceanic domain was poorly deformed and preserves its extensional fabric.

Refraction models only show a limited underthrusting of the oceanic domain under the accretionary wedge (Alvarez-Marron et al. 1997; Fernandez-Viejo et al. 1998). Recent forward gravity modeling results support the occurrence of serpentinized mantle underneath the accretionary prism (Druet et al. 2017), suggesting that a domain of exhumed serpentinized mantle once framed the North Galicia passive margin segment (Tugend et al. 2014, 2015b). The occurrence of a weak layer of serpentinized mantle may have facilitated the localization of the initial stages of subduction at the passive margin (Tugend et al. 2014, 2015b), as also suggested during the amagmatic Alpine subduction (McCarthy et al. 2020, 2021). The Galicia margin and more generally the North Iberian margin therefore correspond to a fossil example where subduction initiated at the passive margin.

The age of rifting in the Galicia margin segment is deduced from the analysis of the dives and dredges collected in the Ortegal Spur area (Boillot et al. 1985; Boillot and Malod 1988). Onset of detrital sedimentation during the Berriasian corresponding to the dismantling of the underlying shallow marine Jurassic platform is interpreted as marking the onset of rifting (Boillot et al. 1985; Boillot and Malod 1988). Berriasian to Aptian sediments are unconformably overlain by Late Aptian to Albian organic rich sandstones like the black shale formation drilled in the Western Approach passive margin (Boillot and Malod 1988) and corresponding to the first post-rift sequence. The identification of a pre-Eocene unconformity on the shelf and the slope suggests that convergence initiated prior to the Eocene. The growth of the accretionary prism is calibrated by Palaeogene sediments deposited in the associated foredeep basin to the north (Figure 6.4).

#### 6.2.2.2. *Cantabrian margin*

The Cantabrian margin links the North Galicia margin to the west with the southern part of the Aquitaine margin to the east. The structure of the Cantabrian margin is complex and variable, including lithospheric-scale structures such as the Ventaniella Fault (Fernandez-Viejo et al. 2015; Cadenas et al. 2018), transfer zones such as the Santander Transfer Zone (Pedreira et al. 2007; Roca et al. 2011) and the thick Asturian Basin (Gallastegui 2000; Cadenas and Fernández-Viejo 2017) separating the Le Danois Bank from the shelf to the south (Figure 6.1).

A relatively dense network of reflection and refraction seismic data covers the margin segment, providing good constraints on its crustal structure, here imaged along the ESCIN 4 seismic reflection profile crossing the Asturian basin (Figure 6.4) (Gallart et al. 1995; Alvarez-Marron et al. 1996, 1997; Gallastegui 2000; Fernandez-Viejo and Gallastegui 2005). Constraints on the velocity structure are available from a coincident seismic refraction profile extending onshore in the Cantabrian Mountains (Fernandez-Viejo et al. 1998). The complementary Marconi reflection and refraction seismic experiment covers the Cantabrian margin up to its eastern linkage with the southern Aquitaine margin (Fernandez-Viejo et al. 2011; Roca et al. 2011; Ruiz et al. 2017). The geology of the Asturian shelf and slope is well constrained because of the available borehole data (Gallastegui 2000; Cadenas and Fernandez-Viejo 2017) and numerous dredges and dives (Malod and Boillot 1980; Malod et al. 1982; Boillot and Malod 1988). One of the striking features of the Cantabrian margin is the presence of the thick Asturian Basin, filled by more than 10-km-thick pre-Aptian sediments according to stratigraphic calibrations (Cadenas and Fernandez-Viejo 2017) and overlying a ~15-km-thick crust (Cadenas et al. 2018). This basin is separated from the abyssal plain by the Le Danois Bank, a structural high whose origin is debated. It either corresponds to a continental bloc shaped during rifting (Cadenas et al. 2018, 2020) or to a structural high tectonically uplifted during convergence (Malod et al. 1982). Dives conducted along the slope combined with dredges revealed the occurrence of imbricated thrusts (Malod and Boillot 1980) that are not easily observed on seismic reflection data (Figure 6.4). Similar to the north Galicia margin segment, the accretionary wedge located at the toe of the Le Danois slope has been interpreted to be formed from the inversion of the former hyper-thinned distal part of the margin (Gallastegui 2000; Tugend et al. 2014, 2015; Cadenas et al. 2018, 2020). However, the wedge is significantly wider than in the north Galicia segment, possibly due to the greater incorporation of sediments (Fernandez-Viejo et al. 2012). The structure of the northern foredeep basin is complex, possibly due to the reactivation of former rift structures and continental blocs during convergence (Figure 6.4; Tugend 2013; see also Cadenas et al. (2020)). The nature of the underthrusted crust is debated, showing atypical velocity structure for oceanic crust (Ruiz et al. 2017), hence possibly corresponding to exhumed mantle with minor magmatism by analogy with the OCT of the North Biscay margin (Figure 6.3).

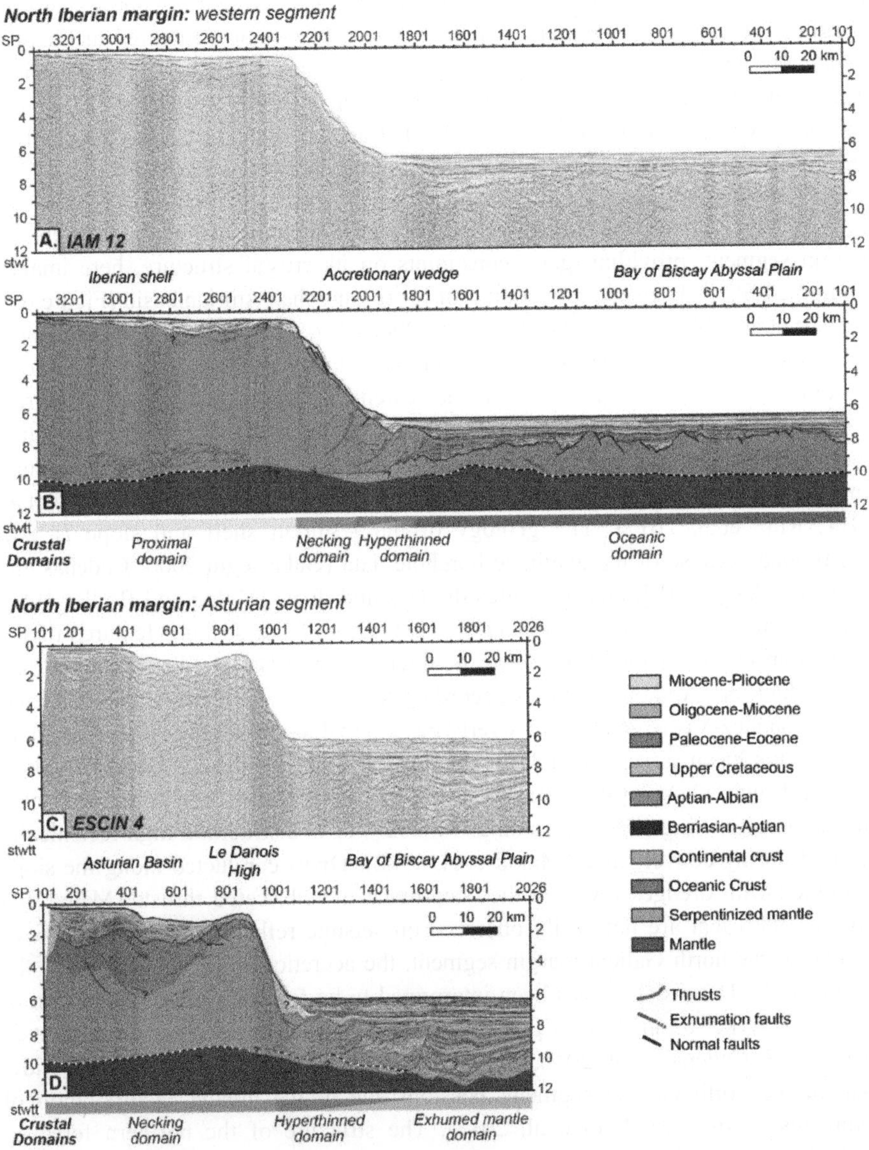


**Figure 6.4.** *Architecture of the North Iberian passive margin*

LEGEND FOR FIGURE 6.4.– *(a) IAM 12 seismic reflection profile across the North Galicia margin (Alvarez-Marron et al. 1997); (b) geological interpretation of the IAM 12 seismic profile (simplified from Tugend et al. (2014)); (c) ESCIN 4 seismic reflection profile across the Cantabrian margin (Alvarez-Marron et al. 1997); (d) geological interpretation of the ESCIN 4 seismic profile (from Tugend (2013)).*

Despite the inversion of the margin during the Pyrenean convergence, different rift events have been identified in Triassic, Late Jurassic and Early Cretaceous (Boillot and Malod 1988), but the geodynamic context and their spatial distribution remains poorly constrained. A recent reassessment notably highlighted the spatial extent of the Triassic rifting and salt deposition (Cadenas et al. 2020). The structural and stratigraphic analysis of the Asturian Basin suggests it formed prior to the Aptian (Figure 6.4); however, the geodynamic context and onset of rifting remains debated. Depending on the interpretation, the Asturian Basin is considered as a basin formed during the Berriasian-Aptian rifting of the Bay of Biscay (Gallastegui 2000; Cadenas and Fernandez-Viejo 2017; Cadenas et al. 2018), or as a strike-slip basin initiated in latest Jurassic time (Cadenas et al. 2020). The analysis of the samples dredged in the Le Danois Bank area (Malod and Boillot 1980; Malod et al. 1982; Boillot and Malod 1988) shows that the main rift event that thinned the crust of the Cantabrian passive margin initiated in Berriasian to Hauterivan time. This event is recorded by a change from platform to detrital sedimentation and ended in late Aptian time with the deposition of black shales. Apatite ages on granulite samples embedded in Aptian-Albian conglomerate dredged along the Le Danois Bank (Capdevila et al. 1974, 1980) range from 138 ± 7 Ma and 120 ± 8 Ma, suggesting that the samples have been exhumed to the surface, eroded and reworked in deep-basin sediment in Berriasian to early Aptian time (Fügenschuh et al. 2003), consistently with the age of rifting deduced from the sedimentary analysis.

### 6.2.3. *Eastern termination of the Bay of Biscay: the Aquitaine margin and Parentis Basin*

The V-shape of the Bay of Biscay terminates to the east in the Aquitaine margin (Figure 6.1). The structural framework of the transition from the Bay of Biscay abyssal plain to the Aquitaine margin is largely unknown, but the segmenting role of N-S trending accommodation zones like the Santander soft transfer zone has recently been emphasized (Roca et al. 2011). East of the Santander accommodation zone, the Aquitaine margin is structured by the Parentis Basin, which corresponds to a roughly 100 km wide rift basin in between the northern Armorican shelf and the Landes High to the south (Figure 6.1). The Landes High is a relatively weakly thinned continental block separating the Parentis from the Basque Cantabrian basin further south (Bois et al. 1997; Roca et al. 2011; Tugend et al. 2014).

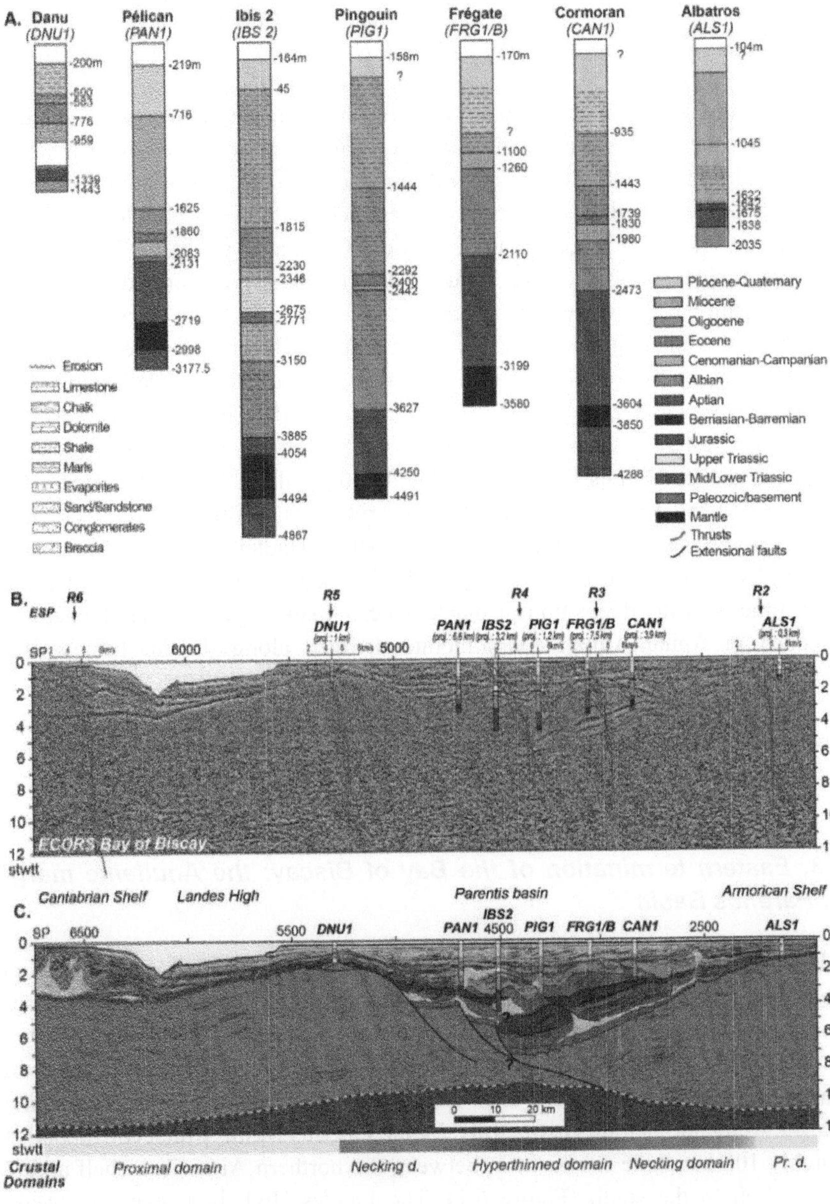


**Figure 6.5.** *Architecture of the Parentis Basin*

LEGEND FOR FIGURE 6.5.– *(a) Drilling results (Bois and Gariel 1994); (b) ECORS Parentis seismic section (Pinet et al. 1987), projected boreholes (Bois et al. 1997; Jammes et al. 2010) and Expending Spread Profiles (ESP) results (Marillier et al. 1988; Tomassino and Marillier 1997); (c) interpretation of the ECORS Parentis seismic profile (from Tugend et al. (2015a)).*

The crustal structure of the Parentis Basin is imaged along the ECORS Bay of Biscay seismic profile, and its sedimentary record is calibrated by borehole data (Figure 6.5; Pinet et al. 1987; Bois and Gariel 1994; Tomassino and Marillier 1997; Bois et al. 1997). The basin is characterized by up to 15 km thick Mesozoic to Cenozoic sediments overlying a severely thinned continental crust, which is less than 7 km thick in the basin center (Pinet et al. 1987; Bois and Gariel 1994; Tomassino and Marillier 1997; Bois et al. 1997; Tugend et al. 2015a). The northern part of the basin is delimited from the southern part by a step in the top basement referred to as the Ibis structure. This structure has been interpreted either as a normal fault (Pinet et al. 1987), a strike slip fault (Jammes et al. 2010) or as related to salt tectonics (Bois and Gariel 1994). The crustal and stratigraphic architecture of the Parentis Basin is largely asymmetric, with the northern part showing a continuous sedimentary cover and progressive crustal thinning from the Armorican shelf to the basin center, while in the southern part, the sedimentary cover is disrupted by salt tectonic activity and a series of steps in the top basement (Pinet et al. 1987; Bois et al. 1994). The stratigraphic architecture of the Parentis Basin is largely controlled by the occurrence of a thick Triassic salt layer (Brunet 1991; Bois and Gariel 1994; Biteau et al. 2006; Ferrer et al. 2008, 2012; Jammes et al. 2010a), which allowed the decoupling of deformation between supra-salt formations and the underlying basement. This decoupling level is at the origin of locally intense salt tectonic activity since the Jurassic. Similar settings are described in other peri-Iberian rift basins (Lagabrielle et al. 2020; Saspiturry et al. 2021), such as the Arzacq (Issautier et al. 2020; Ducoux et al. 2021) or the Columbrets Basin (Etheve et al. 2018).

Different interpretations have been proposed to explain the geometry and asymmetric crustal thinning of the Parentis Basin (Pinet et al. 1987), including a synclinal basin (Bois and Gariel 1994; Bois et al. 1997), half-graben type basin (Ferrer et al. 2008), or supra-detachment basin floored by one (Jammes et al. 2010a, 2010b, 2010c) or by a series of large offset normal faults (Figure 6.5; Tugend et al. 2014, 2015a). The nature of the basement flooring the southern part of the Parentis Basin over the Landes High remains largely unknown, as only the uppermost part has been sampled and included clays and Triassic volcanic rocks under a thin Upper Cretaceous-Triassic sedimentary cover (DANU and TARANIS wells, Figure 6.5). South of the Landes High, toward the Basque Cantabrian Basin, the Meso-Cenozoic cover is affected by Pyrenean deformation overprinting an intense salt tectonics activity (e.g. Ferrer et al. 2008, 2012), while little to no deformation is observed in the Parentis Basin. The analysis of subsidence rates from borehole data suggests that most of the extension of the Parentis Basin occurred from Kimmeridgian to Albian time (Brunet 1984, 1994). Two pulses of tectonic activity are distinguished: from Kimmeridgian to earliest Cretaceous time and from Barremian to Albian time, the fastest subsidence rates being reported in Aptian time (Brunet 1983, 1994). Subsidence results are compatible with the age of rifting reported in the adjacent margins of the Bay of Biscay. Crustal thinning in the Parentis Basin from ~30 km (proximal domain) to less than 10 km thick crust (hyper-thinned domain) occurs within a 30–40 km distance (Figure 6.5). Such localized crustal thinning combined with subsidence analysis results has been interpreted as related to a two-stage formation of the Parentis Basin, which included an initial strike slip movement in latest Jurassic, followed by orthogonal extension in Aptian time (Jammes et al. 2010).

## 6.3. The Bay of Biscay and the Iberian-European plate boundary

The Bay of Biscay oceanic domain and its continental margins are part of a wider rift system that formed at the plate transition between Iberia and Europe in the latest Jurassic to Early Cretaceous time (Figure 6.1(b); Tugend et al. 2015b). The architecture of the Iberian and European plate boundary shows several spatially disconnected rift arms: (1) the Bay of Biscay–Parentis, (2) Pyrenean-Basque-Cantabrian and (3) Central Iberian rift arms (Figure 6.1(b); Salas and Casas 1993; Vergés and García-Senz 2001; Roca et al. 2011; Tugend et al. 2014, 2015), the latter two now being exposed onshore as relics integrated as part of the Pyrenean orogen s.l. and Iberian Range (Figure 6.1). The spatial distribution of the different rift arms and sub-basins reflects the complex partitioning between transtensional and extensional deformation that occurred during rifting (Tugend et al. 2015b; Nirrengarten et al. 2018; King et al. 2021). The detailed investigation of the tectonic and subsidence history of the different sub-basins revealed the diachronous polyphase formation of the rift arms of the Iberian-European plate boundary (Figure 6.6; Tugend et al. 2015b).

The onset of rifting that led to the opening of the Bay of Biscay is difficult to date with precision, but occurred subsequently to the Late Jurassic onset of the left-lateral movement of Iberia relative to Europe (Figure 6.6(a); e.g. Rosenbaum et al. 2002; Nirrengarten et al. 2018). At that time, the formation of the Iberian-European plate boundary was recorded within a wide corridor of transtensional deformation, where different rift arms progressively shaped (Tugend et al. 2015b). The occurrence of faunas typically found in the Atlantic (Durand-Delga 1973) indicate that a marine pathway existed at that time with the Bay of Biscay and Atlantic domain (Figure 6.6(a)), whereas the other arms were connected to the Tethyan domain (Mas et al. 1993; Salas and Casas 1993). First evidence of rift activity in the Bay of Biscay is Berriasian (e.g. Auffret et al. 1979; Derégnaucourt 1981; Thinon et al. 2002), but a Late Jurassic onset is recorded in the Parentis Basin (Brunet 1994) and inferred in the Asturian Basin (Cadenas et al. 2020). Crustal thinning related to the formation of the proximal and necking domain seems to occur prior to Aptian time (Berriasian-Barremian; Thinon et al. 2002; Tugend et al. 2014, 2015b), while deformation focused on the distal parts of the margins in the Aptian, possibly leading to mantle exhumation in the OCT (Thinon et al. 2003; Tugend et al. 2014, 2015b).

Onset of magmatic oceanic accretion in the Bay of Biscay is only indirectly constrained. At the westernmost extremity of the Bay of Biscay, in the Goban Spur margin, the first sediments deposited on top of the oceanic crust are Albian in age (DSDP Leg 80, site 550; De Graciansky and Poag 1985). Drilling results from the DSPD leg 48 and stratigraphic correlations conducted toward the distal margin suggest that the first sediments deposited on top of the oceanic crust bordering the Western Approach margin are Aptian-Albian in age (Figure 6.3; Montadert et al. 1979). At the scale of the Iberian-European plate boundary, a major change in thc subsidence and deformation histories of the different sub-basins of the rift system is recorded in Aptian-Albian time (Figure 6.6(b); Tugend et al. 2015b). In the Central Iberian rift arm, fault activity progressively cessed (Salas and Casas 1993), whereas the onset of significant crustal thinning initiated in the Pyrenean-Basque-Cantabrian rift arm (Jammes et al. 2009; Tugend et al. 2015b). Rifting in the Pyrenean-Basque-Cantabrian rift arm is largely controlled by NNE-SSW to NE-SW transfer zones (Figure 6.6(b)), an observation generally interpreted as marking a change in the movement of Iberia relative to Europe from left-lateral to roughly north-south in late Aptian time (Jammes et al. 2009, 2010; Tugend et al. 2015b). Whether an abrupt or progressive change from strike-slip to orthogonal deformation occurred remains, however, unknown. It should however be noted that larger plate kinematic models require a component of left lateral strike slip motion even after Aptian time in order to avoid compression in the Pyrenean-Basque-Cantabrian rift arm prior to the Santonian (Nirrengarten et al. 2018).

The onset of seafloor spreading occurred at Aptian-Albian time in the western Bay of Biscay (Montadert et al. 1974), but was probably not synchronous throughout the entire Bay of Biscay, with a slightly younger age suggested in the eastern part (Tugend et al. 2015b; Le Maire et al. 2021). The V-shape morphology of the Bay of Biscay and the pattern of magnetic anomalies in its eastern part suggest that the Bay of Biscay tentatively propagated to the east, while seafloor spreading occurred until late Santonian to early Campanian time (Figure 6.6(c)). In that case, the delayed extreme thinning recorded in the Pyrenean-Basque-Cantabrian rift arm in latest Early Cretaceous time to early Late Cretaceous is interpreted as the tentative propagation of the plate boundary between Iberia and Europe (Figures 6.6(b) and (c)).

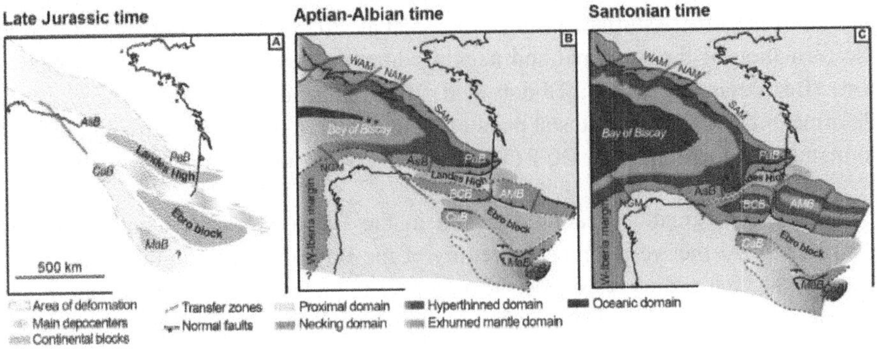


**Figure 6.6.** *Simplified palaeographic maps illustrating the evolution of the Bay of Biscay in the frame of the formation of the Iberia-European plate boundary (from Tugend et al. (2015b))*

LEGEND FOR FIGURE 6.6.– *(a) Onset of the oblique rifting stage (Late Jurassic); (b) onset of seafloor spreading (Aptian–Albian); (c) failed tentative localization of plate boundary (before Santonian). AMB: Arzcaq-Mauléon Basin; AsB: Asturian Basin; BCB: Basque Cantabrian Basin; CaB: Cameros Basin; CoB Columbrets Basin; MaB: Maestrat Basin; NAM North Armorican margin; PaB: Parentis Basin; SAM: South Armorican margin; WAM: Western Approaches margin.*

## 6.4. Future challenges

Despite incredible knowledge improvement on the formation and evolution of the Bay of Biscay, numerous questions remain. They are related to (1) the influence of previous tectonic events, (2) the partitioning of transtensional deformation between the different sub-basins and the structures accommodating the observed segmentation and (3) the plate kinematic context, or the role of the passive margin structure on the distribution of contractional deformation during convergence.

Several tectonic events occurred prior to the Early Cretaceous rifting that structured the lithosphere at crustal scale and deeper. The Variscan orogeny and the post-orogenic evolution was notably related to the formation of major shear zones such as the northern and southern Armorican shear zones that seem to partly control subsequent passive margin segmentation (Tugend et al. 2014; Le Maire et al. 2021; King et al. 2021). Evidence of Triassic rifting is extensively observed along the Iberia-European plate boundary, but the extent and influence of the related structures remain to be explored. Furthermore, Triassic rifting was related to salt deposition that largely controlled the subsidence evolution and stratigraphic architecture of overlying sequences during subsequent deformation (e.g. Columbrets Basin: Etheve et al. 2018; Arzcaq Basin: Issautier et al. 2020; Ducoux et al. 2021; Pyrenean-Cantabrian junction: Lescoutre et al. 2021; Pyrenees: Lagabrielle et al. 2020; Ford et al. 2021; Manatschal et al. 2021; Saspiturry et al. 2021 and cited references).

It is now clear that the opening of the Bay of Biscay is kinematically linked to the rifting and oceanic spreading that occurred between the Western Iberian and Newfoundland passive margin from the Late Jurassic onward (e.g. Nirrengarten et al. 2018), with an oblique rifting phase in the Bay of Biscay as a consequence (Olivet 1996; Tugend et al. 2015b). However, the timing and partitioning of deformation between purely strike-slip, transtensional and extensional deformation remains debated in the different sub-basins during the left lateral movement of Iberia relative to Europe. The Bay of Biscay passive margins are segmented possibly as the consequence of an oblique rifting in a strongly pre-structured lithosphere. The onset of oceanic spreading and later eastward propagation possibly occurred in a slightly oblique context; however, the kinematic context of seafloor spreading onset and the structure and segmentation of the former oceanic ridge are largely unknown.

Though not detailed in this chapter, the architecture of the Iberian-European plate boundary inherited from rifting and oceanic spreading largely controlled the localization of contractional deformation during the Pyrenean orogeny from Late Cretaceous onward (Tugend et al. 2014, 2015b). Onset of contraction seems to have been preferentially accommodated at the boundary between former rift domains, notably at the edge of the continental crust (Figure 6.3; Tugend et al. 2014). The occurrence of exhumed serpentinized mantle in the OCT inferred from seismic observations and gravity inversion results could explain why contractional deformation focused there (Tugend et al. 2014, 2015b), while it seems more diffuse in the oceanic domain (Thinon et al. 2001). With increasing shortening, the distal margin was progressively incorporated in the accretionary wedge framing the incipient subduction stage recorded along the North Iberia margin, while the necking and proximal domains acted as a buttress (Tugend et al. 2014, 2015b; Cadenas et al. 2018). The North Iberian margin represents a key location to investigate the modalities of subduction initiation at magma-poor passive margins (Tugend et al. 2014, 2015b; McCarthy et al. 2020, 2021). Despite a conceptual knowledge of the impact of passive margin during subsequent inversion, the role of their segmentation and lateral variations in the OCT magmatic supply are still poorly addressed.

## 6.5. Acknowledgments

We thank the geophysical data repository of the research group on the Structure and Dynamics of the Earth (http:/geodb.ictja.csic.es) for providing the IAM 12 and ESCIN 4 seismic profiles. JT acknowledges Total SA, now TotalEnergies, for a post-doctoral fellowship to investigate the structure of the Bay of Biscay in the frame of the Orogen project, co-funded by Total, the BRGM and the CNRS-INSU.

## 6.6. References

Alvarez-Marrón, J., Pérez-Estaún, A., Danñobeitia, J.J., Pulgar, J.A., Martínez-Catalán, J.R., Marcos, A., Bastida, F. (1996). Seismic structure of the northern continental margin of Spain from ESCIN deep seismic profiles. *Tectonophysics*, 264(1/4), 153–74 [Online]. Available at: https://doi.org/10.1016/S0040-1951(96)00124-2.

Alvarez-Marrón, J., Rubio, E., Torne, M. (1997). Subduction-related structures in the north Iberian margin. *Journal of Geophysical Research: Solid Earth*, 102(B10), 22497–22511 [Online]. Available at: https://doi.org/10.1029/97JB01425.

Asti, R., Saspiturry, N., Angrand, P. (2022). The Mesozoic Iberia-Eurasia plate boundary: A wide domain of distributed transtensional deformation rather than a single wrench fault. *Earth-Science Reviews*, 230, 104040 [Online]. Available at: https://doi.org/10.1016/j.earscirev.2022.104040.

Auffret, G.A., Pastouret, L., Cassat, G., de Charpal, O., Cravatte, J., Guennoc, P. (1979). Dredged rocks from the armorican and celtic margins. Initial Reports of the Deep Sea Drilling Project, U.S. Government Printing Office, 48(68), 473–491 [Online]. Available at: https://doi.org/10.2973/dsdp.proc.48.152.1979.

Avedik, F. (1975). Seismic refraction survey in the western approaches to the English Channel: Preliminary results. *Philosophical Transactions of the Royal Society of London. Series A, Mathematical and Physical Sciences*, 279(1288), 29–39 [Online]. Available at: https://doi.org/10.1098/rsta.1975.0038.

Avedik, F. and Howard, D. (1979). Preliminary results of a seismic refraction study in the Meriadzek-Trevelyan Area, Bay of Biscay. Initial Reports of the Deep Sea Drilling Project, U.S. Government Printing Office, 48(68), 379–387 [Online]. Available at: https://doi.org/10.2973/dsdp.proc.48.153.1979.

Avedik, F. and Olivet, J.L. (1994). NORGASIS cruise, RV Le Nadir [Online]. Available at: https://doi.org/https://doi.org/10.17600/94000050.

Avedik, F, Camus, A.L., Ginsburg, A., Montadert, L., Roberts, D.G., Whitmarsh, R.B. (1982). A seismic refraction and reflexion study of the continent-ocean transition beneath the North Biscay Margin. *Philosophical Transactions of the Royal Society of London. Series A, Mathematical and Physical Sciences*, 305(1489), 5–25 [Online]. Available at: https://doi.org/10.1098/rsta.1982.0023.

Bacon, M., Gray, F., Matthews, D.H. (1969). Crustal structure studies in the Bay of Biscay. *Earth and Planetary Science Letters*, 6(5), 377–385 [Online]. Available at: https://doi.org/10.1016/0012-821X(69)90188-5.

Ballèvre, M., Bosse, V., Ducassou, C., Pitra, P. (2009). Palaeozoic history of the Armorican Massif: Models for the tectonic evolution of the suture zones. *Comptes Rendus Géoscience*, 341(2/3), 174–201 [Online]. Available at: https://doi.org/10.1016/j.crte.2008.11.009.

Ballèvre, M., Fourcade, S., Capdevila, R., Peucat, J.J., Cocherie, A., Fanning, C.M. (2012). Geochronology and geochemistry of Ordovician felsic volcanism in the Southern Armorican Massif (Variscan Belt, France): Implications for the breakup of Gondwana. *Gondwana Research*, 21(4), 1019–1036 [Online]. Available at: https://doi.org/10.1016/j.gr.2011.07.030.

Barbier, F.l., Duvergé, J., Le Pichon, X. (1986). Structure profonde de la marge Nord-Gascogne. Implications sur le mécanisme de rifting et de formation de la marge continentale. *Bulletin des Centres de Recherches Exploration-Production Elf-Aquitaine*, 10(1), 105–121.

Becker, J.J., Sandwell, D.T., Smith, W.H.F., Braud, J., Binder, B., Depner, J., Fabre, D. (2009). Global bathymetry and elevation data at 30 arc seconds resolution. *SRTM30_PLUS*, 4, 355–371 [Online]. Available at: https://doi.org/10.1080/01490410903297766.

Biteau, J.J., Le Marrec, A., Le Vot, M., Masset, J.M. (2006). The Aquitaine basin. *Petroleum Geoscience*, 12(3), 247–273 [Online]. Available at: https://doi.org/10.1144/1354-079305-674.

Boillot, G. and Malod, J. (1988). The north and north-west Spanish continental margin. *Revista de la Sociedad Geológica de España*, 1(3), 295–316.

Boillot, G., Dupeuble, P.A., Malod, J. (1979). Subduction and tectonics on the continental margin off northern Spain. *Marine Geology*, 32(1/2), 53–70 [Online]. Available at: https://doi.org/10.1016/0025-3227(79)90146-4.

Boillot, G., Temine, D., Malod, J.A., Capdevila, R., Cousin, M., Dupeuble, P.A., Gonzalez-Lodeiro, F. (1985). Exploration par submersible de la marge continentale au Nord-Ouest de l'Espagne. *Bulletin de la Société Géologique de France*, I(1), 89–102 [Online]. Available at: https://doi.org/10.2113/gssgfbull.I.1.89.

Bois, C. and Gariel, O. (1994). Deep seismic investigation in the Parentis Basin (Southwestern France). *Hydrocarbon and Petroleum Geology of France*, 173–186 [Online]. Available at: https://doi.org/10.1007/978-3-642-78849-9_13.

Bois, C., Pinet, B., Gariel, O. (1997). The sedimentary cover along the ECORS Bay of Biscay deep seismic reflection profile. A comparison between the Parentis Basin and other European rifts and basins. *Mémoires de la Société Géologique de France*, 171(1833), 143–165.

Brunet, M.F. (1984). Subsidence history of the Aquitaine basin determined from subsidence curves. *Geological Magazine*, 121(5), 421–428 [Online]. Available at: https://doi.org/10.1017/s0016756800029952.

Brunet, M.F. (1991). Subsidence et géodynamique du bassin d'Aquitaine. Relations avec l'ouverture de l'Atlantique. PhD Thesis, Université Paris 6.

Brunet, M.F. (1994). Subsidence in the Parentis Basin (Aquitaine, France): Implications of the thermal evolution. In *Hydrocarbon and Petroleum Geology of France*, Mascle, A. (ed.). Springer, Berlin/Heidelberg [Online]. Available at: https://doi.org/10.1007/978-3-642-78849-9_14.

Bullock, A.D. and Minshull, T.A. (2005). From continental extension to seafloor spreading: Crustal structure of the Goban spur rifted margin, southwest of the UK. *Geophysical Journal International*, 163(2), 527–546 [Online]. Available at: https://doi.org/10.1111/J.1365-246X.2005.02726.X.

Cadenas, P. and Fernández-Viejo, G. (2017). The Asturian basin within the North Iberian Margin (Bay of Biscay): Seismic characterisation of its geometry and its mesozoic and cenozoic cover. *Basin Research*, 29(4), 521–541 [Online]. Available at: https://doi.org/10.1111/bre.12187.

Cadenas, P., Fernández-Viejo, G., Pulgar, J.A., Tugend, J., Manatschal, G., Minshull, T.A. (2018). Constraints imposed by rift inheritance on the compressional reactivation of a hyperextended margin: Mapping rift domains in the North Iberian margin and in the Cantabrian Mountains. *Tectonics*, 37(3), 758–785 [Online]. Available at: https://doi.org/10.1002/2016TC004454.

Cadenas, P., Manatschal, G., Fernández-Viejo, G. (2020). Unravelling the architecture and evolution of the inverted multi-stage North Iberian-Bay of Biscay rift. *Gondwana Research*, 88(December), 67–87 [Online]. Available at: https://doi.org/10.1016/J.GR.2020.06.026.

Capdevila, R., Lamboy, M., Leprêtre, J.P. (1974). Découverte de granulites, de charnockites et de syénites néphéliniques dans la partie occidentale de la marge continentale Nord-Espagnole. *Comptes Rendus de l'Academie des Sciences, Série D*, 278, 17–20.

Capdevila, R., Boillot, G., Lepvrier, C., Malod, J.A., Mascle, G. (1980). Les formations cristallines du banc Le Danois (marge Nord Ibérique). *Comptes Rendus de l'Academie des Sciences de Paris, Série D*, 291, 317–320.

Carey, S.W. (1958). The tectonic approach to continental drift. In *Continental Drift: A Symposium*. University of Tasmania, Hobart.

Chantraine, J., Autran, A., Cavelier, C. (2003). Carte géologique de la France au 1/1 000000 – 6e édition révisée. BRGM, Orléans.

Charpal, O., Guennoc, P., Montadert, L., Roberts, D.G. (1978). Rifting, crustal attenuation and subsidence in the Bay of Biscay. *Nature*, 275(5682), 706–711 [Online]. Available at: https://doi.org/10.1038/275706a0.

Damotte, B., Grau, G., Gray, F., Limond, W.Q., Patriat, P. (1971). Utilisation conjointe des méthodes sismique réflexion et réfraction pour la détermination des vitesses. In *Histoire structurale du golfe de Gascogne*, Institut Français du Pétrole (ed.). Technip, Paris.

Debyser, J., Le Pichon, X., Montadert, L. (1971). *Histoire structurale du golfe de Gascogne*, vol. 2. Technip, Paris.

Derégnaucourt, D. (1981). Contribution à l'étude géologique du golfe de Gascogne. PhD Thesis, Université Paris VI.

Derégnaucourt, D. and Boillot, G. (1982). Structure géologique du golfe de Gascogne. *Bulletin Bureau de Recherches Géologiques et Minières*, 3, 149–178.

Druet, M., Muñoz-Martín, A., Granja-Bruña, J.L., Carbó-Gorosabel, A., Acosta, J., Llanes, P., Ercilla, G. (2018). Crustal structure and continent-ocean boundary along the Galicia continental margin (NW Iberia): Insights from combined gravity and seismic interpretation. *Tectonics*, 37(5), 1576–1604 [Online]. Available at: https://doi.org/10.1029/2017TC004903.

Ducoux, M., Masini, E., Tugend, J., Gómez-Romeu, J., Calassou, S. (2021). Basement-decoupled hyperextension rifting: The tectono-stratigraphic record of the salt-rich Pyrenean necking zone (Arzacq Basin, SW France). *GSA Bulletin*, July [Online]. Available at: https://doi.org/10.1130/B35974.1.

Durand-Delga, M. (1973). Les calpionelles du golfe de Gascogne, témoin de l'ouverture de l'Atlantique Nord. *Bulletin de la Société Géologique de France*, 7, 22–24.

Etheve, N., Mohn, G., Frizon de Lamotte, D., Roca, E., Tugend, J., Gómez-Romeu, J. (2018). Extreme mesozoic crustal thinning in the eastern Iberia margin: The example of the Columbrets Basin (Valencia Trough). *Tectonics*, 37(2), 636–662 [Online]. Available at: https://doi.org/10.1002/2017TC004613.

Fernández-Viejo, G. and Gallastegui, J. (2005). The ESCI-N project after a decade: A synthesis of the results and open questions. *Trabajos de Geología*, 25, 9–27.

Fernández-Viejo, G., Gallart, J., Pulgar, J.A., Gallastegui, J., Dañobeitia, J.J., Córdoba, D. (1998). Crustal transition between continental and oceanic domains along the north Iberian margin from wide angle seismic and gravity data. *Geophysical Research Letters*, 25(23), 4249–4252 [Online]. Available at: https://doi.org/10.1029/1998GL900149.

Fernández-Viejo, G., Gallastegui, J., Pulgar, J.A., Gallart, J. (2011). The MARCONI reflection seismic data: A view into the eastern part of the Bay of Biscay. *Tectonophysics*, 508(1/4), 34–41 [Online]. Available at: https://doi.org/10.1016/J.TECTO.2010.06.020.

Fernández-Viejo, G., Pulgar, J.A., Gallastegui, J., Quintana, L. (2012). The fossil accretionary wedge of the Bay of Biscay: Critical wedge analysis on depth-migrated seismic sections and geodynamical implications. *The Journal of Geology*, 120(3), 315–331 [Online]. Available at: https://doi.org/10.1086/664789.

Fernández-Viejo, G., Cadenas, P., Acevedo, J., Llana-Fúnez, S. (2021). The unevenness of the north Iberian crustal root, a snapshot of an elusive stage in margin reactivation. *Geology*, 49(12), 1426–1430.

Ferragne, A. (1978). L'ouverture du golfe de Gascogne. Différentes conceptions et connaissances actuelles. *Bulletin Bureau de Recherches Géologiques et Minières*, 4(2), 95–142.

Ferrer, O., Roca, E., Benjumea, B., Muñoz, J.A., Ellouz, N., Gallart, J., Díaz, J. (2008). The deep seismic reflection MARCONI-3 profile: Role of extensional mesozoic structure during the Pyrenean contractional deformation at the eastern part of the Bay of Biscay. *Marine and Petroleum Geology*, 25(8), 714–730 [Online]. Available at: https://doi.org/10.1016/J.MARPETGEO.2008.06.002.

Ferrer, O., Jackson, M.P.A., Roca, E., Rubinat, M. (2012). Evolution of salt structures during extension and inversion of the offshore Parentis Basin (Eastern Bay of Biscay). *Geological Society, London, Special Publications*, 363(1), 361–380 [Online]. Available at: https://doi.org/10.1144/SP363.16.

Fugenschuh, B., Froitzheim, N., Capdevila, R., Boillot, G. (2003). Offshore granulites from the Bay of Biscay margins: Fission tracks constrain a proterozoic to tertiary thermal history. *Terra Nova*, 15(5), 337–342 [Online]. Available at: https://doi.org/10.1046/j.1365-3121.2003.00502.x.

Galdeano, A., Moreau, M.G., Pozzi, J.P., Berthou, P.Y., Malod, J.A. (1989). New paleo-magnetic results from Cretaceous sediments near Lisboa (Portugal) and implications for the rotation of Iberia. *Earth and Planetary Science Letters*, 92(1), 95–106 [Online]. Available at: https://doi.org/10.1016/0012-821X(89)90024-1.

Gallart, J., Fernández-Viejo, G., Díaz, J., Vidal, N., Pulgar, J.A. (1995). Deep structure of the transition between the Cantabrian Mountains and the North Iberian margin from wide-angle ESCI-N data. *Rev. Soc. Geol. España*, 8(4), 365–382.

Gallastegui, J. (2000). Estructura cortical de la cordillera y margen continental Cantábricos: Perfiles ESCI-N. *Trabajos de Geología*, 22, 3–234.

Gallastegui, J., Pulgar, J.A., Gallart, J. (2002). Initiation of an active margin at the North Iberian continent-ocean transition. *Tectonics*, 21(4), 15–14 [Online]. Available at: https://doi.org/10.1029/2001TC901046.

Gillard, M., Tugend, J., Müntener, O., Manatschal, G., Karner, G.D., Autin, J., Sauter, D., Figueredo, P.H., Ulrich, M. (2019). The role of serpentinization and magmatism in the formation of decoupling interfaces at magma-poor rifted margins. *Earth-Science Reviews*, 196(September), 102882 [Online]. Available at: https://doi.org/10.1016/j.earscirev.2019.102882.

Ginzburg, A., Whitmarsh, R.B., Roberts, D.G., Montadert, L., Camus, A. (1985). The deep seismic structure of the northern continental margin of the Bay of Biscay. *Annales Geophysicae*, 3, 499–510.

Graciansky, P.C. and Poag, C.W. (1985). Geologic history of Goban Spur, Northwest Europe continental margin. Initial Reports of the Deep Sea Drilling Project, U.S. Government Printing Office, 80 [Online]. Available at: https://doi.org/10.2973/dsdp.proc.80.158.1985.

Grau, G., Montadert, L., Delteil, R., Winnock, E. (1973). Structure of the European continental margin between Portugal and Ireland, from seismic data. *Tectonophysics*, 20(1/4), 319–339 [Online]. Available at: https://doi.org/10.1016/0040-1951(73)90121-2.

Grimaud, S., Boillot, G., Collette, B.J., Mauffret, A., Miles, P.R., Roberts, D.B. (1982). Western extension of the Iberian-European plate boundary during the Early Cenozoic (Pyrenean) convergence: A new model. *Marine Geology*, 45(1/2), 63–77 [Online]. Available at: https://doi.org/10.1016/0025-3227(82)90180-3.

Groupe CYMOR; Léo, P., Gérard-André, A., Jean-Marie, A., Paul, B., Paul, D., Michel, S., Jacques, S., Jean-René, V. (1981). La marge continentale armoricaine, résultats d'observations en submersible et de dragages dans le canyon Shamrock. *Comptes Rendus de l'Académie des Sciences de Paris*, 292(9), 741–748.

Issautier, B., Saspiturry, N., Serrano, O. (2020). Role of structural inheritance and salt tectonics in the formation of pseudosymmetric continental rifts on the European margin of the hyperextended Mauléon basin (Early Cretaceous Arzacq and Tartas Basins). *Marine and Ptroleum Geology*, 118, 104395 [Online]. Available at: https://doi.org/10.1016/j.marpetgeo.2020.104395.

Jammes, S., Manatschal, G., Lavier, L., Masini, E. (2009). Tectonosedimentary evolution related to extreme crustal thinning ahead of a propagating ocean: Example of the western Pyrenees. *Tectonics*, 28(4), TC4012 [Online]. Available at: https://doi.org/10.1029/2008TC002406.

Jammes, S., Manatschal, G., Lavier, L. (2010a). Interaction between prerift salt and detachment faulting in hyperextended rift systems: The example of the Parentis and Mauléon basins (Bay of Biscay and Western Pyrenees). *AAPG Bulletin*, 94(7), 957–975 [Online]. Available at: https://doi.org/10.1306/12090909116.

Jammes, S., Lavier, L., Manatschal, G. (2010b). Extreme crustal thinning in the Bay of Biscay and the Western Pyrenees: From observations to modeling. *Geochem. Geophys. Geosystems*, 11, Q10016 [Online]. Available at: https://doi.org/10.1029/2010GC003218.

Jammes, S., Tiberi, C., Manatschal, G. (2010c). 3D architecture of a complex transcurrent rift system: The example of the Bay of Biscay-Western Pyrenees. *Tectonophysics*, 489(1/4), 210–226 [Online]. Available at: https://doi.org/10.1016/j.tecto.2010.04.023.

King, M.T., Welford, J.K., Cadenas, P., Tugend, J. (2021). Investigating the plate kinematics of the Bay of Biscay using deformable plate tectonic models. *Tectonics*, 40(7), e2020TC006467 [Online]. Available at: https://doi.org/10.1029/2020TC006467.

Lagabrielle, Y., Asti, R., Duretz, T., Clerc, C., Fourcade, S., Teixell, A., Labaume, P., Corre, B., Saspiturry, N. (2020). A review of Cretaceous smooth-slopes extensional basins along the Iberia-Eurasia plate boundary: How pre-rift salt controls the modes of continental rifting and mantle exhumation. *Earth-Science Reviews*, 201(February), 103071 [Online]. Available at: https://doi.org/10.1016/j.earscirev.2019.103071.

Laughton, A.S. and Berggren, W.A. (1972). Initial Reports of the Deep Sea Drilling Project. U.S. Government Printing Office [Online]. Available at: https://doi.org/10.2973/dsdp.proc.12.1972.

Laughton, A.S., Berggren, W.A., Benson, R.N., Davies, T.A., Franz, U., Musich, L.F., Perch-Nielsen, K., Ruffman, A.S., van Hinte, J.E., Whitmarsh, R.B. et al. (1972a). Site 118. Initial Reports of the Deep Sea Drilling Project, U.S. Government Printing Office, 12 [Online]. Available at: https://doi.org/10.2973/dsdp.proc.12.109.1972.

Laughton, A.S., Berggren, W.A., Benson, R.N., Davies, T.A., Franz, U., Musich, L.F., Perch-Nielsen, K., Ruffman, A.S., van Hinte, J.E., Whitmarsh, R.B. et al. (1972b). Site 119. Initial Reports of the Deep Sea Drilling Project, U.S. Government Printing Office, 12 [Online]. Available at: https://doi.org/10.2973/dsdp.proc.12.110.1972.

Le Borgne, E., Le Mouël, J.L., Le Pichon, X. (1971). Aeromagnetic survey of south-western Europe. *Earth and Planetary Science Letters*, 12(3), 287–99 [Online]. Available at: https://doi.org/10.1016/0012-821X(71)90213-5.

Le Maire, P., Thinon, I., Tugend, J., Issautier, B., Martelet, G., Paquet, F., Proust, J.N., Nalpas, T., Peyrefitte, A., Canva, A. (2021). New magnetic compilation and interpretation of the Bay of Biscay and surrounding continental shelves. *Bulletin de la Société Géologique de France*, 68, 1989–1997.

Le Pichon, X. and Barbier, F. (1987). Passive margin formation by low-angle faulting within the upper crust: The northern Bay of Biscay margin. *Tectonics*, 6(2), 133–150 [Online]. Available at: https://doi.org/10.1029/TC006I002P00133.

Le Pichon, X. and Sibuet, J.C. (1981). Passive margins: A model of formation Xavier. *Journal of Geophysical Research*, 86(B5), 3708 [Online]. Available at: https://doi.org/10.1029/JB086iB05p03708.

Lefort, J.P. (1975). Le socle péri-armoricain. Étude géologique et géophysique du socle submergé à l'ouest de la France. PhD Thesis, Université de Rennes.

Lefort, J.P. (1980). Un "fit" structural de l'Atlantique Nord : arguments géologiques pour corréler les marqueurs géophysiques reconnus sur les deux marges. *Marine Geology*, 37(3/4), 355–69 [Online]. Available at: https://doi.org/10.1016/0025-3227(80)90110-3.

Lescoutre, R., Manatschal, G., Muñoz, J.A. (2021). Nature, origin, and evolution of the Pyrenean-Cantabrian junction. *Tectonics*, 40(5), e2020TC006134 [Online]. Available at: https://doi.org/10.1029/2020TC006134.

Malod, J.A. and Boillot, G. (1980). Campagne CYADANOIS, 9-22 juillet 1979. *Publications du centre national pour l'exploitation des océans. Résultats des campagnes à la mer*, 20.

Malod, J.A. and Mauffret, A. (1990). Iberian plate motions during the Mesozoic. *Tectonophysics*, 184(3/4), 261–278 [Online]. Available at: https://doi.org/10.1016/0040-1951(90)90443-C.

Malod, J.A., Boillot, G., Lepvier, C., Mascle, G., Taugourdeau-Lantz, J., Capdevila, R., Dupeuble, P.A., Müller, C. (1982). Subduction and tectonics on the continental margin off northern Spain: Observations with the submersible Cyana. *Geological Society, London, Special Publications*, 10(1), 309–315 [Online]. Available at: https://doi.org/10.1144/GSL.SP.1982.010.01.20.

Manatschal, G., Chenin, P., Lescoutre, R., Miró, J., Cadenas, P., Saspiturry, N., Masini, E., Chevrot, S., Ford, M., Jolivet, L. et al. (2021). The role of inheritance in forming rifts and rifted margins and building collisional orogens: A Biscay-Pyrenean perspective. *BSGF – Earth Sciences Bulletin*, 192(November), 55 [Online]. Available at: https://doi.org/10.1051/bsgf/2021042.

Marillier, F., Tomassino, A., Patriat, P., Pinet, B. (1988). Deep structure of the Aquitaine shelf: Constraints from expanding spread profiles on the ECORS Bay of Biscay transect. *Marine and Petroleum Geology*, 5(1), 65–74 [Online]. Available at: https://doi.org/10.1016/0264-8172(88)90040-2.

Mas, M., Ramón, J., Alonso, A., Guimerá, J. (1993). Evolución tectonosedimentaria de una cuenca extensional intraplaca: La cuenca finijurásica-eocretácica de Los Cameros (La Rioja-Soria). *Revista de la Sociedad Geológica de España*, 6(3/4), 129–144.

Matthews, D.H. and Williams, C.A. (1968). Linear magnetic anomalies in the Bay of Biscay: A qualitative interpretation. *Earth and Planetary Science Letters*, 4(4), 315–320 [Online]. Available at: https://doi.org/10.1016/0012-821X(68)90094-0.

McCarthy, A., Tugend, J., Mohn, G., Candioti, L., Chelle-Michou, C., Arculus, R., Schmalholz, S.M., Müntener, O. (2020). A case of Ampferer-type subduction and consequences for the Alps and the Pyrenees. *American Journal of Science*, 320(4), 313–372 [Online]. Available at: https://doi.org/10.2475/04.2020.01.

McCarthy, A., Tugend, J., Mohn, G. (2021). Formation of the Alpine orogen by amagmatic convergence and assembly of previously rifted lithosphere. *Elements*, 17(1), 29–34 [Online]. Available at: https://doi.org/10.2138/gselements.17.1.29.

Montadert, L. and Roberts, D.G. (1979). Initial Reports of the Deep Sea Drilling Project. U.S. Government Printing Office, 48 [Online]. Available at: https://doi.org/10.2973/dsdp.proc.48.1979.

Montadert, L. and Winnock, E. (1971). *Histoire structurale du golfe de Gascogne*. Technip, Paris.

Montadert, L., Damotte, B., Fail, J.M., Delteil, J.R., Valery, P. (1971). Structure géologique de la plaine abyssale du golfe de Gascogne. In *Histoire structurale du golfe de Gascogne*, Institut Français du Pétrole (ed.). Technip, Paris.

Montadert, L., Winnock, E., Deltiel, J.R., Grau, G. (1974). Continental margins of Galicia-Portugal and Bay of Biscay. In *The Geology of Continental Margins*, Burk, C.A. and Drake, C.L. (eds). Springer, Berlin/Heidelberg [Online]. Available at: https://doi.org/10.1007/978-3-662-01141-6_24.

Montadert, L., Roberts, D.G., de Charpal, O., Guennoc, P. (1979). Rifting and subsidence of the northern continental margin of the Bay of Biscay. Initial Reports of the Deep Sea Drilling Project, U.S. Government Printing Office, 48, 1025–1060 [Online]. Available at: https://doi.org/10.2973/dsdp.proc.48.154.1979.

Nirrengarten, M., Manatschal, G., Tugend, J., Kusznir, N., Sauter, D. (2018). Kinematic evolution of the Southern North Atlantic: Implications for the formation of hyperextended rift systems. *Tectonics*, 37(1), 89–118 [Online]. Available at: https://doi.org/10.1002/2017TC004495.

Olivet, J.L. (1996). La cinématique de la plaque ibérique. *Bulletin des Centres de Recherches Exploration-Production Elf-Aquitaine*, 20(1), 131–195.

Olivet, J.L., Bonin, J., Beuzart, P., Auzende, J.M. (1984). Cinématique de l'Atlantique Nord et Central. *Publications du Centre National pour l'Exploitation des Océans* (CNEXO), Série Rapports scientifiques et techniques, 54.

Pastouret, L. and Auffret, G.A. (1976). Observations sur les microfaciès des roches sédimentaires prélevées sur la marge armoricaine. *Revue de l'Institut Français du Pétrole*, 31(3), 401–425.

Pastouret, L. and Maury, R.C. (1982). Présence de pillow-lavas tholéitiques dans le canyon Shamrock (marge continentale armoricaine) ; leur place dans l'histoire du golfe de Gascogne. *Comptes Rendus de l'Academie des Sciences Serie II*, 294(11), 669–674.

Pastouret, L., Masse, J.P., Philip, J., Auffret, G.A. (1974). Sur la présence d'Aptien inférieur à Faciès urgonien sur la marge continentale armoricaine. Conséquences paléogéographiques. *Comptes Rendus de l'Académie des Sciences de Paris*, 278(16), 2011–2014.

Pedreira, D., Pulgar, J.A., Gallart, J., Torné, M. (2007). Three-dimensional gravity and magnetic modeling of crustal indentation and wedging in the western Pyrenees-Cantabrian Mountains. *Journal of Geophysical Research*, 112(B12), B12405 [Online]. Available at: https://doi.org/10.1029/2007JB005021.

Pedreira, D., Afonso, J.C., Pulgar, J.A., Gallastegui, J., Carballo, A., Fernàndez, M., Garcia-Castellanos, D., Jiménez-Munt, I., Semprich, J., García-Moreno, O. (2015). Geophysical-petrological modeling of the lithosphere beneath the Cantabrian Mountains and the North-Iberian Margin: Geodynamic implications. *Lithos*, 230(August), 46–68.

Pinet, B., Montadert, L., Curnelle, R., Cazes, M., Marillier, F., Rolet, J., Tomassino, A. (1987). Crustal thinning on the Aquitaine shelf, Bay of Biscay, from deep seismic data. *Nature*, 325(6104), 513–516 [Online]. Available at: https://doi.org/10.1038/325513a0.

Pinet, B., Sibuet, J.C., Lefort, J.P., Schroeder, I., Montadert, L. (1991). Structure profonde de la marge des entrées de la Manche et du plateau continental celtique : le profil WAM. *Mémoires de la Société Géologique de France*, 159, 167–183.

Pulgar, J.A., Gallart, J., Fernández-Viejo, G., Pérez-Estaún, A., Álvarez-Marrón, J. (1996). Seismic image of the Cantabrian Mountains in the western extension of the Pyrenees from integrated ESCIN reflection and refraction data. *Tectonophysics*, 264(1/4), 1–19 [Online]. Available at: https://doi.org/10.1016/S0040-1951(96)00114-X.

Ries, A.C. (1978). The opening of the Bay of Biscay – A review. *Earth-Science Reviews*, 14(1), 35–63 [Online]. Available at: https://doi.org/10.1016/0012-8252(78)90041-7.

Roca, E., Muñoz, J.A., Ferrer, O., Ellouz, N. (2011). The role of the Bay of Biscay mesozoic extensional structure in the configuration of the Pyrenean orogen: Constraints from the MARCONI deep seismic reflection survey. *Tectonics*, 30(2), TC2001 [Online]. Available at: https://doi.org/10.1029/2010TC002735.

Roest, W.R. and Srivastava, S.P. (1991). Kinematics of the plate boundaries between Eurasia, Iberia, and Africa in the North Atlantic from the Late Cretaceous to the present. *Geology*, 19(6), 613 [Online]. Available at: https://doi.org/10.1130/0091-7613(1991)019<0613:KOTPBB>2.3.CO;2.

Rosenbaum, G., Lister, G.S., Duboz, C. (2002). Relative motions of Africa, Iberia and Europe during Alpine orogeny. *Tectonophysics*, 359(1/2), 117–129 [Online]. Available at: https://doi.org/10.1016/S0040-1951(02)00442-0.

Ruffell, A. (1995). Evolution and hydrocarbon prospectivity of the Brittany Basin (Western Approaches Trough), offshore north-west France. *Marine and Petroleum Geology*, 12(4), 387–407 [Online]. Available at: https://doi.org/10.1016/0264-8172(95)96902-3.

Ruiz, M., Díaz, J., Pedreira, D., Gallart, J., Pulgar, J.A. (2017). Crustal structure of the North Iberian Continental margin from seismic refraction/wide-angle reflection profiles. *Tectonophysics*, 717(October), 65–82 [Online]. Available at: https://doi.org/10.1016/j.tecto.2017.07.008.

Salas, R. and Casas, A. (1993). Mesozoic extensional tectonics, stratigraphy and crustal evolution during the Alpine cycle of the eastern Iberian basin. *Tectonophysics*, 228(1/2), 33–55.

Saspiturry, N., Razin, P., Baudin, T., Serrano, O., Issautier, B., Lasseur, E., Leleu, S. (2019). Symmetry vs. asymmetry of a hyper-thinned rift: Example of the Mauléon basin (Western Pyrenees, France). *Mar. Pet. Geol.*, 104, 86–105 [Online]. Available at: https://doi.org/10.1016/j.marpetgeo.2019.03.031.

Saspiturry, N., Issautier, B., Razin, P., Baudin, T., Asti, R., Lagabrielle, Y., Allanic, C., Serrano, O., Duretz, T. (2021). Review of Iberia-Eurasia plate-boundary basins: Role of sedimentary burial and salt tectonics during rifting and continental breakup. *Basin Research*, 33, 1626–1661 [Online]. Available at: https://doi.org/10.1111/bre.12529.

Sibuet, J.C. and Collette, B.J. (1991). Triple junctions of Bay of Biscay and North Atlantic: New constraints on the kinematic evolution. *Geology*, 19(5), 522.

Sibuet, J.C. and Le Pichon, X. (1971). Structure gravimétrique du golfe de Gascogne et le fossé marginal nord-espagnol. In *Histoire structurale du golfe de Gascogne*, Institut Français du Pétrole (ed.). Technip, Paris.

Sibuet, J.C., Monti, S., Réhault, J.P., Durand, C., Gueguen. E. (1993). Quantification de l'extension liée à la phase pyrénéenne et géométrie de la frontière de plaques dans la partie ouest du golfe de Gascogne. *Comptes Rendus de l'Académie des Sciences. Série 2, Mécanique, Physique, Chimie, Sciences de l'Univers, Sciences de la Terre*, 317(9), 1207–1214.

The Shipboard Scientific Party (1979a). Site 401. Initial Reports of the Deep Sea Drilling Project, U.S. Government Printing Office, 48 [Online]. Available at: https://doi.org/10.2973/dsdp.proc.48.104.1979.

The Shipboard Scientific Party (1979b). Site 402/Hole 402A. Initial Reports of the Deep Sea Drilling Project, U.S. Government Printing Office, 48 [Online]. Available at: https://doi.org/10.2973/dsdp.proc.48.105.1979.

The Shipboard Scientific Party (1979c). Sites 399, 400, and Hole 440A. Initial Reports of the Deep Sea Drilling Project, U.S. Government Printing Office, 48 [Online]. Available at: https://doi.org/10.2973/dsdp.proc.48.103.1979.

Thinon, I. (1999). Structure profonde de la Marge Nord Gascogne et du Bassin Armoricain. PhD Thesis, Université de Bretagne Occidentale, Brest.

Thinon, I., Fidalgo-González, L., Réhault, J.P., Olivet, J.L. (2001). Déformations pyrénéennes dans le golfe de Gascogne. *Comptes Rendus de l'Académie des Sciences – Series IIA – Earth and Planetary Science*, 332(9), 561–568 [Online]. Available at: https://doi.org/10.1016/S1251-8050(01)01576-2.

Thinon, I., Réhault, J.P., Fidalgo-González, L. (2002). La couverture sédimentaire SynRift de la marge Nord-Gascogne et du bassin armoricain (golfe de Gascogne) : à partir de nouvelles données de sismique réflexion. *Bulletin de la Société Géologique de France*, 173(6), 515–522.

Thinon, I., Matias, L., Réhault, J.P., Hirn, A., Fidalgo-González, L., Avedik, F. (2003). Deep structure of the Armorican basin (Bay of Biscay): A review of norgasis seismic reflection and refraction data. *Journal of the Geological Society*, 160(1), 99–116 [Online]. Available at: https://doi.org/10.1144/0016-764901-103.

Tomassino, A. and Marillier, F. (1997). Processing and interpretation in the tau-p domain of the ECORS Bay of Biscay expanding spread profiles. *Mémoires de la Société Géologique de France (1833)*, 171, 31–43.

Tugend, J. (2013). Role of hyperextension for the formation of rift systems and its implication for reactivation processes and Orogen formation: The example of the Bay of Biscay and Pyrenees. PhD Thesis, Université de Strasbourg.

Tugend, J., Manatschal, G., Kusznir, N.J., Masini, E., Mohn, G., Thinon, I. (2014). Formation and deformation of hyperextended rift systems: Insights from rift domain mapping in the Bay of Biscay-Pyrenees. *Tectonics*, 33(7), 1239–1276 [Online]. Available at: https://doi.org/10.1002/2014TC003529.

Tugend, J., Manatschal, G., Kusznir, N.J., Masini, E. (2015a). Characterizing and identifying structural domains at rifted continental margins: Application to the Bay of Biscay margins and its Western Pyrenean fossil remnants. *Geological Society, London, Special Publications*, 413(1), 171–203 [Online]. Available at: https://doi.org/10.1144/SP413.3.

Tugend, J., Manatschal, G., Kusznir, N.J. (2015b). Spatial and temporal evolution of hyperextended rift systems: Implication for the nature, kinematics, and timing of the Iberian-European plate boundary. *Geology*, 43(1), 15–18 [Online]. Available at: https://doi.org/10.1130/G36072.1.

Vaillant, P. (1988). L'extrémité occidentale de la marge Nord Gascogne : contexte stratigraphique, structural et cinématique, implications géodynamiques. PhD Thesis, Université de Bretagne Occidentale, Brest.

Vergés, J. and García-Senz, J. (2001). Mesozoic evolution and cainozoic inversion of the Pyrenean rift. *Mémoires du Muséum National d'Histoire Naturelle (1993)*, 186, 187–212.

Vissers, R.L.M. and Meijer, P. (2012). Mesozoic rotation of Iberia: Subduction in the Pyrenees? *Earth-Science Reviews*, 110(1/4), 93–110.

Whitmarsh, R.B, Avedik, F., Saunders, M.R. (1986). The seismic structure of thinned continental crust in the Northern Bay of Biscay. *Geophysical Journal International*, 86(2), 589–602.

Yang, P., Welford, J.K., Peace, A.L., Hobbs, R. (2020). Investigating the Goban Spur rifted continental margin, offshore Ireland, through integration of new seismic reflection and potential field data. *Tectonophysics*, 777(February), 228364.

# 7

# The Trimouns Talc-Chlorite Deposit

**Michel DE SAINT BLANQUAT[1], Philippe DE PARSEVAL[1], Alexandre BOUTIN[1], Benoit QUESNEL[2] and Marc POUJOL[3]**

[1] *GET, CNRS, Université Toulouse-III-Paul-Sabatier, IRD, CNES, OMP, Toulouse, France*

[2] *CREGU, GeoRessources, CNRS, Université de Lorraine, Nancy, France*

[3] *Géosciences Rennes, CNRS, Université de Rennes 1, France*

## 7.1. The central place of Trimouns in the geology of the Pyrenees

The Trimouns talc-chlorite deposit (Figure 7.1) is located in the heart of a North Pyrenean Massif (NPM), the Montagne de Tabe (MdT), often called the Saint-Barthélémy massif, which is a Variscan basement massif. Until the end of the 1980s, researchers interpreted NPMs as massifs free from any trace of post-Variscan deformation, apart from the fact that they are thrusted over the North Pyrenean foreland during the Pyrenean orogeny (Souquet and Peybernès 1987). The Trimouns talc deposit was therefore considered as a witness of the Variscan orogeny (see the overview by Fortuné et al. (1980)).

The recent results on metasomatism and deformation at Trimouns, which are the subject of this chapter, show unambiguously that the formation of talc-chlorite

---

For a color version of all the figures in this chapter, see www.iste.co.uk/saspiturry/evolution1.zip.

*Evolution of the Pyrenees during the Variscan and Alpine Cycles 1*,
coordinated by Nicolas SASPITURRY, Jessica UZEL, Alexandre ORTIZ. © ISTE Ltd 2024.

mineralization is associated with the rifting phase at the beginning of the Alpine cycle (250 Ma current), and that the structure inherited from the Variscan cycle in the Pyrenees (about 330–260 Ma) plays a significant role.

These results are to be contrasted with the numerous recent works on the Mesozoic series and peridotites, interpreting the North Pyrenean Zone (NPZ) as a hot, hyperextended passive continental paleomargin with mantle locally exhumed up to the surface (Lagabrielle and Bodinier 2008; Jammes et al. 2009; Lagabrielle et al. 2010, 2020; Clerc et al. 2012; Lagabrielle et al. 2016; de Saint Blanquat et al. 2016; Saspiturry et al. 2019, 2021a, 2021b). Formed in the Early Cretaceous, this margin was inverted during the Pyrenean orogeny initiating at Late Cretaceous-Paleogene time (see overview in Teixell et al. (2018)). The new metasomatism and deformation data at Trimouns allow us to replace the antemesozoic basement within the Early Cretaceous hyperextension context, but also to understand the tectono-metamorphic processes involved under the pre-rift cover, when the continental crust thins until it breaks, in association with mantle exhumation.

**Figure 7.1.** *Aerial view of the Trimouns quarry. From west to east (from left to right), we can distinguish: the wall formations (beige), the deposit (white) and the roof formations (gray). Note the Chateau de Montségur in the background (magnifying glass not provided) (image by MSB 2016)*

Trimouns is therefore central to Pyrenean geology, since it is possible to observe and study a geological record that covers the entire Variscan and Alpine cycles, and particularly the relationships between their different pre-, syn-, late- and postorogenic phases. Trimouns also means we can study specific geological processes, such as the relationship between partial melting, metamorphism and deformation of the continental crust (Variscan cycle), Early Cretaceous continental crust thinning and the role of Variscan structural inheritance during hyperextension, as well as the role of the Early Cretaceous structural template during the subsequent Upper Cretaceous and Tertiary orogenic phases. The geology of Trimouns thus opens up perspectives on the study of paleorifts elsewhere on earth, particularly with respect to the structural behavior of the basement, fluid circulation and associated mineralization.

## 7.2. Pyrenean talc and its exploitation

Numerous talc occurrences can be observed in the Pyrenees (Figure 7.2), most often associated with tectonic accidents, detachments, thrusts or faults. Most of them are derived from the transformation of Ante-Mesozoic metasediments and are located in the NPM and the Axial Zone (AZ), sometimes in contact with magmatic intrusions (e.g. Quérigut pluton). They are generally characterized by the association between dolomitic limestones and silico-aluminous rocks (micaschistes, migmatites and granites), and almost always correspond to associations of talc and chlorite.

The most important talc deposit in the Pyrenees is that of Trimouns, in the MdT, which also contains the formerly exploited Porteille deposit, above Mont-Ferrier. The other main deposits are located in the AZ, in the massifs of the Roque de France (Les Salines and La Vajol on the Spanish side, Reynes, Mas Parer and Montner on the French side), Canigou (Corneilla-de-Conflent), at the contact of the Quérigut-Millas pluton (Col de Jau, Counozouls), and in a few other NPM, for example, Trois-Seigneurs in Ariège (Rabat) or the small massif of Espezel in the Aude department (Figure 7.2). Rare occurrences of talc have been observed in the "Chaînons Béarnais" (Saraillé) (Lagabrielle et al. 2010; Corre et al. 2018).

The populations who lived in or near the Pyrenees or who crossed them have always known this white and soft rock, with a very soft touch (talc is the least hard mineral on the Mohs scale, and the word talc also designates the rock composed only of talc, sometimes also called soapstone when it is massive). The talc pencils found on the *Dame à la capuche* site in Brassempouy in the Landes region, dating back more than 20,000 years, the talc mixed with the pigments in the 11,500-year-old paintings in the Cave of Niaux and the 12th-century talc-chlorite statues discovered in Montségur all bear witness to this.

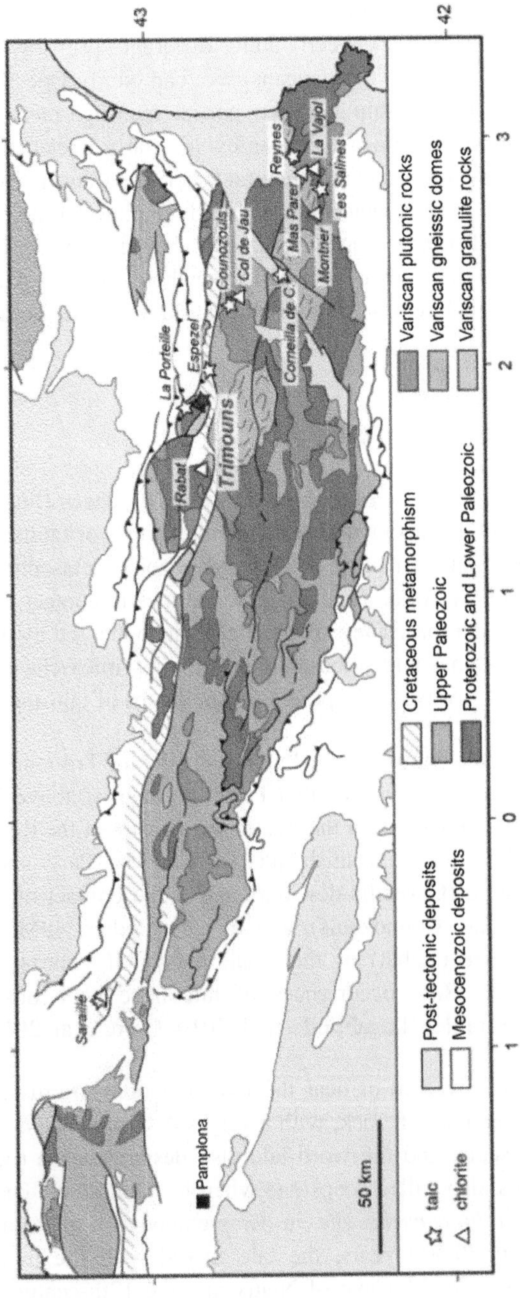


**Figure 7.2.** *Simplified geological map of the Pyrenees with location of the main talc-chlorite deposits*

The Trimouns quarry is located on the eastern slopes of the Soularac peak, the highest point of the MdT (2,368 m). The name Trimouns comes from “Tres Monts”, the name of the pass, now disappeared, located at an altitude of 1,840 m and about 1 km south of the Peyre pass, where talc was found before any exploitation. These two passes constituted one of the passages to go from the high valley of the Ariège to Montségur and to the plain of Lavelanet. However, the “three mounts” have long since disappeared as a result of the exploitation of this talc layer, which sinks eastward under the mountain. The mining concession is shared between the communes of Montségur, Bestiac, Lordat and Vernaux. The administrative buildings and the processing plant are located in Luzenac, hence the common name of Luzenac talc. Founded in 1905, the first company of exploitation, called Société des talcs de Luzenac, was built around the exploitation of the quarries of Porteille (exploited until 1969), and especially of Trimouns. Currently, the deposit is exploited by the group Imerys talc, a subsidiary of Imerys.

The very specific properties of talc have led to its use in many industrial sectors. It is an inert mineral, chemically and biologically stable, which can be mixed with most mineral and organic products without degrading. It is a tough phyllosilicate, with each sheet resistant to ordinary chemical attack and to heat up to 1,000°C. Its layered structure, linked by Van der Walls bonds, gives it a preferential orientation, a soft touch and a lubricating effect. Finally, it is adsorptive and hydrophobic, and easily binds organic molecules and releases them gradually without altering them.

The Trimouns deposit supplies many industries where talc is used as a mineral filler: paper, polymers, paints, ceramics, food, fertilizers, wastewater treatment, cosmetics and pharmaceuticals. A car, for example, contains more than 10 kg of talc (plastics in bumpers and dashboards, tires, body paint, etc.). The current use of this mineral is also part of a sustainable development logic, as it is a natural mineral that does not require chemical treatment, and in some cases replaces substances considered harmful.

The open pit deposit (Figure 7.1) is located between 1,650 and 1,800 m above sea level and is only mined from May to October due to winter snow cover. The talc is mined along 5-m-high steps that run parallel to the elongation of the deposit (Figure 7.3). The wide variety of ores, from talc to chlorite and all the intermediates, as well as the presence of barren blocks, require highly selective extraction and sorting. The use of small mechanical shovels allows this selection and leads to more than a dozen qualities of ore classified directly on the deposit, according to the whiter to darker color, and thus the more or less talc or chlorite mineralogy (Figure 7.3(b)). Each ton of talc extracted requires the evacuation of about 10 tons of waste rock (Figures 7.3(c) and (d)). Stored in silos, the ores are then brought down to

Luzenac via a cable car that links the quarry to the plant in the valley in less than 30 min. The 5-km-long cable car operates at a rate of 180 tons per hour, with about 100 skips, each of which can hold almost 1.5 ton of ore. The braking is applied to the cable when the skips are lowered, this system produces the electricity necessary for its own operation and also provides power for other equipment on the site. At Luzenac, a large portion of the ore then passes through a fully automated optical sorting station. A high-resolution camera analyzes the color of each piece of talc and directs its fall by a jet of compressed air. This technology optimizes the recovery of the purest white ore and meets the growing market demand for this type of product. The ore is finally dried, crushed and shipped worldwide. This is a far cry from the manual sorting of ore by "piquette" that existed until the late 1990s (Figure 7.3(a)), the transport by oxen that was replaced in 1904 by the first cable car, and the manual filling of skips that ended in 1962. There were 800 workers in 1950 and 400 in 1980. Today, there are only 100 or so, half of whom are seasonal workers, to extract 450,000 tons of talc per year, that is, about 10% of the world's production, which makes Trimouns the largest talc mine in the world. Since the beginning of the industrial exploitation of the site, more than 20 million tons of talc have been extracted.

**Figure 7.3.** *The exploitation of talc in Trimouns. (a) Extraction and manual sorting of talc in 1985; (b) extraction and sorting with a mechanical shovel in 2015; (c) overburden at work; (d) roof steps and overburden*

## 7.3. History of the work

Since the first works listed in the "*Aperçu sur l'étude géologique de l'Ariège*" by François (1841) until today, about 50 scientific publications and 15 PhD theses have been published on the MdT and the Trimouns site.

The first PhD thesis is that of Zwart (1954), who drew the first geological map of the massif. He shows the presence of catazonal rocks, the first described in the Pyrenees. The Paleozoic was then studied more specifically by Mangin (1967) for the northern edge, and by Arnoux and Majza (1986) for the eastern edge. Guchereau (1975) studied migmatites and gneisses and defined the geometry of isograds centered around a migmatitic dome. He also highlights an increasing deformation toward the upper part of the gneisses. All these authors interpret the structure of the massif as resulting from Variscan orogenic deformation.

At the end of the last century, works based on petrostructural studies (Passchier 1982; de Saint Blanquat 1985, 1989) reinterpreted the structure of the MdT and proposed that it is largely the result of extensive south-verging tectonics, possibly Cretaceous in age (Passchier 1984), and related to extension and strike-slip deformation, exhumation of the sub-continental mantle associated with a passive margin (de Saint Blanquat et al. 1986), late Variscan extensional context (de Saint Blanquat et al. 1990) or both (de Saint Blanquat 1993). The difficulty of interpretation lies in the strong structural ambiguity related to the remarkable constancy of the orientation for the stretching lineation in the gneisses and migmatites and the systematic top to south sense of shear, and this from high (800°C, ductile) to low (<300°C, brittle) temperature.

The consequence at Trimouns, located in the eastern part of the MdT, was that its formation context could also be questioned and associated with mid-Cretaceous extensive events. Indeed, until this work, the formation of the Trimouns mineralization was assumed to be associated with a westward thrust (Fortuné et al. 1980). More recent work (de Saint Blanquat 1985, 1989) has shown that this thrust does not explain the strong reduction of the lithologic pile observed at Trimouns (Upper Paleozoic on migmatites). Moreover, the lineation synchroneous of the talc formation is N-S and horizontal, carried by a N-S schistosity with a variable dip to the east. The absence of structural markers showing E-W displacements and the abundance of markers showing N-S displacements have therefore led to the interpretation of syn- to late-migmatite deformation as ante-Pyrenean N-S shear, with a dexterous strike-slip at Trimouns (de Saint Blanquat 1985, 1989).

Geochronological data obtained on rare earth minerals from Trimouns then confirmed this hypothesis by showing that talc was formed in the Cretaceous (around 100 Ma), that is, during the operation of the transtensive rift created by the opening of the Bay of Biscay and the rotation of the Iberian plate (Schärrer et al. 1999).

Supported by the study of the relationship between fluids and mineralization (de Parseval 1992), all these results have recently been taken up in the framework of several academic projects (CNRS-INSU, ANR-Pyramid) or in partnership with the operator (SA des talcs de Luzenac, Rio Tinto, then Imerys) or other industrial partners (TotalEnergies, Convergence/Fluids project). This work has allowed us to specify this new extensional and cretaceous model and clearly establish its relevance (see details below). Analytical developments (in situ U-Th-Pb dating on titaniferous minerals) have allowed us to date the metasomatism at the origin of the Trimouns talc between the Aptian and Turonian more precisely (125–90 Ma) (Boutin et al. 2016) and show that it was contemporary with a regional metasomatism that affected the northern Pyrenean basement and the northern axial zone (Boulvais et al. 2007; Poujol et al. 2010; Fallourd et al. 2013). This shows the existence in the Middle Cretaceous of a long and intense fluid circulation event of regional extent.

Studies on the transformation of dolomites into talc ores and of the fluids responsible for the mineralization began with Mengaud (1908), and continued notably by Capdecomme (1950), and then Zwart (1953), Thiebaut et al. (1965), Aranitis (1967), Fortuné (1971), Moine (1982), Moine et al. (1979, 1982, 1989), de Parseval (1992), Boiron et al. (2005) and Boulvais et al. (2006).

This body of work has made it possible to characterize the metasomatic transformations at the origin of the deposit, metasomatosis corresponding to a metamorphic transformation under the action of a fluid inducing major changes in chemical and mineralogical composition between the starting rocks and those transformed (dolomite versus talc).

For the majority of authors (Capdecomme, Thiebaut, Fortuné and Moine), talc results from the metasomatosis of dolomites through salty and siliceous solutions. The talcification is accompanied by a release of Ca, which can precipitate in the form of calcite. Other authors (Mengaud and Aranitis) reject the talc-dolomite linkage and appeal to magnesian solutions, which would have reacted on the muscovite-rich horizons and on the limestones, giving clinochlore rocks and dolomites. The silica released by the formation of clinochlore and the dissolution of quartz would have combined in solution with magnesia to give talc. The conditions of formation of the mineralization have been estimated at Trimouns at 200–300°C and less than 2 kbar (de Parseval 1992; Boiron et al. 2005).

## 7.4. Geological context

The MdT is an NPM, that is, a massif located north of the North Pyrenean Fault (NPF), and consisting of Ante-Mesozoic basement, uprooted and transported on Meso- and Cenozoic sediments during the Upper Cretaceous and Eocene orogenic phases. It is therefore composed of Ante-Mesozoic rocks (Figure 7.4). The lithologic units that compose it have undergone a complex, polyorogenic, polymetamorphic and poly-metasomatic history.

To the north, the MdT oberlaps the Nalzen Basin, a sandstone-pelite basin (the "black flysch") of Albian to Campanian age, via a south dipping thrust (Mangin 1967; Baby 1988). To the east, it dips beneath the brecciated complex of the Inner Metamorphic Zone (IMZ), containing peridotite-element, and its Mesozoic cover (Sault Plateau) via a weakly east-dipping polyphased tectonic contact, first extensional and related to Cretaceous rifting, and then compressional and related to the orogenic inversion (de Saint Blanquat et al. 2016). To the south, this same contact becomes subvertical or even steeply dipping to the north and the MdT overlaps the IMZ and the Tarascon Basin to the south. Finally, to the west, the MdT is in contact with the Arize NPM via the Croquié fault (Figure 7.4).

The MdT contains, from bottom to top: (1) a gneissic series, the lower part of which can be related to the low-pressure high-temperature (LP-HP) granulite facies and the upper part to the amphibolite facies (Zwart 1954; Vielzeuf 1984; de Saint Blanquat 1989; de Saint Blanquat et al. 1990; Lemirre 2018); this series contains mainly paragneisses, but also marbles, amphibolites and orthogneiss lenses. All these lithologies are derived from the transformation of sedimentary rocks interlayed with effusive and intrusive magmatic rocks, all of Upper Proterozoic to Lower Paleozoic age; (2) a heterogeneous migmatite assemblage containing various, mainly leucogranitic granitoids, which gradually grades upward to micaschists and schists derived from the transformation of Cambrian-Ordovician sediments; (3) black shales rich in organic matter and Silurian carbonates (marbles and dolomites) in which an important polyphase detachment is located, which played a role in the Variscan and Alpine periods; (4) a folded and schistosed Devono-carboniferous sedimentary series, sometimes sheared and foliated. Some Permian rocks are trapped in the northern thrust of the massif.

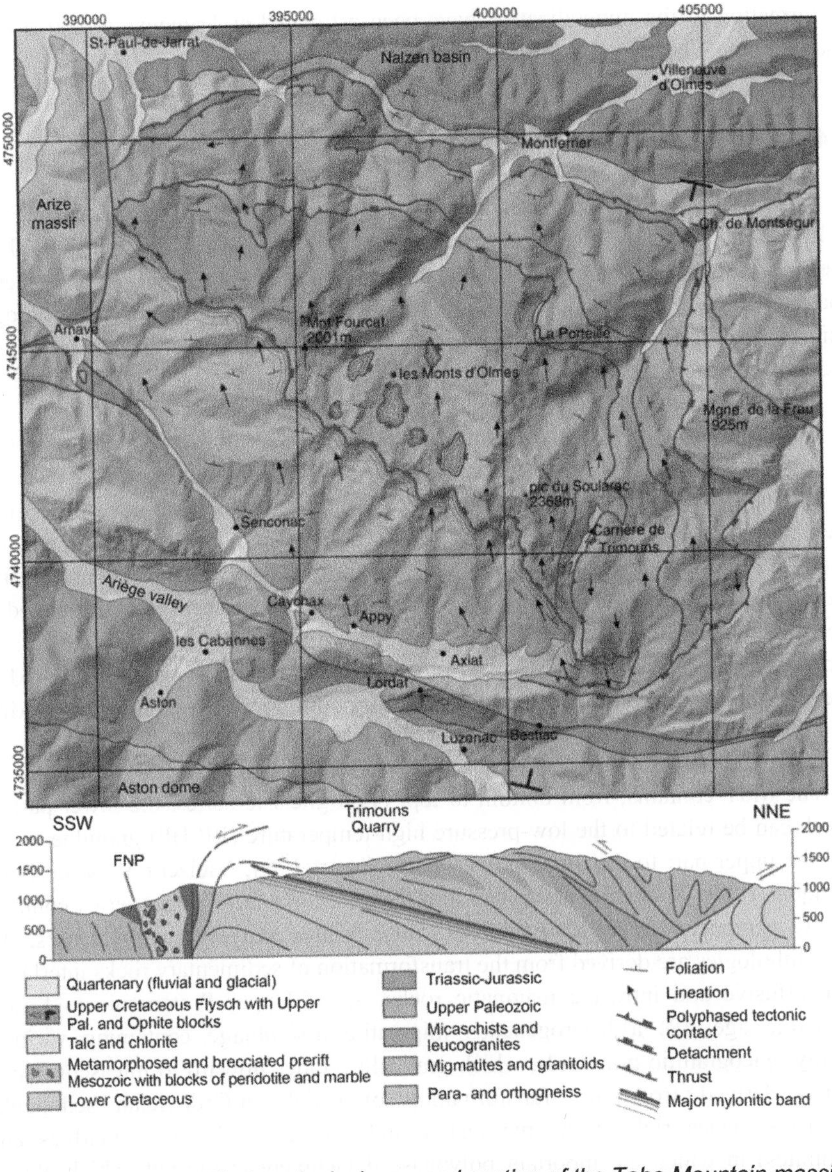


**Figure 7.4.** *C simplified geological map and section of the Tabe Mountain massif*

The structure of the MdT is that of a north- and east-dipping dome fragment. Two main structural discontinuities structure the massif: (1) a large, shallowly dipping ductile shear zone (the Main Mylonitic Band (MMB)), oriented

290–300°/15–20°, which separates the gneisses and migmatites. It carries an NNW-SSE-oriented lineation and shows a top to the south shear direction. The gneiss series is affected by a ductile deformation in retrograde condition, with increasing intensity from the granulites at the base of the massif to the top of the MMB, which is marked by a plurimetric layer of ultramylonite (Passchier 1982, 1985; de Saint Blanquat 1989), and then decreases very abruptly in the migmatitic ensemble, which is not very penetratively deformed, but still affected by localized shear zones. This asymmetric pattern of finite deformation on either side of the MMB is characteristic of detachments. The petrostructural data allow it to be interpreted as a late-Varictican normal shear-zone, possibly replayed in the Cretaceous, and tilted to the position of an apparent reverse shear-zone at the time of the Upper Cretaceous-Eocene inversion (Passchier 1982, 1984; de Saint Blanquat 1985, 1989, 1993; de Saint Blanquat et al. 1990); (2) a polyphase kinetic accident located above the MMB at the Ordovician-Silurian transition, an accident known throughout the Variscan Pyrenees. In the MdT, it has the particularity of being associated with talc, whose formation fossilized the Variscan deformations, which is itself deformed. From northwest to southeast, we find along this contact a succession of talc-chlorite mineralizations of very variable dimensions, la Porteille, Reboule, le Pradas, Trimouns, the peak of Ourlès and the forest of Salvanières. Only the site of Trimouns is currently exploited.

The mapping of metamorphic zones shows very tight isogrades: the transition from epizone (weak metamorphism) to catazone (strong metamorphism, partial melting of the crust) occurs in less than 2–3 km of structural thickness, and a shortfall of about 10 km of thickness is to be noted in comparison with the neighboring NPM (Arize and Trois-Seigneurs). Retromorphosis is present throughout the crystalline series (gneiss and migmatites) and is contemporaneous with the major ductile deformation phase that led to the formation of the MMB (Passchier 1982, 1985; de Saint Blanquat 1989; de Saint Blanquat et al. 1990). The metamorphic gradient observed today is apparent as tectonic in origin, controlled essentially by the two main strongly subtractive shear-zones described above, but also by retromorphic penetrative deformation. Geochronological data on high-temperature low-pressure (HT-LP) metamorphism and associated magmatism show unambiguously that these HT episodes belong to the Variscan cycle (Delaperriere et al. 1994; Lemirre 2018). Ar-Ar data on micas from granulite, gneiss, and migmatite samples yield ages of 110–100 Ma, regardless of rock texture, mylonitic or not (Costa and Maluski 1988). In situ UPb dating of apatites from the crystalline basement of the massif (gneisses, mylonites and migmatites) has yielded ages between 299 and 98 Ma; lower Cretaceous ages are interpreted in terms of ductile deformation ages (Odlum and Stöckli 2020).

## 7.5. Lithology, structure and microstructures

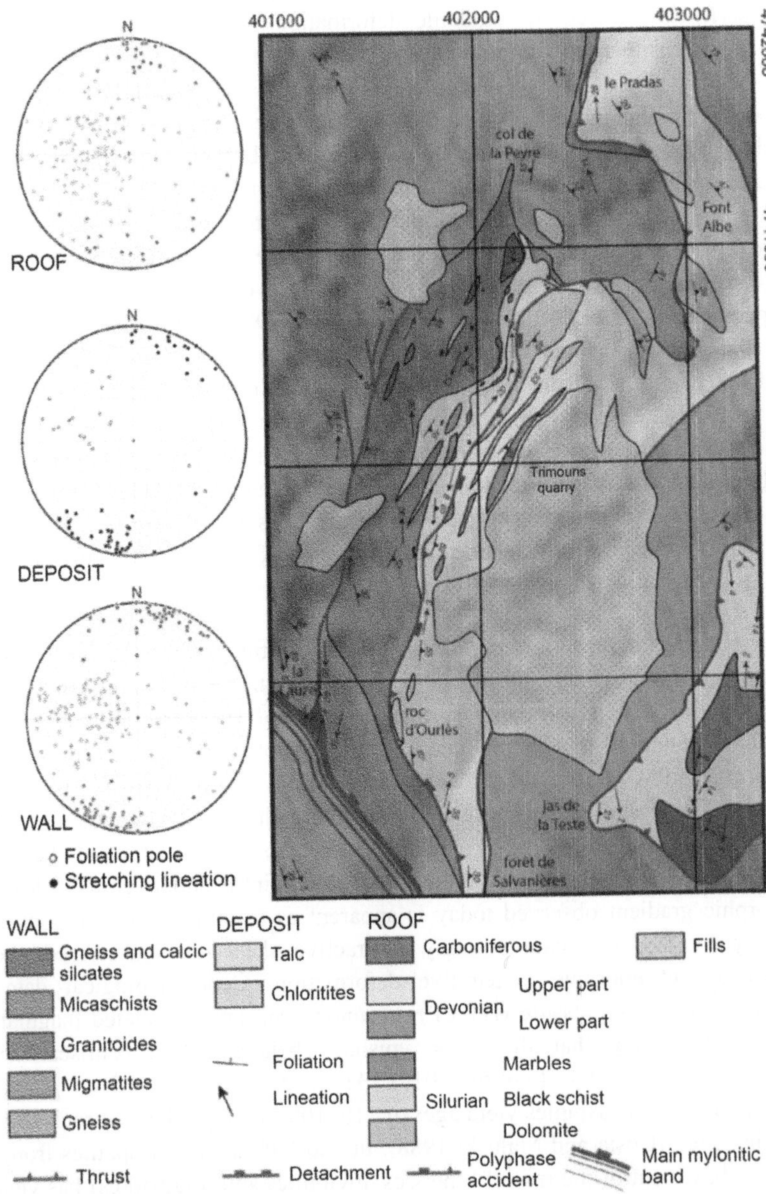


**Figure 7.5.** *Simplified geological map and structural measurements of the Trimouns quarry and its surroundings. Schmidt's canvas, lower hemisphere*

The Trimouns deposit is oriented NNE-SSW and dips variably to the east (Figures 7.5 and 7.6). It is hosted between Precambrian to Cambro-Ordovician formations (migmatites and micaschists) at the footwall and an upper Paleozoic cover (dolomites, limestones, shales and calcchists) at the hanging wall of the deposit. Talc and chlorite are, respectively, derived from the metasomatic alteration of dolomites in the roof and micaschists in the wall.

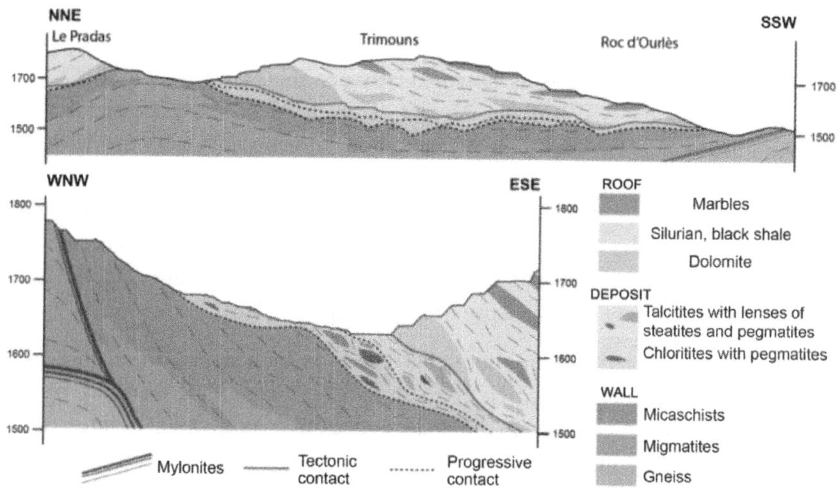


**Figure 7.6.** *Simplified geological sections of the Trimouns quarry, built from field and borehole data. Note that the two sections are not at the same scale*

### 7.5.1. *The wall*

The main lithologies encountered in the wall are (Figure 7.7):

1) Migmatites that contain various granitoids (granites, diorites, leucogranites, pegmatites) in metric to decametric lenses that are intrusive and concordant or in veins. The migmatitic bedding is on average N-S with a variable dip toward the east (Figure 7.7(a)). An N-S-oriented HT foliation is locally observed posterior to the bedding, which carries a subhorizontal N-S stretching lineation. The shear direction is top-to-south.

2) Micaschists that are located above the migmatites and of reduced thickness in the quarry (300 m maximum) compared to the rest of the massif (up to 1,500 m). These are composed of quartz, biotite, muscovite and plagioclase, and sometimes andalusite, rarely garnet, and accessory minerals. Sillimanite and cordierite are

observed in the transition zone with migmatites. Their bedding, of sedimentary or metasedimentary origin (S0-1), is affected by folds of various scales (metric to centimetric) whose axial plane is constituted by the major schistosity (S2), which is NNE-SSW with a dip of about 30° to the east, and carries a subhorizontal lineation. From bottom to top, we observe micaschists containing more or less quartz and not chloritized, and then chloritized micaschists. The transition between migmatites and micaschists is gradual in the northern part of the deposit (anatexis front), whereas a ductile shear zone (the structure called "F1 fault" by the operator) about 10 m thick brings these two lithologies into contact in the southern part of the deposit (Figure 7.7(b)). A structural continuity is obtained between this shear zone and the MMB, which connects to the SSW of the deposit at the level of the garages (at a place called La Lauze). The F1 fault thus appears to be contemporary with the MMB. It is oriented 10E60 and carries a subhorizontal or shallowly south-dipping stretching lineation. The sense of shear is dextral, that is, consistent with the general top-to-south movement of the MMB roof, and probably associated with a change in the geometry of the MMB. Above this shear zone, the micaschists are affected by C/S structures that also show dextral sense of shear (Figure 7.7(c)).

3) A number of varied lithologies that occur as inclusions in the micaschists: (i) calcium silicate gneisses (CSGs), which show a centimetric banding of green (epidote, clinozoisite, actinote, diopside, accessory minerals), white (quartz, plagioclase, accessory minerals) and red (biotite, garnet) levels (Figure 7.7(d)). The CSGs form metric to multi-decametric lenses. They show marked hot deformation with subhorizontal folds parallel to the N-S lineation, and boudinage; (ii) rare lenses of banded dolomitic marbles (calcite, dolomite ± diopside, epidote, prismatic actinote, accessory minerals); (iii) pegmatites in lenses or in N-S veins co-structured with micaschists. As the deposit approaches, the veins become more and more boudinaged and the pegmatites form discontinuous lenses of variable size.

### 7.5.2. *The deposit*

The deposit consists of talc and chlorite mineralization, but also pegmatites and waste rock inclusions belonging to the wall or roof (Figure 7.8). At the scale of the quarry, it presents a relatively simple geometry: lithological contacts, foliation and schistosity have an average N10 orientation with low eastward dips (<40°) to the north, and increasingly high to the south, up to 70°. The stretching mineral lineation carried by these planes has an average orientation parallel to the deposit and is therefore subhorizontal.

The thickness of the mineralized layer is low in the north, more important in the center where it can reach 80 m, and increasingly low toward the south. The deposit contains the following components:

1) Chloritites are composed of clinochlore and some accessory minerals (apatite, titanite, rutile). They are observed in the form of a main layer of variable thickness (from a few meters to a few tens of meters) and anastomosing veins located under the main layer on the wall side. In thin sections, several generations of chlorite are observed: a first generation does not show a marked prefabricated orientation and seems to have crystallized statically. A second generation seems to intersect the first and defines a schistosity (Figure 7.9(a)). Occasionally, C/S structures are observed and formed synchronously with this second generation (Figure 7.8(a)). The different generations of chlorites have globally similar compositions. The micaschist-chlorite contact is progressive and is materialized by a more or less diffuse metasomatic front. It does not correspond to an ancient tectonic structure or to a lithological discontinuity.

2) Talc-chlorites are a mixture of talc and chlorites, which mark the contact between chloritites and talcitites. These are observed as transition zones of metric scale between talcitites and chloritites. Under the microscope, an equivalent amount of talc and chlorite is observed, and the talc and chlorite lamellae are interlocked. The contact between chloritites and talcitites is sometimes diffuse when the transition is gradual, or sometimes very rapid, when it is taken up by the syn- to post-mineralization schistosity (Figure 7.8(b)). It shows no particular deformation and does not correspond to a mechanical discontinuity. Before metasomatism, however, it should correspond to the contact between the base of the Silurian and the underlying basement, in this case the micaschists, that is, a Variscan tectonic contact. The fact that it was not mechanically reused during the syn- or post-mineralization deformation episode(s) can be explained by the fact that metasomatosis begins with a rather static episode (which is suggested by the microstructures, see below) and deformation is then distributed throughout the thickness of the talcitites and chloritites, rocks of equivalent hardness and very easily deformed.

3) The talcitites form a main vein located between the main chorite vein and the Silurian. They contain more than 90% talc and accessory minerals such as apatites, sulfides (pyrite) and inherited graphite. The talcitites most commonly show a marked planar flow, which corresponds to either foliation (Figure 7.8(c)) or schistosity (Figure 7.8(d)), but massive talcitites, or steatites, also occur (Figure 7.8(e)). Inherited deformation structures fossilized by mineralization are often observed in steatite (Figure 7.8(f)).

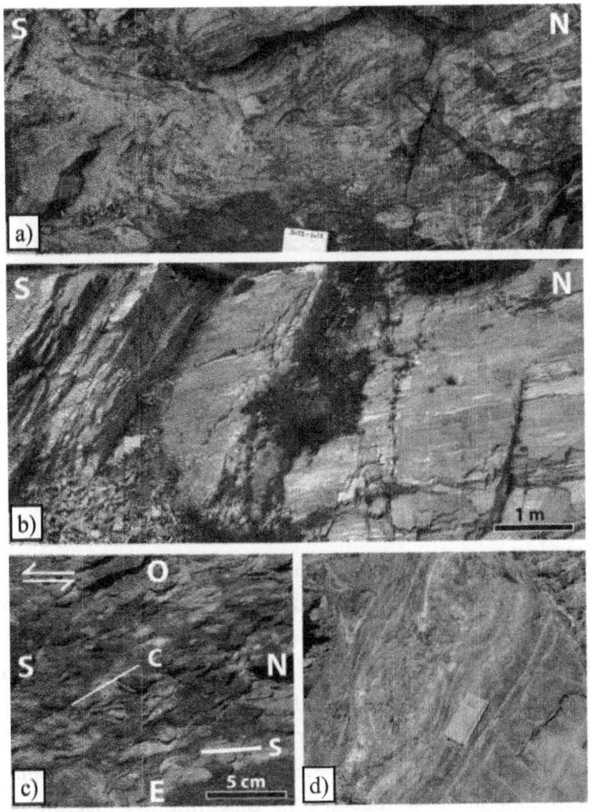


**Figure 7.7.** *Major wall facies. (a) Migmatites, note deformed bedding showing top to the south and postanatexis shear; (b) destral "F1" shear zone, note the stretching lineation weakly plunging to the south; (c) dextral sheared micaschist above F1; (d) boudinized calcium silicate gneiss*

The steatites form lenses of variable size, metric to decametric, elongated parallel to the main orientation of the deposit and essentially located in the center of the deposit over an area about 250 m long and a few tens of meters thick. Thin section observation shows talc lamellae generally smaller than 50 μm without preferential orientation, which suggests a static crystallization of the ore (Figure 7.9(b)). The foliated talcitite is relatively massive, it shows a banding and a preferential orientation of the talc lamellae that form a foliation, which is confirmed by the thin section observation (Figure 7.9(c)). The most common type of talcitite contains a relatively regular schistosity, but which sometimes shows C/S structures with apparent top to the south movement. It may also define tight folds or more complex figures. In thin sections, the schistosed talcitite shows a heterogeneous texture with schistosity planes marked by fine talc flakes and impurities that more or less crosscut a foliated texture or a structure without preferential orientation (Figure 7.9(d)). Foliation and schistosity bear a mineral stretching lineation averaging NNE-SSW, and are most often subparallel, or make an angle of less than 30°. Some samples show several generations of talc, a massive and/or foliated talc 1 intersected by talc 2 veins. These veins are sometimes folded and the axial plane of the folds is constituted by the schistosity.

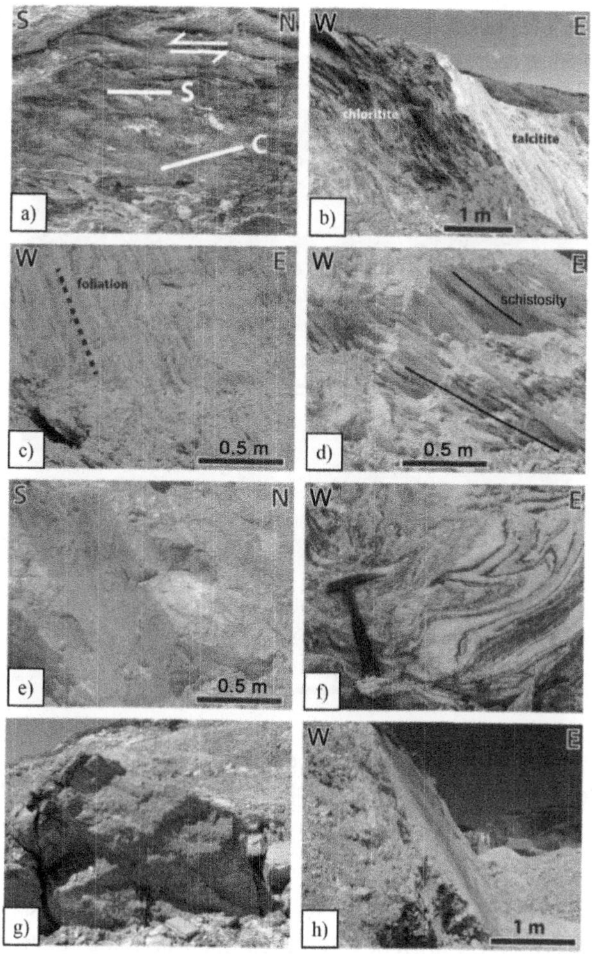


**Figure 7.8.** *Main facies of the deposit. (a) Chloritites with top to the south C/S texture; (b) schistose chloritite-talcitite contact; (c) foliated talcitite; (d) schistose talcitite; (e) steatite; (f) steatite with inherited structure fossilized by talcification; (g) banded marble boulder in the southern part of the deposit; (h) late fault intersecting talc*

Like the other underlying units, the talcitites are crossed by normal faults parallel to the elongation of the deposit, marked by perfectly smooth fault mirrors bearing 90° pitch striations (Figure 7.8(h)). These structures intersect all others and are therefore the latest. In thin sections, we observe that these faults induce talc (re)crystallization: we observe that all the talc lamellae are oriented parallel to the fault, but also that at the level of the fault mirror the talc has the same extinction angles over a thickness of about 30 μm. This suggests either that the fault was formed during the emplacement of the mineralization, or – as shown by the field relationships – that it operated after the mineralization under conditions that allowed talc crystallization or recrystallization.

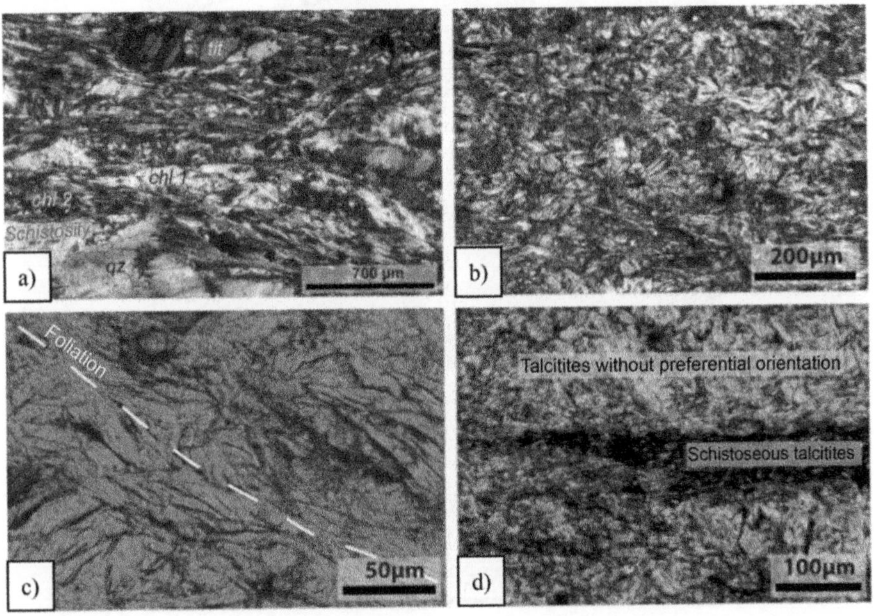


**Figure 7.9.** *Microstructures: (a) the two generations of chlorite, LPA; (b) steatite, LPA; (c) foliated talcitite, LP; (d) schistosed talcitite, LPA; chl: chlorite, tit: titanite, qz: quartz*

4) The deposit contains lenses of wall and roof rocks, micaschists, dolomitic marbles, marbles (Figure 7.8(g)) or GSC. These inclusions are co-structured with the enclosing chloritites and talcitites, suggesting that they are related to the syn- to late-mineral deformation that formed the main schistosity. However, they may also be lenses that are not completely metasomatized.

5) Pegmatites are found in all mineralized levels, chlorites and talcitites. They have the same standard mineralogy as the wall pegmatites, but they differ in several respects. First, their shape can be spherical or in the form of ovoid lenses of no more than a few meters (Figure 7.10(a)). Second, their edge, which is altered on the surface, shows a metasomatic alteration sequence from unaltered pegmatite to chlorite (Figure 7.10(b)). They are observed in the form of strings arranged parallel to the main schistosity, with a long axis parallel to the N-S subhorizontal lineation of the deposit. These rounded pegmatites appear clearly molded by the schistosity of the chloritites and talcitites. In the cross-section, the chlorites are oriented parallel to the schistosity at the edge of the pegmatite and are not oriented in the strain shadows on either side. The fact that they are metasomatized with a rounded shape and the presence of strain shadows chlorite zones show the contemporaneity between metasomatism and deformation. If metasomatisis was prior to deformation, we should indeed observe uniformly transformed rounded pegmatites, without the alteration zonation parallel to the edges of the rounded pegmatite. The strain shadows chlorite zones show that the deformation is clearly syn-mineralization.

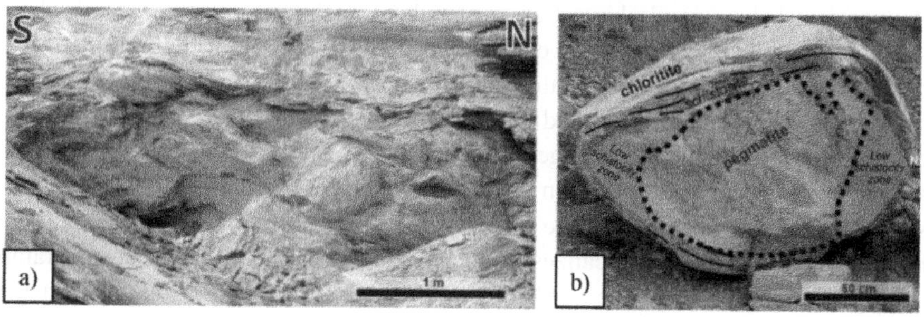


**Figure 7.10.** *(a) Rounded block of pegmatite packed in talc; (b) detail of the internal structure with its metasomatized periphery and its preserved core*

To summarize, the structures and microstructures of the deposit show the following relative chronology: (1) one or more phases of ante-mineralization deformation now fossilized by metasomatosis; a large part of the present structure of the deposit is probably associated with it; (2) a phase of static mineralization, which formed the steatite and massive chloritites; (3) a phase of dynamic mineralization associated with a phase of penetrative deformation that forms the foliated talc. The kinematics associated with these deformations is N-S (lineation), but its direction is not determined; (4) a late- to post-mineralization deformation phase during which the major schistosity of the deposit is formed, which deforms chloritite and talcitite and folds the talc veins of the roof under the conditions of chlorite and talc formation. The kinematics is always N-S, with probably a top to the south shear direction; (5) a late brittle phase, always under the conditions of formation or recrystallization of chlorite and talc, with an E-W extensional kinematics.

### 7.5.3. *The deposit-roof contact*

From a structural point of view, in some places, a continuous transition between talcitites and massive dolomites can be observed, and thus a progressive deposit-roof contact. Most often, this contact corresponds to a mechanical discontinuity linked to the gradient of rigidity between the talcitites and the untransformed dolomite, a gradient far more marked than that between talcitites and chloritites, or between chloritites and micaschists. It thus constitutes a mechanical discontinuity that will easily localize the deformation. However, we observe that in general, the syn- to post-talc deformation is distributed in a heterogeneous way in all the thickness of the deposit, chloritites and talcitites, of equivalent deformability. This contact corresponds to the localized tectonic discontinuity in the Silurian black shales. It extends northward (Porteille quarry), westward (Fumat rock) and then southward (Salvanière, Fabre wood above Caussou), where it shows top to the south shear sense. It is marked by small talco-chlorite lenses along its entire length. It intersects the MMB (Figure 7.4). Toward the south, it descends down into the series, as it contacts the granulite gneisses and the lower Devonian. It is therefore a subtractive contact. We also observe that its geometry is variable and forms successions of flats and ramps: it is subhorizontal in the northern part of Trimouns, then becomes strongly sloped toward the east to horizontalize again to the north of the Caussou buttonhole, and finally verticalizes to the south of the latter. This flat and ramp geometry, also visible to the north of Trimouns at Le Pradas, is reflected in the geometry of the various roof units.

### 7.5.4. *The roof*

The roof is formed by the Upper Paleozoic series, mainly dolomite, black shale, marble and limestone (Figure 7.11). It is deformed and affected by a N-S schistosity with a variable dip to the east, which carries a N-S stretching lineation.

1) The dolomites constitute the protolith of the talcitites. They occur in the form of tectonic lenses packed in the Silurian black shales, generally metric to decametric in size, but up to 80-m thick north of the quarry (Figure 7.11(a)). They are gray, white or banded.

The dolomite-talcite contact marks the hydrothermal alteration front of the dolomite into talc. This front is relatively straightforward, and in less than 1 m we can observe the transition from a 100% dolomitic rock to a 100% talc ore. This transition zone can be marked by a graphite layer. Generally, this mineralogical and thus rheological transition was used by the deformation (see previously). To the north of the quarry, numerous geodes are observed in the dolomites (Figure 7.11(b)), in which we find talc in sheets called talc flower, but also calcite or quartz, as well as a large panel of rare earth minerals ranging from silicates (allanite, dissakisite, etc.), carbonates (bastnaesite, synchisite, etc.) to phosphates (monazite, xenotime, etc.) (Figures 7.11(c) and (d)). These automorphic rare-earth minerals have been associated with talcification-related metasomatic episodes (de Parseval 1992; de Parseval et al. 1997; Schärer et al. 1999) and have been the subject of analyses that have dated Mesozoic metasomatic episodes (Schärer et al. 1999; Boutin et al. 2016). At la Porteille, above the talc-chlorite accident, Mangin (1967) describes a dolomite-marble transition, that is, a dolomitization front, an observation confirmed by Boulvais et al. (2006). The dolomites of the Trimouns roof are therefore probably secondary dolomites resulting from the transformation of Silurian carbonates.

2) The black shales represent most of the Silurian material present at Trimouns (Figure 7.11(e)). They are in contact with the carbonates and dolomites through thin talco-chlorite levels, testifying to the influence of metasomatism within the roof rocks. They are made up of quartz, chlorites, muscovites and some oxides, and are very rich in graphite; they are thus sometimes qualified of ampelites. By going up in the shale series, we note an enrichment in organic matter marked by an increasingly dark and greasy appearance. The whole of these shales is extremely tectonized.

3) Rare pegmatite veins are observable in the roof (Figure 7.11(f)). They are altered to a green chlorite with occasional relics of preserved rock. These pegmatites have the same major element composition as those in the deposit, but higher rare earth contents, and notable niobium, tantalum and beryllium contents. If we place both types of pegmatites in the same lineage (same source), the roof pegmatites are more evolved/differentiated than the deposit pegmatites, which is consistent with their structural position further from their potential source zone, that is, migmatites.

4) Several types of non-dolomitic marmorized carbonates are present in the Silurian. They do not exceed 10 m in thickness. Three types can be distinguished, from bottom to top: (i) at the contact of the talcitite vein, bluish-gray banded marbles, which contain calcite, micas, quartz, prismatic actinotes and titanites. They are chloritized on their edge (over a few centimeters). They are present only in the south of the quarry; (ii) pink to white banded marbles, essentially composed of calcites and phlogopite; and (iii) dark quartz marbles. All these marbles are like the dolomites of tectonic lenses and show boudinage, the space between the boudins being filled with chlorite mineralization containing titanites and automorph calcites.

**Figure 7.11.** *Main roof facies. (a) General view of the wall structure north of the quarry with dolomite-black shale alternations; (b) dolomite geode with talc flower; (c) geode with calcite and quartz; (d) monazite microcrystal in a geode; (e) Silurian black shale rich in organic matter; (f) pegmatite vein intersecting the dolomites and completely chloritized*

5) The Devonian series appears east of the quarry. From bottom to top, calcite, chlorite, muscovite and biotite calcschists are observed, and are considered to be

Lower to Middle Devonian. It is noted that the calcschists at the Silurian contact are partly chloritized (Zwart 1954; Mangin 1967; Arnoux and Majza 1986). Further east, the Upper Devonian facies are more massive, more carbonated (sometimes marmorized) and form thick banks of hot deformed marbles. They are mainly composed of calcite, quartz, chlorite, mica and some biotite. Scapolites were also observed (de Saint Blanquat 1985).

### 7.5.5. *Quaternary formations*

In addition to the fills constituted by the accumulation of waste rock extracted from the deposit, the quaternary formations are essentially glaciar moraines, containing rounded blocks of very variable size. A large part of the talwegs in the northern (Pradas, Basqui, Lasset) and southern (Salvanière and Ourlès) sectors are made up of moraine blocks resulting from the dismantling of the migmatitic ensemble under the influence of the glaciers. The numerous small lakes installed in ancient glacial cirques also testify to the imprint of the quaternary glaciations.

## 7.6. Ore, mineralogy and fluids

### 7.6.1. *Ore*

Over the whole deposit, the ore is formed by alternating layers of dominantly talcous or chloritic mineralogy. This succession is repeated at all scales, from the microsequence in the mixed ores to metric to plurimetric alternations. The talc-dominant ores are located near the dolomites from which they are derived, and the chlorite and talco-chlorite ores are in contact with the wall formations. Finally, the ores from the pegmatites occur in fully transformed rounded lenses, or sometimes in plurimetric aureoles around the residues of healthy pegmatites.

#### 7.6.1.1. *Talcous ore*

It is possible to define qualities of talc ranging from black to white with pearly reflections. The latter are the purest of the deposit. The intermediate qualities are banded talcs and gray talcs that are more or less dark, with the coloring essentially due to graphite in punctuations in very small quantities in the ore. The banded talcs occur as alternating dark and light centimetric bands found in the unprocessed dolomitic levels. Small crystals of pyrite and pyrrhotite, sometimes deformed, are frequent.

The conservation of the structures of the banded dolomites in certain talc samples implies a constant volume dolomite-talc transformation. An equilibrium reaction of the type is given as:

$$3CaMg(CO_3)_2 + 4SiO_{2(aq)} + 6H^+ \rightarrow Mg_3Si_4O_{10}(OH)_2 + 3Ca^{2+} + 6CO_2 + 2H_2O$$

This reaction imposes a volume decrease of about 30%, which can be compensated by the direct precipitation of talc from the percolating solution following the equation:

$$4SiO_{2(aq)} + 3Mg^{2+} + 4H_2O \rightarrow Mg_3Si_4O_{10}(OH)_2 + 6H^+$$

| | Dolomite | Talc |
|---|---|---|
| | Infiltration | |
| $SiO_2$ | *0.60* | *61.67* |
| $Al_2O_3$ | Tr | 0.19 |
| $Fe_2O_3t$ | 0.29 | 0.68 |
| MnO | 0.03 | Tr |
| MgO | 20.41 | 31.52 |
| CaO | *32.53* | *0.02* |
| $Na_2O$ | Tr | Tr |
| $K_2O$ | Tr | Tr |
| $TiO_2$ | Tr | Tr |
| $P_2O_5$ | 0.07 | 0.11 |
| L.O.I. | *46.40* | *4.84* |
| Total | 100.33 | 99.03 |

**Table 7.1.** *Whole rock composition in %wt of a dolomite and a talc ore. The numbers in italics show the main changes related to the metasomatic transformation (L.O.I. (Loss On Ignition): a loss of weight which corresponds mainly to $CO_2$ for the dolomite and $H_2O$ for the talcous ore)*

Let the balance equation (globally realized at constant volume) be written as:

$$2CaMg(CO_3)_2 + Mg^{2+} + SiO_{2(aq)} + 2H^+ \rightarrow Mg_3Si_4O_{10}(OH)_2 + 2Ca^{2+} + 4CO_2$$

In both cases (Table 7.1), silica ($SiO_2$) is provided in solution. Ca and $CO_2$ are leached out and are partly found as calcite in the dolomite geodes. The carbon represented by the inherited graphite and disseminated in the talcous matrix or in the banding seems stable during this transformation.

#### 7.6.1.2. *Chlorite ore*

##### 7.6.1.2.1. Micaschist series

These are located at the wall as well as in tectonic slices inside the deposit. We can observe two main types of protolith: (1) a relatively massive and homogeneous clear one where millimetric sulphides can be observed; (2) a more schistosed rock with alternating light (quartzo-felds-pathic) and dark (mica-graphite association, etc.) beds, the latter often being partly chloritized (scratchable with the fingernail). In the case of the clearest micaschists, we can observe the development of reddish zones with phlogopite comparable to those of the pegmatitic series. Those where light and dark beds alternate show only slight reddish nuances masked by the dark color of the rock. The ore follows the zones described above and is more or less schistosed.

##### 7.6.1.2.2. Pegmatitic series

| | Pegmatite | Phlogopite zone | Quartz-chlorite zone | Chlorite ore |
|---|---|---|---|---|
| | | *Infiltration* | | |
| $SiO_2$ | *71.7* | 62.0 | 46.0 | *35.8* |
| $Al_2O_3$ | 14.0 | 13.9 | 12.6 | 17.6 |
| $Fe_2O_3t$ | 0.6 | 0.9 | 1.2 | 1.5 |
| MnO | Tr | Tr | Tr | Tr |
| MgO | *4.0* | 14.0 | 30.4 | *33.0* |
| CaO | 0.6 | 0.4 | 0.3 | 0.4 |
| $Na_2O$ | *3.2* | 1.8 | tr | *Tr* |
| $K_2O$ | *1.8* | 1.6 | 0.4 | *0.1* |
| $TiO_2$ | 0.1 | 0.0 | 0.0 | 0.0 |
| $P_2O_5$ | 0.2 | 0.2 | 0.2 | 0.3 |
| L.O.I. | *2.7* | 5.3 | 9.0 | *11.3* |
| Total | 98.9 | 100.1 | 100.1 | 100.1 |

**Table 7.2.** *Whole rock composition in %wt of an example of pegmatitic alteration line into a talc ore. The text and values in italics show the main changes related to the metasomatic transformation (L.O.I. (Loss On Ignition): a loss of weight corresponding mainly to $H_2O$)*

Pegmatites are present throughout the deposit. It is common to observe all the stages between the original and metasomatized rocks that constitute the ore in the same sample (Figure 7.10(b) and Table 7.2). The initial pegmatite is represented by a white gray rock with individual black tourmaline crystals. This zone, which constitutes the core of the pegmatitic rounded block, is generally surrounded by a centimetric reddish-brown zone of phlogopite where the tourmaline crystals progressively disappear. The following zone directly in contact with the previous one stands out by its characteristic green color, but is not yet completely transformed, as evidenced by its rough touch related to the presence of residual quartz. This centimetric zone is gradually changing to chlorite ore. The mineralized part in this type of transformation is identifiable by its green color, but also because it can be easily separated from the non-mineralized part. As for talc, this type of ore can present itself in a very schistose form with a sheeted flow, or in a more compact mass where structures of the initial pegmatite seem to persist. The color of the ore varies from pale to dark green.

#### 7.6.1.3. *Interpretation and synthesis*

The three series described above, dolomite-talc, micaschist-chlorite and pegmatite-chlorite, are the most common and most easily observed on the deposit. The talcitites and chloritites are derived from the metasomatic alteration of dolomitic rocks and silico-aluminous rocks (micaschist and pegmatitic series), respectively.

De Parseval (1992) speaks of coupled reactions between the processes of talcification and chloritization (Figure 7.12). Therefore, the fluid at the origin of talc, enriched in $SiO_2(aq)$ during chloritization, acquires a silica undersaturation during talcification and will be enriched in $CO_2$. This fluid by its composition will then chloritize the micaschists and be loaded again in silica to transform the dolomites into talc. Such a system would require closed-circuit fluid circulation mechanisms on the scale of the deposit, a hypothesis that is difficult to defend in the current state of knowledge of the deposit.

The talc-chlorite ores from micaschists and pegmatites are mainly constituted by a chlorite of the clinochlore type and associated with lower or higher proportions of talc. Note that they differ essentially by (1) their color, light to dark green for ores from pegmatites, greenish gray to dark gray for those associated with micaschists; (2) the nature of accessory trace minerals: titanite (up to 2 or 3% of the rock) and/or rutile (1%) only for ores from micaschists. These titanium minerals are derived from the release of Ti during the chloritization of micas (biotites) and will be suitable for geochronological dating.

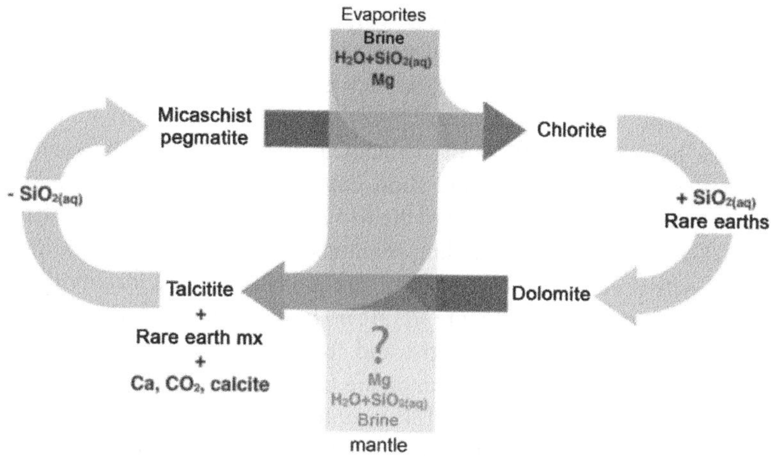


**Figure 7.12.** *Simplified diagram of metasomatosis at the Trimouns scale*

## 7.6.2. *Mineralogy*

**Figure 7.13.** *Some examples of rare earth microminerals present at Trimouns. (a) Dissakisite-(Ce); (b) astnaesite-(Ce); (c) gatelite-(Ce); (d) trimounsite-(Y). Scale bar: 1 mm (collection and images of Guy Bernadi)*

In addition to talc ($Mg_3Si_4O_{10}(OH)_2$), chlorite ($(Mg,Fe,Al)_6(Si,Al)_4O_{10}(OH)_8$) and associated accessory minerals (apatite, titanite), Trimouns contains other mineralogical specificity (Figure 7.13). They are also the result of metasomatosis and mainly of the alteration of micaschists into chlorite ore. This transformation leads to the leaching of rare earths contained in micaschists. These rare earths diluted in the residual metasomatic fluid can be incorporated in the rare earth minerals of the dolomite geodes. More than 15 rare-earth mineral species have been described (Gatel et al. 2002), including trimounsite-(Y) (Figure 7.13(d)), a titanium silicate with yttrium and heavy rare earths. Trimouns is the type locality (Piret et al. 1990), but also gatelite-(Ce) (Figure 7.13(c)) (Bonazzi et al. 2003, 2019), named after P. Gatel, an amateur mineralogist and founder of the French Association of Micromonitors. Among the most common are bastnaesite-(Ce) (Figure 7.13(b)), and especially dissakisite-(Ce) (Figure 7.13(a)), within which fluid inclusions that have led to an understanding of the nature and origin of fluids are observed (de Parseval 1992).

### 7.6.3. *Origin of the fluids and formation conditions of the deposit*

Moine et al. (1989) demonstrated that the mineralizing fluids were mainly brines (very salty fluids), having circulated under mesothermal conditions (~300°C, ~1.5 kbar). These authors reached these conclusions by analyzing fluid inclusions (Figure 7.14) in quartz and apatites associated with chlorite ores (de Parseval 1992). These apatites are contemporaneous with the formation of the ores and have revealed the presence of $CO_2$ in the inclusions. Studies of fluid inclusions were continued with those of inclusions trapped in automorph quartz from geodes located in the dolomitic levels at the top of the deposit (see Figure 7.14). In addition to quartz, these geodes contain rare-earth minerals (dissakisite-(Ce), bastnaesite-(Ce), etc.), calcite, dolomite, as well as talcum flower, attesting to a common origin with the mineralizations of the deposit. These studies have confirmed the highly saline character of the main fluid, while refining the estimate of the P/T conditions of the deposit formation to establish them at $250°C < T < 320°C$, $P < 2.5$ kbar, and identifying traces of $CO_2$, $CH_4$, $N_2$ in the gas phase of the fluid inclusions (de Parseval et al. 2004; Boiron et al. 2005, 2007). In order to determine the nature and origin of the fluids that formed the Trimouns talc deposit, some of these studies have used the isotopic composition of metasomatic minerals (talc, chlorite, dolomite, calcite, quartz, rare earth minerals) to suggest a link between these highly saline fluids and the Triassic evaporitic series present in the Pyrenees (Boulvais et al. 2006).

Subsequently, other works have refined the understanding of the modalities of formation of the deposit by suggesting that (1) the magnesium necessary for the formation of the deposit initially came from the brines and did not result from their interactions with the percolated rocks (Boiron et al. 2007), (2) the dolomitization of the calcitic marbles and subsequent talcification of the dolomites resulted from the same fluid circulation system and not from separate hydrothermal events (Boulvais et al. 2006; Quesnel et al. 2019), and finally (3) the brines were evaporated seawater (Boiron et al. 2007; Leisen et al. 2012; Quesnel et al. 2019), and were directly expelled from the Triassic evaporitic series (Quesnel et al. 2019).

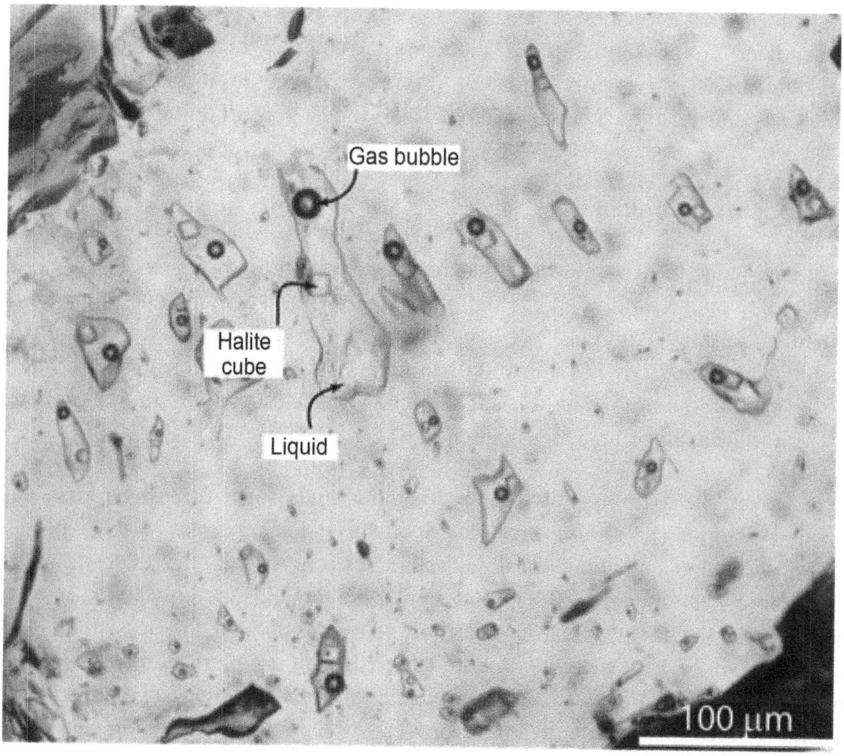


**Figure 7.14.** *Typical three-phase fluid inclusion fields of those observed at Trimouns in automorphic quartz*

Recently, it has been suggested that an alternative source of brine may have been involved in the formation of the deposit. Indeed, the serpentinization process of ultramafic rocks could be at the origin of brine formation in different geological contexts (Scribano et al. 2017; Hovland et al. 2018). The process of serpentinization involves the hydration of ultramafic rocks in contact with a fluid. Therefore, this process consumes water and can therefore induce the increase in the salinity of the residual fluid. Around the MdT, the numerous serpentinized peridotite relics (Figure 7.4) are evidence of mantle exhumation and serpentinization during the Albo-Cenomanian hyperextension (de Saint Blanquat et al. 2016).

Serpentinization and the formation of the Trimouns deposit may be contemporary events, and it is possible to imagine a link between the two. If the hypothesis of a mantle origin for the brines involved in the formation of the Trimouns deposit has been suggested, the current lack of knowledge on the geochemical characteristics of these atypical brines prevents discussion of their real involvement in the formation of the deposit (Quesnel et al. 2019). Moreover, serpentinization is currently undated, whereas the age of the formation of the Trimouns deposit is now known on the basis of robust geochronological data.

## 7.7. Geochronological data

In the late 1990s, the first geochronological study (Schärer et al. 1999), conducted on xenotime-(Y) and monazite-(Ce), demonstrated that talc mineralization at Trimouns was anchored in the Alpine cycle (precollision, ca. 100 Ma). This study has allowed us to relate the Paleozoic terrains of Trimouns to the crustal thinning and to place the SiMg metasomatism in the Pyrenean Albo-Ceno-Manian rifting (see Chapter 5).

However, the dating of talc required another study for two reasons. First, since talc and chlorite cannot (yet) be dated directly, Schärer et al. (1999) use two rare-earth minerals contemporary with talcification to date the deposit, namely xenotime-(Y) and monazite-(Ce). However, these minerals are only found in a very specific zone of the deposit in the roof dolomites, which are too small to be representative of a structurally complicated deposit of plurikilometre size. The second reason is that since 1999, dating techniques have evolved and several other mineral species such as titanites have proven to be good geochronological markers (Frost et al. 2000).

The study by Boutin et al. (2016) extends the work of Schärer et al. (1999) with the aim of applying proven dating techniques on a wider panel of minerals distributed over the different formations of the whole site. With this in mind, a more complete sampling of rare-earth minerals (xenotime, monazite, allanite) and titaniferous minerals (titanite, rutile) was carried out, essentially at Trimouns, but also on talc-chlorite deposits in the AZ. This geochronological study U-Th-Pb geochronology was carried out by laser ablation ICP-MS (LA-ICP-MS).

Data from the Trimouns samples show that the deposit was formed during at least two successive hydrothermal events. The first was revealed by titanites contained in the chlorite vein near the wall. This event is dated at around 166 Ma (Middle Jurassic). The second event is the one highlighted by Schärer et al. (1999). Data on monazites, xenotimes, allanites and titanites contained in talc-filled geodes and titanites contained in chlorite veins, geodes and veins in the roof dolomites, confirm its Middle Cretaceous age, and allow us to attribute a duration of at least 25 million years (between ca. 122 Ma and 96 Ma, between the end of the Aptian and the Cenomanian).

The history of talco-chlorite metasomatism at Trimouns can thus be described by the following sequence of events (Figure 7.15):

1) Jurassic chlorite formation episodes (170–140 Ma), possibly associated with the diffuse rifting described in Western Europe (Cathelineau et al. 2012), which could be associated with marble dolomitization.

2) A major talc and chlorite formation event in the Middle Cretaceous. It may itself have been formed by several successive hydrothermal pulses. This latter event is spatially and temporally associated with the sodium and calcosodium metasomatisms observed throughout the Pyrenees (Boulvais et al. 2007; Poujol et al. 2010; Fallourd et al. 2014; Clerc et al. 2015; Chelalou et al. 2016; Corre et al. 2016).

Available data show that Cretaceous metasomatism (lato sensu: SiMg, Na and CaNa) begins before HT metamorphism and North Pyrenean alkaline magmatism, and it stops before the end of the rifting phase. The age of onset of metasomatism probably dates the onset of crustal thinning, which logically begins before contact of the pre-rift series with the exhumed mantle. Metasomatism is low to medium temperature, between 200 and 350°C for talc-chlorite (de Parseval 1992), less than 450°C for sodium metasomatism (Boulvais et al. 2007; Poujol et al. 2010) and between 450 and 550°C for calcosodium metasomatism (Fallourd et al. 2014). These temperatures are reached at depths in the order of about 10–12 km. North Pyrenean metamorphism shows higher temperatures, with maxima around 600°C in the pre-rift cover and at the base of synrift basins (Golberg and Leyreloup 1990; Clerc et al. 2015; Chelalou et al. 2016), that is, at shallow depths of less than 5 km, showing the presence of very high gradients, estimated at 60–80°C/km (Vacherat et al. 2014; Hart et al. 2017; Saspiturry et al. 2020). This difference in thermal intensity suggests that the North Pyrenean basement was not – or no longer – located beneath the pre-rift cover during its metamorphism.

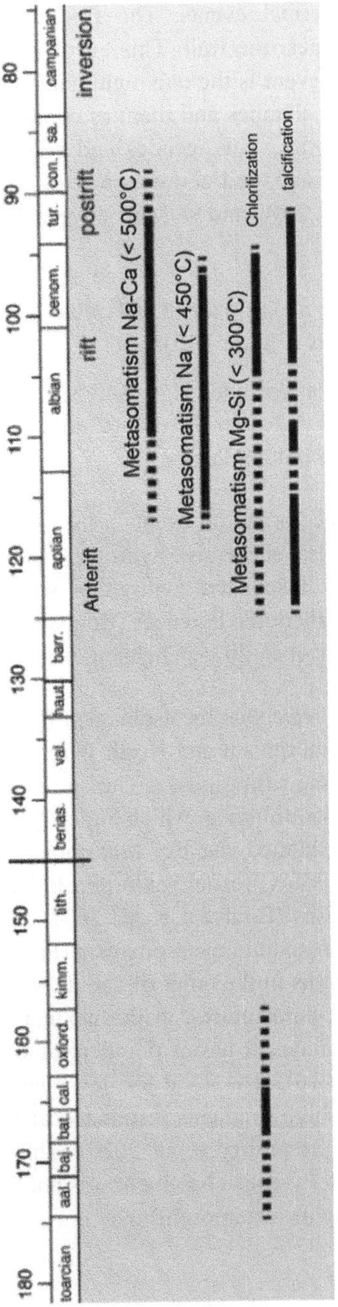


**Figure 7.15.** *Synthesis of geochronological data on metasomatism in the North Pyrenean and Pyrenean axial zones (data from Poujol et al. (2010); Fallourd et al. (2014); Boutin et al. (2016))*

## 7.8. Synthesis and formation model

The presented model takes into account all the data summarized above, and is modified from Boutin (2016), Boutin et al. (2016) and Quesnel et al. (2019) (Figure 7.16). It places the metasomatism and formation of the Trimouns deposit in the context of Cretaceous transtension associated with the rotation of the Iberian plate (120–85 Ma), a context that resulted in the breakup of the continental crust and the exhumation of the mantle to the surface. The simplified geological history is as follows:

1) A tectonic event responsible for the compositional bedding of the migmatites, folds and S0-S1 schistosity in the micaschists and roof rocks is evidence of the "early" Variscan D1 phase (see Chapter 1), which is considered to be of Late Carboniferous age in the Pyrenees (de Saint Blanquat et al. 2019).

2) During the major Variscan phase, the development of HT-LP metamorphism allows the partial melting and formation of the migmatites, which is accompanied by the intrusion of magmas of various compositions, diorite, granodiorite, granite, leucogranite in the migmatites and micaschists, and pegmatite probably extracted from the migmatites in the whole of the Trimouns series. Penetrative deformation (D2) synchronous with or subsequent to the metamorphic peak is responsible for the development of foliation in the migmatites and schistosity in the micaschists, folding and boudinage of the CSGs and roof carbonates, which are marmorized. It results in a vertical thinning and thus a reduction of the series. The deformation gradually becomes localized with the formation of the MMB and the F1 structure. From the beginning to the end of this episode, the kinematics is always top to the SSE and thus dextral at Trimouns.

3) Around 165 Ma, a first chlorite mineralization is formed, which is the consequence of a metasomatism related to the initiation of a diffuse extensive context in the whole western European continent (Cathelineau et al. 2012).

4a) There follows a probable phase of dolomitization of Silurian marbles, undated and whose traces are not visible at Trimouns, since no dolomitization front is observed there.

4b) From about 122 Ma (at the beginning of the Aptian), the main metasomatical phase begins. This begins with a static mineralization phase that forms steatite and massive chlorites under the action of brines.

4c) Mineralization continues in association with penetrative deformation and forms the chloritites and foliated talcitites (Figure 7.14). This deformation is NNE-SW, but its kinematics (dextral or sinistral) is not known.

4d) A late to post-mineralization deformation phase forms the major schistosity of the deposit, still under the conditions of chlorite and talc formation. It is responsible for the folding of the talc veins of the roof. This deformation is top to the NNE-SSW/destral (C/S, some sheared folds). The metasomatic event seems to end at the end of the Cenomanian.

5a) Late normal fault deformations showing E-W extension affect the Trimouns deposit, always in conditions allowing talc and chlorite crystallization. The age of this deformation, ante- or post-inversion, is not constrained.

5b) Inversion begins everywhere in the Pyrenees around 85 Ma (Teixell et al. 2018; Andrieu et al. 2021; Issautier et al. 2022). It is likely responsible for the cessation of metasomatism following mechanical closure of the system.

6) Erosion induced by exhumation, successive glaciations and then mining activity created the current landscape.

## 7.9. Conclusions and perspectives

The geology of the Trimouns site shows:

– A contrast between a geological history associated with high and medium temperatures (>450°C) and crustal metamorphic and magmatic fluids of deep origin, in a context of middle crust (around 10 km depth), testifying to the Variscan cycle, and a colder history (<300 °C) associated with superficial fluids (brines from Mesozoic evaporites, see Fontes and Matray (1993), and from serpentinization of the mantle in the process of exhumation), in a context of upper crust (<6 km), testifying to the Alpine cycle.

– The formation of polyphase mineralization, with a Jurassic phase (170–160 Ma) and a Cretaceous main phase (122–96 Ma in the present state of our data), each of these phases being potentially polyphased.

– The initiation of crustal thinning and thus of pre-orogenic deformation: talco-chlorite metasomatism is part of the extensive pre-orogenic Alpine context. The different episodes of mineralization, each associated with a more or less penetrative deformation event, probably mark the different stages of the extension that will lead to the breakoff of the continental crust and the exhumation of the mantle to the surface, or close to it.

– A metasomatism which seems to stop before the end of the rifting phase, and fluid circulations which stop at the beginning of the tectonic inversion by mechanical closure of the system.

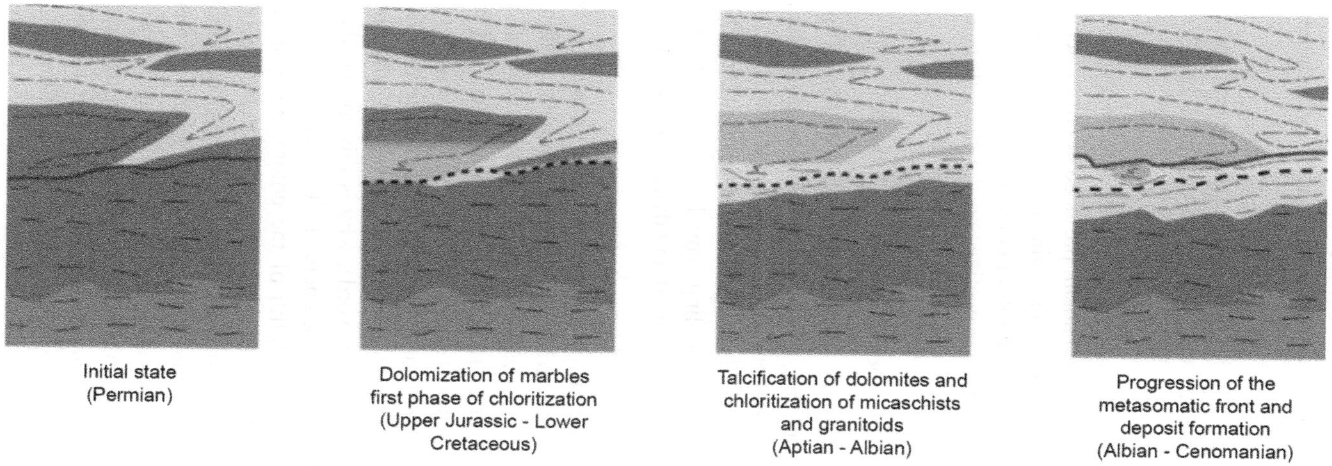


**Figure 7.16.** *Geological model of the genesis of the talc-chlorite deposit of Trimouns. The lithologies of little significance in terms of their volume (pegmatites, calcium silicate gneiss) have not been represented for the sake of clarity. Same legend as in Figure 7.5*

– The fundamental role played by the Variscan structural heritage on the localization of the Cretaceous deformation, and thus by extension on the localization of the fluid circulations at that time.

– The present structure of the Tabe Mountain massif is a fossilized syn-hyperextensional structure that was superimposed on a late Variscan structure, which is itself extensive.

– The context of the passive margin is a favorable geodynamic context for the development of this type of metasomatism: presence of crustal scale faults, inherited from the Variscan chain, which serve as drains for fluid circulation, an adequate PT window, and a long event whose duration could be estimated around 20 Ma. The geology of Trimouns thus opens perspectives on the study of paleo-rifts elsewhere on Earth, particularly with respect to basement behavior.

The questions that remain to be explored are as follows:

– Our estimates on the age and duration of talc-chlorite metasomatism are constrained both by our sampling and by existing methodology (in situ U-Th-Pb on minerals neoformed by metasomatism and containing uranium). While waiting for the perfect method that would allow us to directly date talc or chlorite, we cannot exclude that a more complete sampling would give us perhaps slightly different results on the timing (beginning and end) and dynamics (one or more pulses) of the fluid circulations.

– It is probable that the driving force of fluid circulation is an association between sedimentation, deformation and thermal gradient. Sedimentation through overloading that crushes the evaporites and expels interstitial fluids, extensive deformation (main vertical stress) that thins and shears them and favors the descent of supracrustal fluids (Siebenaller et al. 2006), but also the ascent of deep fluids, and the thermal gradient through its role in the formation of convective loops that hold the system together. Moreover, how do they arrive and why are they located at Trimouns or La Porteille? In other words, why is this spectacular accumulation of talc-chlorite at Trimouns and not elsewhere? Is it the effect of geology (presence of lenses of marble or potential protolith of the dolomites) of a heterogeneous and localized fluid circulation (more fluid than elsewhere) in connection with the flat and ramp geometry of the main structural discontinuity or deformation (mechanical accumulation by tectonics), or the effect of a combination of these three possible causes? The answer to this question probably requires a more complete characterization of the relationship between deformation and metasomatism. The transition between extension (synmineralization) and inversion (postmineralization) and the possible effects of the latter also remain to be identified, if indeed it has been recorded at Trimouns. In any case, it is clear that deformation, fluid circulation and metasomatism are very closely linked;

– How do fluids flow downwards (brines from the Mesozoic series) or upwards (possible mantle "brines") in a strike-slip shear zone? Although much work has shown that strike-slip faults are commonly associated with fluid flow, hydrothermalism, and mineralization as normal or reverse kinematic faults (e.g. (Nemcok et al. 2002; Wibberley and Shimmamoto 2003; Ford et al. 2009; Hensen et al. 2015), the relationships between fluid flow and deformation remain to be clarified.

– Theoretical calculations based on solution compositions estimate the amount of fluid needed to make 1 kg of talc at 1,000 L (B. Moine, pers. com., 2021). It would be interesting to make a quantitative balance of the quantity and flows of fluids that have circulated through the system.

– Dolomitization has not been studied, neither at Trimouns nor elsewhere. Which fluid(s), which context(s), which age(s)? Contemporary with talcification, as suggested by Boulvais et al. (2006) and Quesnel et al. (2019), or not? This is however an important question in Pyrenean geology, which goes beyond the question of talc formation. The Mesozoic pre-rift series of the IMZ, whether brecciated or not, are indeed dolomitized in many places.

– It cannot be excluded that there was an episode of talco-chlorite mineralization at Trimouns during the Permian, as in the other Pyrenean deposits of the AZ (Boutin et al. 2016). A finer understanding of the interactions between the hyperextensive context and fluid circulations would require a better determination of the precise timing of the emplacement of talc-chlorite mineralization in the different Pyrenean deposits, and thus a more systematic geochronological study, which will have to be carried out after the characterization of the structural history of these deposits and its relationship with mineralization.

– Finally, the geological history documented at Trimouns still needs to be integrated into a schema on the scale of the Pyrenean hyperextended rift.

## 7.10. Acknowledgments

The recent works presented in this overview were obtained under the framework of the following projects: Alexandre Boutin's PhD (Cifre Imerys-ANRT), ANR PYRAMID (PI M. Ford), TotalEnergies Fluids project (M.C. Boiron, M. Cathelineau, L. Truche, E. Gaucher), cartographic surveys for the geological map of Ax-les-Thermes (BRGM), and the various studies in association with Talcs de Luzenac, Rio Tinto, and then Imerys. We would particularly like to thank the different quarry managers at Trimouns for their warm welcome. We would also like to thank Bernard Moine, Philippe Boulvais, Bernard Gavoille and Jean-François Robert for the always stimulating discussions on the geology and origins of this deposit.

## 7.11. References

Andrieu, S., Saspiturry, N., Lartigau, M., Issautier, B., Lasseur, E. (2021). Sedimentary evolution and vertical movements during the Cenomanian to Santonian interval on the Iberian margin of the Pyrenees: Comparison with the southern Aquitaine Basin. *BSGF – Earth Sciences Bulletin*, 192, 19 [Online]. Available at: https://doi.org/10.1051/bsgf/2021011.

Aranitis, S. (1967). Les gisements de talc pyrénéens. Description. Essais d'interprétation de leur genèse. PhD Thesis, BRGM.

Arnoux, F. and Majza, B. (1986). Contribution à l'étude géologique des gisements de talc du massif du Saint Barthélémy (Ariège). PhD Thesis, Université de Franche-Comté, Besançon.

Baby, P., Crouzet, G., Specht, M., Deramond, J., Bilotte, M., Debroas, E.J. (1988). Rôle des paléostructures albo-cénomanienne dans la géométrie des chevauchements frontaux Nord-Pyrénéens. *C. R. Acad. Sc. Paris*, 306, 307–313.

Boiron, M.C., Boulvais, P., Cathelineau, M., Banks, D., Calvayrac, N., Hubert, G. (2005). Fluid circulation at the origin of the Trimouns talc deposit (Pyrenees, France). *European Current Research on Fluid Inclusions (ECROFI XVIII)*, Siena.

Boiron, M.C., Cathelineau, M., Dubessy, J., Fabre, C., Boulvais, P., Banks, D. (2007). Na Ca-Mg rich brines and talc formation in the giant talc deposit of Trimouns (Pyrénées): Fluid inclusion chemistry and stable isotope study. *European Current Research on Fluid Inclusions (ECROFI-XIX)*, Université de Bern.

Bonazzi, P., Bindi, L., Parodi, G. (2003). Gatelite-(Ce), a new REE-bearing mineral from Trimouns, French Pyrenees: Crystal structure and polysomatic relationships with epidote and törnebohmite-(Ce). *Am. Mineral.*, 88, 223–228.

Bonazzi, P., Holstam, D., Bindi, L. (2019). Gatelite-supergroup minerals: Recommended nomenclature and review. *European Journal of Mineralogy*, 31, 173–181.

Boulvais, P., de Parseval, P., D'Hulst, A., Paris, P. (2006). Carbonate alteration associated with talc-chlorite mineralization in the eastern Pyrenees, with emphasis on the St. Barthelemy Massif. *Mineralogy and Petrology*, 88(3/4), 499–526.

Boulvais, P., Ruffet, G., Cornichet, J., Mermet, M. (2007). Cretaceous albitization and dequartzification of Hercynian peraluminous granite in the Salvezines Massif (French Pyrénées). *Lithos*, 93(1/2), 89–106.

Boutin, A. (2016). Contexte géologique du gisement de talc de Trimouns. PhD Thesis, Université de Toulouse.

Boutin, A., de Saint Blanquat, M., Poujol, M., Boulvais, P., de Parseval, P., Robert, J.F. (2016). Succession of metasomatic events in the eastern Pyrenees with emphasis on the Trimouns talc-chlorite deposit. *Int. J. Earth Sci.*, 105(747), 747–770.

Capdecomme, L. (1950). Sur la genèse des talcs pyrénéens. *Bulletin de la Société d'Histoire Naturelle de Toulouse*, 85(313), 313–316.

Cathelineau, M., Boiron, M.-C., Fourcade, S., Ruffet, G., Clauer, N., Belcourt, O., Coulibaly, Y., Banks, D.A., Guillocheau, F. (2012). A major Late Jurassic fluid event at the basin/basement unconformity in western France: 40Ar/39Ar and K–Ar dating, fluid chemistry, and related geodynamic context. *Chemical Geology*, 322/323, 99–120.

Chelalou, R., Nalpas, T., Bousquet, R., Prevost, M., Lahfid, A., Poujol, M., Ringenbach, J.C., Ballard, J.F. (2016). New sedimentological, structural and paleo-thermicity data in the Boucheville Basin (eastern North Pyrenean Zone, France). *C. R. Geosciences*, 348(3/4), 312–321.

Clerc, C., Lagabrielle, Y., Neumaier, M., Reynaud, J.-Y., de Saint Blanquat, M. (2012). Exhumation of subcontinental mantle rocks: Evidence from ultramafic-bearing clastic deposits nearby the Lherz peridotite body, French Pyrenees. *Bulletin Société Géologique de France*, 183(5), 443–459.

Clerc, C., Lahfid, A., Monié, P., Lagabrielle, Y., Chopin, C., Poujol, M., Boulvais, P., Ringenbach, J.-C., Masini, E., de Saint Blanquat, M. (2015). High-temperature metamorphism during extreme thinning of the continental crust: A reappraisal of the north Pyrenean paleo-passive margin. *Solid Earth Discussions*, 7(1), 797–857.

Corre, B., Lagabrielle, Y., Labaume, P., Fourcade, S., Clerc, C., Ballèvre, M. (2016). Deformation associated with mantle exhumation in a distal, hot passive margin environment: New constraints from the Saraillé Massif (Chaînons Béarnais, North-Pyrenean Zone). *Comptes Rendus Geoscience*, 348(3–4), 279–289.

Corre, B., Boulvais, P., Boiron, M.C., Lagabrielle, Y., Marasi, L., Clerc, C. (2018). Fluid circulations in response to mantle exhumation at the passive margin setting in the north Pyrenean zone, France. *Mineralogy and Petrology*, 112, 647–670.

Costa, S. and Maluski, H. (1988). Use the 40Ar-39Ar stepwise heating method for dating mylonite zones: An example from the St Barthélémy massif (Northern Pyrénées, France). *Chem. Geol.*, 72, 127–144.

Delaperriere, E., de Saint Blanquat, M., Brunel, M., Lancelot, J.R. (1994). Géochronologie U-Pb sur zircons et monazites dans le massif du Saint Barthélémy (Pyrénées, France) ; discussion des âges des événements varisques et pré-varisques. *Bull. Soc. Géol. Fr.*, 165(2), 101–112.

Fallourd, S., Poujol, M., Boulvais, P., Paquette, J.-L., de Saint Blanquat, M., Rémy, P. (2014). In situ LA-ICP-MS U–Pb titanite dating of Na–Ca metasomatism in orogenic belts: The North Pyrenean example. *International Journal of Earth Sciences*, 103, 667–682.

Fontes, J.C. and Matray, J.M. (1993). Geochemistry and origin of formation brines from the Paris Basin, France: 1. Brines associated with Triassic salts. *Chemical Geology*, 109(1/4), 149–175.

Ford, A., Blekinsop, T.G., McLellan, J.G. (2009). Factors affecting fluid flow in strike-slip fault systems: Coupled deformation and fluid flow modelling with application to the western Mount Isa Inlier, Australia. *Geofluids*, 9, 2–23.

Fortuné, J.-P. (1971). Contribution à l'étude minéralogique et génétique des talcs pyrénéens. PhD Thesis, Université Paul Sabatier, Toulouse.

Fortuné, J.-P., Gavoille, B., Thiébaut, J. (1980). Le gisement de talc de Trimouns près Luzenac (Ariège). *26e C.G.I., Fasc.*, E10.

François, J. (1841). Aperçu sur l'étude géologique de l'Ariège. *Annales agricoles, littéraires et industrielles de l'Ariège*, 3, 56–64.

Frost, B.R., Chamberlain, K.R., Schumacher, J.C. (2000). Sphene (titanite): Phase relations and role as a geochronometer. *Chemical Geology*, 172(1/2), 131–148.

Gatel, P., Parodi, G.C., de Parseval, P., Fontan, F. (2002). Les minéraux de terres rares à Trimouns. *Le Règne Minéral (Hors-série)*, VIII, 1–76.

Golberg, J.-M. and Leyreloup, A.-F. (1990). High temperature-low pressure Cretaceous metamorphism related to crustal thinning (Eastern North Pyrenean Zone, France). *Contributions to Mineralogy and Petrology*, 104, 194–207.

Guchereau, J.-Y. (1975). Le Saint Barthélémy métamorphique (Pyrénées ariégeoises). Pétrographie et structures. PhD Thesis, Université Paul Sabatier, Toulouse.

Hart, N.R., Stockli, D.F., Lavier, L.L., Hayman, N.W. (2017). Thermal evolution of a hyperextended rift basin, Mauléon Basin, western Pyrenees: Thermal evolution of hyperextended rift. *Tectonics*, 36(6), 1103–1128.

Hensen, C., Scholz, F., Nuzzo, M., Valadares, V., Gràcia, E., Terrinha, P., Liebetrau, V., Kaul, N., Silva, S., Martínez-Loriente, S. et al. (2015). Strike-slip faults mediate the rise of crustal-derived fluids and mud volcanism in the deep sea. *Geology*, 43(4), 339–342.

Hovland, M., Rueslatten, H., Johnsen, H.K. (2018). Large salt accumulations as a consequence of hydrothermal processes associated with "Wilson cycles": A review Part 1: Towards a new understanding. *Marine and Petroleum Geology*, 92, 987–1009.

Issautier, B., Lasseur, E., Saspiturry, N., Andrieu, S., Angrand, A., Serrano, O. (2022). Onset of the Iberian-European plates convergence: Late Cretaceous flexural response of a hot lithosphere. *Tectonophysics*, 843, 229504 [Online]. Available at: https://doi.org/10.1016/j.tecto.2022.229504.

Jammes, S., Manatschal, G., Lavier, L., Masini, E. (2009). Tectonosedimentary evolution related to extreme crustal thinning ahead of a propagating ocean: Example of the western Pyrenees. *Tectonics*, 28(4), 1–24.

Lagabrielle, Y. and Bodinier, J.-L. (2008). Submarine reworking of exhumed subcontinental mantle rocks: Field evidence from the Lherz peridotites, French Pyrenees. *Terra Nova*, 20(1), 11–21.

Lagabrielle, Y., Labaume, P., de Saint Blanquat, M. (2010). Mantle exhumation, crustal denudation, and gravity tectonics during Cretaceous rifting in the Pyrenean realm (SW Europe): Insights from the geological setting of the lherzolite bodies. *Tectonics*, 29(4), TC4012.

Lagabrielle, Y., Clerc, C., Vauchez, A., Lahfid, A., Labaume, P., Azambre, B., Fourcade, F., Dautria, J.M. (2016). Very high geothermal gradient during mantle exhumation recorded in mylonitic marbles and carbonate breccias from a Mesozoic Pyrenean paleo-margin (Lherz area, North Pyrenean Zone, France). *Comptes Rendus Geoscience, Special Issue*, 348(3/4), 257–267.

Lagabrielle, Y., Asti, R., Duretz, T., Clerc, C., Fourcade, S., Teixell, A., Labaume, P., Corre, B., Saspiturry, N. (2020). A review of cretaceous smooth-slopes extensional basins along the Iberia-Eurasia plate boundary: How pre-rift salt controls the modes of continental rifting and mantle exhumation. *Earth-Science Reviews*, 201, 103071 [Online]. Available at: https://doi.org/10.1016/j.earscirev.2019.103071.

Leisen, M., Boiron, M.-C., Richard, A., Dubessy, J. (2012). Determination of Cl and Br concentrations in individual fluid inclusions by combining microthermometry and LAICPMS analysis: Implications for the origin of salinity in crustal fluids. *Chemical Geology*, 330/331, 197–206.

Lemirre, B. (2018). Origine et développement de la thermicité dans les Pyrénées varisques, conséquences sur l'évolution du segment ouest-européen de l'orogène. PhD Thesis, Université Paul-Sabatier, Toulouse.

Mangin, A. (1967). Étude géologique de la partie septentrionale du massif du Saint Barthélémy. PhD Thesis, Université Paul Sabatier, Toulouse.

Mengaud, M. (1909). Les gisements de talc du Massif du Saint Barthélémy. *Bulletin de la Société d'Histoire Naturelle de Toulouse*, 41(3), 71–98.

Moine, B. (1982). Géochimie des transformations métasomatiques à l'origine du gisement de talc et chlorite de Trimouns – II. Approche des conditions physico-chimiques de la métasomatose. *Bulletin de Minéralogie*, 105, 76–88.

Moine, B., Gavoille, B., Thiébaut, J. (1979). Géochimie des transformations métasomatiques à l'origine du talc et chloritites de Trimouns (Luzenac, Pyrénées, France). *7[e] Réunion des Sciences de la Terre*, Lyon.

Moine, B., Gavoille, B., Thiébaut, J. (1982). Géochimie des transformations métasomatiques à l'origine du gisement de talc et chlorite de Trimouns – I. Mobilité des éléments et zonalités. *Bulletin de Minéralogie*, 105, 62–75.

Moine, B., Fortuné, J.-P., Moreau, P., Viguier, F. (1989). Comparative mineralogy, geochemistry and conditions of formation of two metasomatic talc and chlorite deposits: Trimouns (Pyrenees, France) and Rabenwald (Eastern Alps, Austria). *Economic Geology*, 84(5), 1398–1416.

Nemcok, M., Henk, A., Gayer, R.A., Vandycke, S., Hathaway, T.M. (2002). Strike-slip fault bridge pumping mechanism: Insights from field-based palaeostress analysis and numerical modelling. *Journal of Struc. Geol.*, 24, 1885–1901.

Odlum, M.D. and Stockli, D.F. (2020). Geochronologic constraints on deformation and metasomatism along an exhumed mylonitic shear zone using apatite U-Pb, geochemistry, and microtextural analysis. *Earth and Planetary Science Letters*, 538, 116177.

de Parseval, P. (1992). Étude minéralogique et géochimique du gisement de talc et chlorite de Trimouns. PhD Thesis, Université Paul Sabatier, Toulouse.

de Parseval, P., Moine, B., Fortuné, J.-P., Ferret, J. (1993). Fluid-mineral interactions at the origin of the Trimouns talc and chlorite deposit (Pyrénées, France). *Current Research in Geology Applied to Ore Deposits*, 12, 205–208.

de Parseval, P., Fontan, F., Agouy, T. (1997). Composition chimique des minéraux de terres rares de Trimouns (Ariège, France). *Comptes Rendus de l'Académie des Sciences de Paris*, 324, 625–630.

de Parseval, P., Jiang, S., Fontan, F., Wang, R., Martin, F., Ferret, J. (2004). Geology and ore genesis of the Trimouns talc chlorite ore deposit. *Acta Petrologica Sinica*, 20(4), 877–886.

Passchier, C.W. (1982). Mylonitic deformation in the St-Barthélémy Massif, French Pyrénées, with emphasis on the generation relationship between ultramylonite and pseudotachylite. *G.U.A. Papers of Geology*, 1(16), 1–173.

Passchier, C.W. (1984). Mylonite dominated footwall geometry in a shear zone. *Geological Magazine*, 121, 429–436.

Passchier, C.W. (1985). Water-deficient mylonite zones – An example from the Pyrénées. *Lithos*, 18, 115–127.

Piret, P., Deliens, M., Pinet, M. (1990). La trimounsite-(Y), nouveau silicotitanate de terres rares de Trimouns, Ariege, France. *European Journal of Mineralogy*, 2(5), 725–729.

Poujol, M., Boulvais, P., Kosler, J. (2010). Regional-scale Cretaceous albitization in the Pyrenees: Evidence from in situ U-Th-Pb dating of monazite, titanite and zircon. *Journal of the Geological Society*, 167, 751–767.

Quesnel, B., Boiron, M-C., Cathelineau, M., Truche, L., Rigaudier, T., Bardoux, G., Agrinier, P., de Saint Blanquat, M., Masini, E., Gaucher, E. (2019). Nature and origin of mineralizing fluids in hyperextensional systems: The case of Cretaceous Mg metasomatism in the Pyrenees. *Geofluids*, 7213050.

de Saint Blanquat, M. (1985). Recherche sur la tectonique distensive contemporaine du métamorphisme d'âge crétacé dans le massif Nord-pyrénéen du Saint Barthélémy et sa bordure. Thesis, Université de Montpellier II.

de Saint Blanquat, M. (1989). La faille normale ductile du massif du Saint Barthélémy. Âge et signification de l'extension crustale dans la zone nord-pyrénéenne. PhD Thesis, Université de Montpellier II.

de Saint Blanquat, M. (1993). La faille normale ductile du massif du Saint Barthélémy. Évolution hercynienne des massifs nord-pyrénéens catazonaux considérée du point de vue de leur histoire thermique. *Géod. Acta*, 6, 59–77.

de Saint Blanquat, M., Brunel, M., Mattauer, M. (1986). Les zones de cisaillements du massif Nord Pyrénéen du Saint Barthélémy, témoins probables de l'extension crustale d'âge crétacé. *C. R. Acad. Sci. Série II*, 303, 1339–1344.

de Saint Blanquat, M., Lardeaux, J.M., Brunel, M. (1990). Petrological arguments for high temperature extensional deformation in the Pyrenean variscan crust (Saint Barthélémy massif, Ariège). *Tectonophysics*, 177(1–3), 245–262.

de Saint Blanquat, M., Bajolet, F., Grand'Homme, A., Proietti, A., Zanti, M., Boutin, A., Clerc, C., Lagabrielle, Y., Labaume, P. (2016). Cretaceous mantle exhumation in the central Pyrenees: New constraints from the peridotites in eastern Ariege (North Pyrenean zone, France). *Comptes Rendus Geoscience, Special Issue*, 348(3/4), 268–278.

de Saint Blanquat, M., Cochelin, B., Lemirre, B., Duchêne, S., Denèle, Y. (2019). Le cycle varisque dans les Pyrénées, tectonique et thermicité. *Géochroniques*, 152(December), 16–20.

Saspiturry, N., Razin, P., Baudin, T., Serrano, O., Issautier, B., Lasseur, E., Leleu, S. (2019). Symmetry vs. asymmetry of a hyper-thinned rift: Example of the Mauléon Basin (Western Pyrenees, France). *Mar. Pet. Geol.*, 104, 86–105 [Online]. Available at: https://doi.org/10.1016/j.marpetgeo.2019.03.031.

Saspiturry, N., Lahfid, A., Baudin, T., Guillou-Frottier, G., Razin, P., Issautier, B., Le Bayon, B., Serrano, O., Lagabrielle, Y., Corre, B. (2020). Paleogeothermal gradients across an inverted hyperextended rift system: Example of the Mauléon fossil rift (Western Pyrenees). *Tectonics*, 39(10), 1–36 [Online]. Available at: https://doi.org/10.1029/2020TC006206.

Saspiturry, N., Issautier, B., Razin, P., Andrieu, S., Lasseur, E., Allanic, C., Serrano, O., Baudin, T., Cochelin, B. (2021a). Review of the syn-rift to early post-rift depositional systems of the Cretaceous Mauléon rift: Sedimentary record of continental crust hyperextension and mantle denudation (Western Pyrenees). *Bulletin de la Société Géologique de France*, 1–33 [Online]. Available at: https://doi.org/10.1051/bsgf/2021044.

Saspiturry, N., Issautier, B., Razin, P., Baudin, T., Asti, R., Lagabrielle, Y., Allanic, C., Serrano, O. (2021b). Review of Iberia-Eurasia plate-boundary basins: Role of sedimentary burial on depth-dependent continental crust ductile thinning during rifting and continental breakup. *Basin Research*, 33(2), 1626–1661 [Online]. Available at: https://doi.org/10.1111/bre.12529.

Schärer, U., de Parseval, P., Polvé, M., de Saint Blanquat, M. (1999). Formation of the Trimouns talc-chlorite deposit (Pyrenees) from persistant hydrothermal activity between 112 and 97 Ma. *Terra Nova*, 11, 30–37.

Scribano, V., Carbone, S., Manuella, F.C., Hovland, M., Rueslatten, H., Johnsen, H.-K. (2017). Origin of salt giants in Geofluids abyssal serpentinite systems. *Int. Journal of Earth Sci.*, 106(7), 2595–2608.

Siebenaller, L., Boiron, M.C., Vanderhaeghe, O., Hibsch, C., Jessell, M.W., Andre-Mayer, A.S., France-Lanord, C., Photiades, A. (2013). Fluid record of rock exhumation across the brittle-ductile transition during formation of a Metamorphic Core Complex (Naxos Island, Cyclades, Greece). *J. Metamorphic Geol.*, 31, 313–338.

Souquet, P. and Peybernès, B. (1987). Allochtonie des massifs primaires nord-pyrénéens des Pyrénées centrales. *C. R. Acad. Sc. Paris*, 305, 733–739.

Teixell, A., Labaume, P., Ayarza, P., Espurt, N., de Saint Blanquat, M., Lagabrielle, Y. (2018). The present-day and past crustal structure of the Pyrenean-Cantabrian belt: A review and recent developments from new concepts and data. *Tectonophysics*, 724/725, 146–170.

Thiebaut, J., Debeaux, M., Morre, N. (1965). Sur la genèse du gisement de talc de Trimouns (Ariège). *C. R. Soc. Géol. Fr.*, 78–79.

Vacherat, A., Mouthereau, F., Pik, R., Bernet, M., Gautheron, C., Masini, E., Le Pourhiet, L., Tibari, B., Lahfid, A. (2014). Thermal imprint of rift-related processes in orogens as recorded in the Pyrenees. *Earth Planet. Sci. Lett.*, 408, 296–306.

Vielzeuf, D. (1984). Relations de phases dans le faciès granulite et implications géodynamiques. L'exemple des granulites des Pyrénées. PhD Thesis, Université de Clermont-Ferrand, Besançon.

Wibberley, C.A.J. and Shimamoto, T. (2003). Internal structure and permeability of major strike-slip fault zones: The Median Tectonic Line in Mie Prefecture, Southwest Japan. *Journal of Structural Geology*, 25, 59–78.

Zwart, H.J. (1954). La géologie du massif du Saint Barthélémy. *Leidse Geol. Meded.*, 18, 1–229.

# List of Authors

Paul ANGRAND
GET
CNRS
Université Toulouse-III-Paul-Sabatier
IRD
CNES
OMP
Toulouse
France

Thierry BAUDIN
BRGM
Orléans
France

Romain BOUSQUET
Géosciences Rennes
CNRS
Université de Rennes 1
France
and
Institute of Geosciences
Christian-Albrechts-Universität zu Kiel
Germany

Alexandre BOUTIN
GET
CNRS
Université Toulouse-III-Paul-Sabatier
IRD
CNES
OMP
Toulouse
France

Bénédicte CENKI
Géosciences Montpellier
CNRS
Université de Montpellier
France

Roman CHELALOU
Géosciences Rennes
CNRS
Université de Rennes 1
France

Bryan COCHELIN
State Key Laboratory for
Mineral Deposits Research
University of Nankin
China

Alexandre CUGERONE
Géosciences Montpellier
CNRS
Université de Montpellier
France
and
Department of Earth Sciences
University of Geneva
Switzerland

Yoann DENÈLE
GET
CNRS
Université Toulouse-III-Paul-Sabatier
IRD
CNES
OMP
Toulouse
France

Charles GUMIAUX
ISTO
CNRS
Université d'Orléans
and
BRGM
Orléans
France

Tarik KERNIF
Géosciences Rennes
CNRS
Université de Rennes 1
France

Yves LAGABRIELLE
Géosciences Rennes
CNRS
Université de Rennes 1
France

Benjamin LE BAYON
BRGM
Orléans
France

Frédéric MOUTHEREAU
GET
CNRS
Université Toulouse-III-Paul-Sabatier
IRD
CNES
OMP
Toulouse
France

Manuel MUÑOZ
Géosciences Montpellier
CNRS
Université de Montpellier
France

Thierry NALPAS
Géosciences Rennes
CNRS
Université de Rennes 1
France

Émilien OLIOT
Géosciences Montpellier
CNRS
Université de Montpellier
France

Alexandre ORTIZ
Géosciences Rennes
CNRS
Université de Rennes 1
France

Philippe DE PARSEVAL
GET
CNRS
Université Toulouse-III-Paul-Sabatier
IRD
CNES
OMP
Toulouse
France

Marc POUJOL
Géosciences Rennes
CNRS
Université de Rennes 1
France

Benoit QUESNEL
CREGU
GeoRessources
CNRS
Université de Lorraine
Nancy
France

Michel DE SAINT BLANQUAT
GET
CNRS
Université Toulouse-III-Paul-Sabatier
IRD
CNES
OMP
Toulouse
France

Nicolas SASPITURRY
Géosciences Montpellier
CNRS
Université de Montpellier
France

Isabelle THINON
BRGM
Orléans
France

Julie TUGEND
ISTeP
CNRS
INSU
Sorbonne Université
and
GEC
CY Cergy Paris Université
France

Jessica UZEL
Géosciences Rennes
CNRS
Université de Rennes 1
France

# Index

## A, B



## C, D

## E, F



## I, K, L



## M, N, O



## P, R



## S, T



## U, W

Printed and bound by CPI Group (UK) Ltd, Croydon, CR0 4YY
13/09/2026
14960999-0006